AF477137

Microbiology Quiz
(A Handbook for Competitive Exam)

Microbiology
Quiz
(A Handbook for Competitive Exam)

Jayababu Mudili

Head, Department of Microbiology
V.S. Lakshmi Women's Degree & P.G. College
Kakinada (A.P.)

PharmaMed Press
An imprint of Pharma Book Syndicate

A Unit of BSP Books Pvt., Ltd.

4-4-309/316, Giriraj Lane,
Sultan Bazar, Hyderabad - 500 095.

© 2014, *by Publisher*

All rights reserved. No part of this book or parts thereof may be reproduced, stored in a retrieval system or transmitted in any language or by any means, electronic, mechanical, photocopying, recording or otherwise without the prior written permission of the publishers.

Published by

PharmaMed Press
An imprint of Pharma Book Syndicate

A unit of BSP Books Pvt., Ltd.

4-4-309/316, Giriraj Lane, Sultan Bazar, Hyderabad - 500 095.
Phone: 040-23445605, 23445688; Fax: 91+40-23445611
E-mail: info@pharmamedpress.com

ISBN : 978-93-85433-21-4 (HB)

Preface

Success in Competitive Entrance Examinations of Microbiology requires a broad overview of the subject both at the intermediate and Degree level of examination. The knowledge of the fundamental principles and exhaustive practice in solving Objective Exercises is most mandatory for the students.

The present book Microbiology Quiz is intended at introducing the learner with all the theories, examples, concepts and principles ought to be known and also provides thorough practice by way of numerous objective exercises.

The First Chapter serves as an escort for the learner towards taking Competitive Entrance Tests. The Do's and Do Not's in preparing and appearing for competitive tests have been scrupulously discussed. Further, certain instructions and suggestions have been provided for the teachers and question paper setters regarding putting forth clear and uncomplicated questions.

The realm of the subject matter has been distributed in the next Twenty Chapters (Chapters 2 to 21) in order to facilitate comfortable learning under key titles.

Each title has been justified with precise introductory study-material comprising of significant terminology and examples. As far as possible the ideas and concepts are illustrated in tabular forms to give a rapid bird's eye view.

The objective exercises provided in each Chapter have been written conscientiously to cover questions of easy, difficult, analytical as well as reasoning type. Care has also been taken to include the commonly asked questions in many competitive entrance exams conducted by various Universities.

The Answer Key to the various Exercises is provided at the end of each Chapter. Explanations to the answers have been endowed with, wherever felt necessary.

The Appendix comprises of three Grand Tests and a Pictorial Quiz. The Grand Test Papers have been set to incorporate questions of increasing difficulty levels ranging from preliminary - (Grant Test 1), advanced – (Grand Test 2) and reasoning and analytical - (Grand Test 3). These Grand Tests will allow the learner to comprehend his stand in taking the tests. The Pictorial Quiz is a visual test of identifying Scientists, Organisms, Experiments, etc.

It is sincerely hoped that the present book will serve its purpose to both teachers and students alike. Comments and criticisms are welcome for further improvement.

Author

Acknowledgement

I am gratified by numerous individuals for their priceless professional and technical support that has made the present book possible. Even as I am unable to acknowledge each one in person, I wish to mention the names of a few noteworthy cohorts.

In the first place, I acknowledge the generosity of the Management & Principal **Smt. V. Sandhya** of V.S.Lakshmi Women's Degree & P.G. College, Kakinada, who provided the required cooperation and encouragement during the process of writing.

Many thanks are due, to my colleagues and seniors, **Dr. A. Srinivasulu, Smt. M. Usha Rani and Shri N. Sharief Md.**, for their valuable suggestions made towards writing certain Chapters.

I owe a special debt of gratitude to my beloved students **Ms. M. Subhashini, Ms. K. Revati, Ms. G. Neelima** and **Mrs. Karimunisa Begum**, all Lecturers in VSL College, for their invaluable help in compiling and proof-reading of the final text.

It is with great pride that I mention the cooperation and understanding of my wife and kids that was very much needed and readily given.

My sincere acknowledgements to all those authors whose valuable writings served as reference material for preparation of the various objective exercises.

Last but not the least; I am most thankful to the Publisher for the continuous interactions and a patient wait until completion of this project.

Author

Contents

APPENDIX

1

Stimulation Towards Competitive Entrance Examinations

Introduction, The Purpose of Competetitive Entrance tests, Gear up to take the Challenge, Be mentally fit for the competitive test, Be physically fit for the competitive test, Adopt the right procedure of study, Adopt the right approach for answering, Major types of MCQs, The Teacher

Introduction

In contemporaneous times, almost every student seeking higher education encounters some or the other entrance tests in his educational career. Entrance examinations are conducted by prestigious Universities, deemed Universities, Research Centres, Autonomous Institutions, State and National level academic bodies. An organization considers the entrance test as a perfect tool to select the best applicants when there is large number of students seeking admission, as against a limited number of available seats.

The Purpose

The prime mission of an entrance test is to select candidates from among a large number of applicants. However, there may be several ancillary purposes behind holding such tests.

- The result of these tests give an idea of what percent of the applicants are actually qualified enough to join the higher courses. Hence, the organizing body usually proposes the qualifying marks for the test. Only the qualified individuals are invited for counseling where further selection is easier.

- Some institutions or Research Centres that maintain very high standards of academic profile do not confide upon the standards of the lower degrees of other Universities as up to the mark. Hence they conduct their own screening tests instead of selecting the so called toppers from other University or institution. Such tests are indeed challenging and have a flavor of the organizing institute. The student needs to prepare himself after considering the kind of research or academic activities undertaken by the organizing institute as many questions are likely to be churned out of their own areas of interests.

- Some PG courses are open to graduates from other subjects. For instance a graduate of Botany or Zoology may be eligible to take M.Sc. course in Microbiology. In such a case the University demands that the applicant should have some fundamental knowledge of the new subject. Moreover, it is not fair to select meritorious candidates from different streams into one course purely on the basis of their UG marks. In such cases the University conducts a simple entrance test and fills the seats as per the rank obtained, irrespective of their ground subject. Such tests are the easiest but require a clear mind of perception and a strong foundation.

- Entrance tests are also conducted to provide added benefits proposed by the organization on a limited basis. For Eg. The first ten toppers will be exempted from tution fee. In such a case an entrance test becomes essential.

Apart from the above mentioned reasons there may be other purposes for conducting competitive tests but whatever the reason may be a competitive test is intended to check the individual's knowledge, analytical thinking ability and clarity of a concept.

Gear up to take the challenge

- Both fundamental knowledge as well as a winning spirit is required to make it to the top of entrance tests. Even as the student is too occupied preparing himself for the test, somewhere he tends to lack the boost to achieve success. Most of the students are found with extreme negativity of mindset during the preparation phase. Fear of the test and underestimation of ones own capability and "loser" attitude from the beginning, all add to their failure.

- The applicant needs to be in a psychologically sound state, seeming himself that he knows he will be successful if he prepares effectively. Just say to yourself that "it's my area of study and all I need to do is a thorough brush up of my knowledge".

- A few highly focused students unfortunately come under the spell of extreme stress. They are so tensed about the final outcome that they only think about the test even as they take a break to relax. They are even unable to concentrate on what they are reading because the mind is too stressed about *Will I get the seat finally?*

- *What if I don't get through?*

- *What if I am offered a seat in a remote town?* And such other botherations.

- These thoughts lead to serious stress and may result in acute depression and obvious failure.

- Hence, it is necessary to keep any kind of stress at bay and just relax during regular breaks. A cup of coffee, a short walk, a brief nap, are all refreshing. Watching a comic animation or even a movie is a must when you are too worked up.

- Quite contrary to this, there are some students who are rather "over confident" about themselves. "Only I can do it, no one else can", its after all a matter of attempting 100 questions by choosing the correct answer. Do I need to actually prepare for a test which will have simple questions in my own subject. These thoughts I am afraid are too bold and definitely injurious. Such a student will tend to neglect sincere preparation and repeated practice. Do not pretend to be confident just because you are lazy to put in genuine effort.

- In short, gear up to take the challenge with the right spirit , dispel stress and concentrate on the present chapter that you are studying instead of distracting your mind about the ultimate result and finally, avoid being too confident.

The style of competitive entrance tests

- Majority of entrance tests are of multiple choice questions (MCQ) type, while a few of them include subjective/comprehensive questions. It is beyond the scope of this book to elaborate on the essay type questions. Hence, focus will be laid on those tests that pose purely objective questions.

- Usually, the paper is set to include fundamental questions with a view to test your clarity of the concept. Yet a few other categories may also be asked.

The following constructive and detrimental approaches are intended to give guidance to the candidates to gain the right spirit of success.

Be mentally fit for the competitive test

Constructive Approach	Detrimental Approach
<ul><li>Assure yourself of your positive ability</li><li>Set your goal and determine to achieve it</li><li>Be confident that you can outshine the others</li><li>Get ready for the kind of hard-work you had never done before</li><li>Identify your inner strengths and sharpen the same</li><li>You know you can do it and that you will…</li></ul>	<ul><li>Never take the test as an impossible task</li><li>Do not lose confidence and at the same time never be overconfident of the test</li><li>Avoid stressing your brain thinking of the events of failure in the test</li><li>Do not underestimate your knowledge and capacity</li><li>Prevent any delay in reaching the exam hall. Last minute rush will leave the mind disturbed.</li></ul>

Be physically fit for the competitive test

Constructive Approach	Detrimental Approach
• Maintain a stress free mind and body on the day of the examination • Seek medical assistance in case of any ill health • Have a sound sleep on the previous night to wake up fresh. • Have a filling breakfast and boost your energy with fresh juices or milk that keeps the system far from hunger • If possible take a 30min meditation session prior to the test to refresh the brain. Regular meditation improves the concentration prowess.	• Never come to the examination on an empty stomach • Avoid severe work-outs and exercises on the day of the test that may exert the body and mind to discomfort. • Avoid night-outs as this will only leave the body tired and irritable. • Do not get anxious of the test

Adopt the right procedure of study

Constructive Approach	Detrimental Approach
• Give a justified review to each and every topic so as to have a clear knowledge of the entire subject. • Give extreme care to study illustrations and examples for every sub-topic, as many MCQs are based on examples. • Observe finer details of diagrams, structures, pathways, tables and flow-sheets from standard textbooks. • An objective approach must be given when preparing for the test. Learning voluminous print matter and writing essays will not be of much help. • Refer to as many textbooks as possible to comprehend each topic. On a daily basis, it is absolutely necessary to practice as many objective exercises as possible on every topic to gain expertise in taking the test.	• Do not learn the topics by-heart without understanding the principles. • Avoid getting too deep into the details of the topic. It is just sufficient to have a fair and clear idea of the concept. • At the same time, do not give a casual browse over the topic. It is essential to study the nuance of the subject matter. • Do not simply attempt various multiple choice exercises without reading about the topic. • Never learn MCQ's with the answer key. Attempt all the questions and then check with the answers.

Adopt the right approach of answering

Constructive Approach	Detrimental Approach
• Read each word of the question carefully to understand the intention of testing. • Wherever applicable try to tell the answer to yourself without looking at the options. A direct glance at the options may otherwise create unnecessary confusion • If you are unable to get the correct answer, at least try to rule out the wrong answers. You will end up with fewer options to choose from. • Start from the first question and go ahead with answering all the easy questions. In case you come across a question where you can get the answer but some time is required for reasoning, just mark (@ or any other symbol) with a pencil next to that question and go ahead.	• Do not attempt the questions in a hurry. • Before you mark the correct answer, just justify to yourself why the other options are incorrect or not appropriate. • Avoid attempting the questions non-stop. Take a simple deep breath, relax for a while and move on the next question especially after answering every 6-10 questions. • It is reported that even the most alert brains get a little careless after attempting 10 questions at a stretch. Hence, a trivial break is essential.

Table Contd…

Constructive Approach	Detrimental Approach
• Similarly if a particular question is totally new to you, mark it X with a pencil and proceed to the next question. • After attempting all the easy questions, now go ahead with the reasoning/calculation type problems marked by you and tackle them carefully. • Finally get back to the questions you have marked X. Even the best of experts have said that guessing is not wrong in MCQ type tests. Simply guessing the correct answer for 5 questions (out of 100) should not do much harm even if there is negative marking. Therefore, feel safe to simply guess the correct answer for up to 5 questions instead of leaving them un-attempted.	• After finishing the paper, do not develop a tendency to change answers just out of doubt in your second reading of the full paper. It is often most likely that the first choice is the correct one. The more you think of the problem, the more you tend to get misled. Hence, change a particular answer only if you are doubly sure, or if you had not read the question properly earlier.

Major types of MCQs

Question	Remarks
1. Cells of *Corynebacterium diphtheriae* are known to show a. Safety pin morphology b. Chinese letter configuration c. Budding yeast morphology d. Queen's necklace morphology *Ans:* b 2. Bacteria with prosthecae and holdfast a. *Caulobacter* b. *Sphaerotilus* c. *Pseudomonas* d. *Leucothrix* *Ans:* a 3. The temp/time employed in UHT processing of milk is a. 137.8^0C / 2 Seconds b. 148.9^0C / 3 Seconds c. 161^0 F / 5 Seconds d. 145^0F / 3 Seconds *Ans:* a	Questions 1, 2 & 3 are typical **knowledge based questions**. Unless you know the fact, you will not be able to answer such questions. Such questions are posed to test your fundamental subject knowledge.
4. The viral particle may be called as a. Nucleocapsid b. Protomer c. Capsid d. Viroid *Ans: a* 5. The internal organelles can be appreciated in the cells by a. Bright field microscopy b. Dark field microscopy c. Fluorescence microscopy d. Phase contrast microscopy *Ans: d*	Questions 4 & 5 are typical **logic based questions**. Even if you might have no idea of the answer, thought process will drive you to the correct answer. Example, in the 4th problem, if you know that a virus is made up of nucleic acid and capsid, it is easy to deduce the answer. Similarly in the 5th problem, you have to think logically that unless a contrast is created within the cell structure, we cannot observe the inner details.

Table *Contd...*

Question	Remarks

6. Arrange the steps A, B, C & D in the correct sequence for Manufacture of Beer
 	
A. Boiling the wort
B. Malting
C. Mashing
D. Adding Hops
 	
  a. A, D, C, B
  b. C, B, D, A
  c. B, C, A, D
  d. B, D, A, C
 	
Ans: c
	Questions 6, 7 & 8 are **Analytical and reasoning type questions**. These types of questions are relatively few in the exams. However, they usually serve as the Rank deciders. Only those candidates who have a sound subject foundation supported by strong analytical skill can answer them correctly. One must read the problem very cautiously, giving importance to each word in order to approach to the correct answer.

7. Which of these is FALSE statement with respect to killed vaccines
 	
  a. They must be administered by intravascular injection
  b. They are more effective against systemic viruses than against viruses that replicate in local mucosal sites
  c. They are more stable under environmental conditions when compared to attenuated vaccines
  d. Relatively smaller amounts of Ag is required for injection before induction of immunity
 	
Ans: d
	Some analytical & logic based questions are included in all the Chapters of this book. Particularly Grand Test 3 in the Appendix has been set with the intention of checking the analytical skill of the learner.

8. A 12y old boy complains of nausea, mild diarrhea and faint abdominal pain after 8h of consuming ladoo bought from a roadside vender. The suspected pathogen could be
 	
  a. *Clostridium botulinum*
  b. *Shigella dysenteriae*
  c. *Staphylococcus aureus*
  d. *Brucella*
 	
Ans: c	

The Teacher

- A certain amount of preparation is required in order to give guidance/coaching to the competitive tests.
- It is necessary that the teacher structures his lecture differently from the routine class work.
- The topic needs to be discussed very constructively with the most critical facts and by including several hardcore examples. It is a waste of time to teach the entire concept in detail.
- Every class must end with a lot of discussion of examples and illustrations based on the topic. A few oral exercises may be posed by the tutor to cross-check the concept base.
- The pace of the class should not be too fast which may otherwise escape the interest of the learner.
- Daily tests must be conducted comprising of choicest questions on the topic discussed the previous day.
- The MCQs must be framed with proper choice of words, in a simple manner.

Examples of badly framed questions	Remarks	An improved version of the same question

1. Which of these is FALSE
  a. Eutrophication causes Decrease in plant, algal and animal biomass
  b. Anoxic conditions may develop in the lake due to eutrophication
  c. Turbidity of the lake increases due to eutrophication
  d. Eutrophication shortens the life span of the lake
	The purpose of the problem must be focused in the main statement itself. The option must be clear and precise with minimum words.	1. What is FALSE about lake-eutrophication
		  a. Decrease in biomass
		  b. Causes anoxic conditions
		  c. Increases turbidity
		  d. Shortens the life span

2. MacConkey agar which is a differential as well as a selective medium is differential due to ______ and selective due to ________ in the medium.
  a. Bile salt, Lactose
  b. Lactose, Bile salt
  c. lactose, neutral red
  d. Bile salt, Neutral red
	The main statement should be short, clear and must not give a comprehension type picture	2. The differential and selective nature of Mac Conkey agar is due to ______ and ______ respectively.
		  a. Bile salt, Lactose
		  b. Lactose, Bile salt
		  c. lactose, neutral red
		  d. Bile salt, Neutral red

3. The autoclave is used for
  a. Sterilizing glassware
  b. Sterilizing aprons
  c. Sterilizing culture media
  d. All
	The contents of the options need not be repetitive. In such cases, the repeated word must be incorporated into the main statement.	3. Which of these material can be sterilized using an autoclave
		  a. Glassware
		  b. Aprons
		  c. Culture media
		  d. All

4. Jaundice is primarily a disease of?
  a. Children
  b. Liver
  c. North Asia
  d. High mortality
	As far as possible, the options must be revolving around the same theme. They must not cover multiple areas.	5. Jaundice is primarily a disease of?
		  a. Lungs
		  b. Liver
		  c. Intestine
		  d. Brain

5. The building block of proteins is an
  a. Amino acid
  b. Nucleic acid
  c. Monosaccharide
  d. Fatty acid
	The use of article "an" in the question may suggest the correct answer automatically. This must be avoided.	5. The building block of proteins is an/a
		  a. Amino acid
		  b. Nucleic acid
		  c. Monosaccharide
		  d. Fatty acid

6. Stanley crystallized TMV in
  a. 1935
  b. 1936
  c. 1937
  d. 1938
	The options given in the question must not be too confusing and similar to the correct answer. For example, in this case the intention is only to know the period of crystallization of viruses and not the exact date. Hence the options must cover a wider range of periods.	6. Stanley crystallized TMV in
		  a. 1935
		  b. 1836
		  c. 1985
		  d. 1947

2

Origin and Scope of Microbiology

Retrace your Subject...

Discovery of Microorganisms, Debate over spontaneous generation, The Golden Age of Microbiology, Significant historical events in the development of Microbiology, Significant Titles of Microbiologists.

- **Discovery of microorganisms** – Antoni van Leeuwenhoek, Father of Microbiology, first observed live microbes (*animalcules)* through his magnifying lenses between 1673 and 1723 and described bacterial motility and morphology.

- **Debate over Spontaneous Generation** – the faulty **Theory of Abiogenesis** hypothesized that *living things emerged spontaneously from non-living objects.* This was supported by biologists such as Aristotle, Rudolf Virchow & John Needham. Beginning with 1668, several scientists opposed the theory of abiogenesis. A new theory called the **Theory of Biogenesis** evolved. The scientists who conducted experiments to disprove the theory of abiogenesis include: Francesco Redi (three jar meat experiment), Lazzaro Spallanzani (heating broth in sealed flasks), Louis Pasteur (heating broth in open end long swan necked flasks), John Tyndall (intermittent heating of broth to kill spores – tyndallization).

- **The Golden Age of Microbiology** – 1857-1914: The causative agents of several disease causing microbes were discovered during this period. It also witnessed the advances in laying the foundations of immunology and microbial metabolism.

- **Significant discoveries of the Golden Age:** Fermentation Technology (Pasteur), Pasteurization (Pasteur), Germ Theory of Diseases (Robert Koch), Aseptic Surgical techniques (Joseph Lister), Causative agent of Anthrax (*Bacillus anthracis* By Robert Koch), Koch's Postulates – A systematic approach to prove microbial causative agent (Robert Koch), Vaccination (Edward Jenner), Discovery of the wonder drug penicillin (Alexander Fleming), First use of Chemotherapeutic agent Salvarsan for treatment of syphilis (Paul Ehrlich).

Significant historical events in the development of Microbiology

NAME OF THE DISCOVERER	YEAR	MICROBIOLOGICAL EVENT
Girolamo Fracastoro	1546	Disease is caused by an invisible creature
Jansen	1519-1608	Developed first useful compound microscope
Antony von leeuwenhoeck	1676	Discovered animalcules.
Franciscoredi	1688	Worked on theory of spontaneous generation
Lazaro spallanzani	1765-1776	Worked against spontaneous generation theory.

Table *Contd...*

NAME OF THE DISCOVERER	YEAR	MICROBIOLOGICAL EVENT
Edward jenner	1798	Introduced cowpox vaccine for small pox.
Theodar schwann and schleiden	1838-1839	Cell theory.
Augustino bassi	1835-1844	Discovered that a fungus causes silk worm disease.
Louis Pasteur	1857	Lactic acid fermentation is due to microorganisms.
	1861	Disproved spontaneous generation
	1881	Developed anthrax vaccine
	1885	Developed rabies vaccine.
Joseph Lister	1867	Published his work on antiseptic surgery.
Joseph Lister	1878	Pure culture of bacteria-lactics
Robert Koch	1876-1877	Discovered causative agent of anthrax as *Bacillus anthracis*. Identified *Vibrio cholerae* cause cholera
	1881	Introduced gelatin for culture media.
	1882	Discovered tubercle bacillus
	1884	Proposed Staining techniques.
Meister	1876-1877	Discovered nucleic acids.
Petri Richard	1887	Proposed use of Petridishes.
Winogradsky	1887-1890	Studied sulfur, nitrifying bacteria
Martinus willem and Beijerinck	1888	Isolated rhizobium.
Beijerinck	1899	Proved that virus causes Tobacco Mosaic disease
Von behring	1890	Antitoxin for diphtheria and tetanus.
Ivanowsky	1892	Virus causation of TMV
Bordet	1895	Complement
Buchner	1897	Prepared Yeast extracts.
Ronald Ross	1897	Showed malaria transmission by mosquito.
Land steiner	1902	Blood groups.
Von behring kitasato	1890	Described specific humurol factors[antibodies]
Wright	1903	Discovered antibodies in the immunized animal blood.
Schaudinn and Hoffman	1906	Syphilis caused by Treponema pallidum
Wassermann	1906	Complement fixation test for syphilis.
Ricketts	1906	Rocky mountain spotted fever
Paul Ehrlich	1910	Chemotherapetic agent for syphilis
D'herelle and Twort	1915-1917	Discovered bacteriophages.
Alexander Fleming	1921	Lysozyme
Alexander Fleming	1929	Penicillin
Bergey's	1923	Bacterial classification (First edition)
Fred Griffith	1928	Bacterial transformation
Ruska	1933	Transmission electron microscope.

NAME OF THE DISCOVERER	YEAR	MICROBIOLOGICAL EVENT
Stanley	1935	Crystallized TMV.
Beadle and tatum	1941	One gene-one enzyme hypothesis
Avery	1944	Showed DNA carries information during transformation.
Selman A.waksman	1944	Streptomycin
Lederberg and tatum	1946	Described bacterial conjugation.
Lwoff	1950	Lysogenic bacteriophages.
Harshey and chase	1952	Showed that bacteriophages inject DNA into host cell.
Zinder and lederberg	1952	Generalized transduction.
d.waston and H.C crick	1953	Proposed double helix of DNA
Jacob and wollman	1955	Discovered F factor is Plasmid
Jerne and burnet	1955	Proposed clonal selection theory
Yallow	1959	Radio immune assay technique.
Jacob and monad	1961	Operon model of gene regulation.
Nirenberg and Khorana	1961-66	Elucidated genetic code.
Arber and smith	1970	Discovered restriction Endo nucleases.
Temin and Baltimore	1970	Reverse transcriptase and retro viruses
Ames	1973	Developed bacterial assay for detection of mutagens and carcinogens.
Kohler and milstein	1975	Monoclonal antibodies.
Gallo and montagnier	1983-84	Isolated HIV
Mullis	1983-84	PCR

Significant Titles for Microbiologists

Antony von leeuwenhoeck (1632 -1723)	Father of Microbiology
Louis Pasteur (1822 -1895)	Father of Modern Microbiology
Robert Koch (1843-1910)	Father of bacteriology
Edward Jenner 1796	Father of Vaccination
Alexander Fleming (1929)	Father of Antibiotics
Paul Ehrlich (1908)	Father of Chemotherapeutics
Sergei Nikolayevich Winogradsky	Father of Soil Microbiology
Matinus Beijerinck (1851-1931)	Father of Virology
Joseph Lister	Father of Antisepsis
Elias Fries (1794-1878)	Father of Mycology (Study of Fungi)
Dr M.O.P. Iyengar (1886-1963)	Father of Indian Phycology (Study of Algae)
Max Perutz (1914-2002)	Father of Molecular Biology
Jaques Miller (1931)	Father of Modern Immunology

Exercises

1. India's contribution to the development of ancient Microbiology is evident from   [   ]
   a. Septic tanks in Mohenjodaro and Harappan regions
   b. Mention of seeds of disease in the Vedas
   c. Description of fish pickle fermentation in Indian scriptures
   d. Mention of 'wine' in the Bible

2. Swan neck flasks were designed by   [   ]
   a. F. Redi   b. Louis Pasteur   c. Robert Koch   d. Aristotle

3. Opposed the theory of Biogenesis   [   ]
   a. Spallanzani   b. F. Redi   c. H. Schrodder   d. John Needham

4. Intermittent sterilization was introduced by   [   ]
   a. Schwann   b. Liebig   c. John Tyndall   d. E. Jenner

5. Agastino Bassi is known for his proposal of   [   ]
   a. Anthrax is caused by bacteria
   b. Causative agent of silkworm disease is a fungus
   c. Germ theory of disease
   d. Some bacteria can form resistant spores

6. Edward Jenner introduced   [   ]
   a. Vaccination against small pox   b. Disinfection of tables in labor rooms
   c. Antibiotic against fungi   d. Use of Agar-Agar in culture media

7. Golden era of Microbiology was established between   [   ]
   a. 1950 – 1970   b. 1760 – 1810   c. 1860 – 1910   d. 1945 – 1966

8. Father of modern bacteriology   [   ]
   a. Robert Koch   b. Pasteur   c. Leeuwenhoek   d. F. P. Rous

9. Father of modern Microbiology   [   ]
   a. Robert Koch   b. Pasteur   c. Leeuwenhoek   d. F. P. Rous

10. Edelman and Porter are associated with   [   ]
    a. Discovery of Ribozyme
    b. Development of crystallographics in EM
    c. Discovery of Restriction enzymes
    d. Research on structure of Antibodies

11. Sulfa drugs were discovered by   [   ]
    a. Tatum   b. Ruska   c. Domagk   d. J. M. Bishop

12. Animalcules were reported by   [   ]
    a. Koch   b. Jenner   c. Leeuwenhoek   d. Pasteur

13. The discovery of Penicillin and its therapeutic value was studied by   [   ]
    a. Fleming   b. Chain   c. Florey   d. All

14. Enders, Robin and Robbins were known for   [   ]
    a. Discovery of Sarcoma virus   b. Cultivation of polio virus in tissue culture
    c. Discovery of prions   d. Discovery of histocampatability Ags

15. Ruska is known for development of   [   ]
    a. Rabies vaccine   b. Recombinant Hepatitis B vaccine
    c. TEM   d. Vaccine against yellow fever

16. Which of these is a Koch Postulate   [   ]
    a. The causative agent of a disease must be isolated in the lab culture media
    b. All bacteria are prokaryotes
    c. Viruses cannot grow on culture media
    d. Living things do not arise spontaneously from non-living things

17. Pasteur was NOT associated with   [   ]
    a. Timed heating of milk   b. Canning of cooked food
    c. Development of vaccine   d. Lactic acid fermentation

18. Robert Koch was associated with   [   ]
    a. gelatin based solid culture media
    b. etiological role of bacteria in Anthrax
    c. Isolation of microbes on slices of boiled potato
    d. a, b, and c

19. As per the current knowledge of Microbial diseases, Koch's postulates are   [   ]
    a. Acceptable for all diseases
    b. Not acceptable for any disease
    c. Mostly acceptable with slight conditional changes

20. Beijerinck is associated with   [   ]
    a. Soil Microbiology   b. Virology
    c. Food Microbiology   d. a & b only

21. The germ theory was further supported by   [   ]
    a. Louis Pasteur   b. Leeuwenhoek   c. S. Ochoa   d. J Monod

22. Father of Modern Microbiology   [   ]
    a. Leeuwenhoek   b. Pasteur   c. Koch   d. Jenner

23. Role of microorganisms in biogeochemical cycles was elucidated by   [   ]
    a. S. E. Luria   b. S. B. Prussiner
    c. Sergei Winogradsky   d. S. A. Waksman

24. Buchner is known for the description of   [   ]
    a. Aerobic and anaerobic metabolism   b. Process of canning
    c. Cell free fermentation   d. Acetone-Butanol Pathway

25. Which of these is NOT a inference of Pasteur   [   ]
    a. Crystallization of TMV
    b. Sheep has immunity against anthrax
    c. Presence of microorganisms in air
    d. Separation of tartaric acid crystals into two types

26. The leprosy bacterium (*M. leproae*) was first discovered by   [   ]
    a. Hansen   b. Alexander Fleming
    c. M Gast   d. Thomas Spackman

27. Chick embryo technique to cultivate viruses was developed by   [   ]
    a. Conrat & Williams   b. W M Stanley
    c. E W Goodpasteur   d. Holmea

28. Klebs-Loeffler's Bacterium   [   ]
    a. *Klebsiella pneumoniae*   b. *Bacillus anthracis*
    c. *Corynebacterium diphtheriae*   d. *Haemophilus influenzae*

29. Von Behring is known for his work on   [   ]
    a. Purification of antibiotics
    b. Oral polio vaccine
    c. Mode of action of streptomycin
    d. Immunization of tetanus and diphtheria.

30. Kitasato is related with the isolation of   [   ]
    a. *Streptococcus pyogenes*
    b. *Clostridium tetani*
    c. *Mycobacterium tuberculosis*
    d. *Vibrio cholerae*

31. Adolf Mayer first reported   [   ]
    a. Mosaic disease in tobacco
    b. Smut disease of maize
    c. Blast disease of rice
    d. Rust disease of wheat

32. Father of Virology   [   ]
    a. W. M. Stanley
    b. Dimitri Ivanowsksy
    c. Beijerinck Martinus
    d. F. H. d'Herelle

33. Father of Bacteriology   [   ]
    a. Lazzaro Spallanzani
    b. Leeuwenhoek
    c. Robert Koch
    d. Louis Pasteur

34. Bacillary forms in the blood of patients infected with Rocky mountain spotted fever first revealed   [   ]
    a. Clamydiae   b. Rickettsiae   c. Arthrobacter   d. Actinobacteria

35. *Staphylococcus*, the pathogen of abscesses was forst isolated by   [   ]
    a. Ogston   b. Edward Jenner   c. John Tyndal   d. Agastino Bassi

36. Gonococci were discovered by   [   ]
    a. Loius Laveran
    b. Stanislaus Prowazek
    c. Neisser
    d. Domagk

37. Which of these is NOT related to early virology research   [   ]
    a. F. Loeffler
    b. P. Frosch
    c. Alexander Yersin
    d. Peyton Rous

38. Paul Ehrlich is associated with   [   ]
    a. Methods of standardizing toxins and antitoxins
    b. Side chain theory of immunity
    c. Both a & b
    d. Neither a nor b

39. Complement factors in the serum were first described by   [   ]
    a. J. Bordet   b. Von Behring   c. N. K. Jerne   d. J. Murray

40. Elie Metchnikoff   [   ]
    a. Demonstrated anaphylaxis
    b. Produced antitoxins by immunizing horses
    c. Observed the role of phagocytes in immunity
    d. a, b & c

41. Calmette & Guerin   [   ]
    a. Developed vaccine against tuberculosis
    b. Discovered Rh factor
    c. Worked on antihistamines
    d. Described precipitation reaction

42. Pick out the FALSE pair   [   ]
    a. A. E Wright – Identified opsonic activity to phagocytosis
    b. Mac Farlane Burnet – Proposed Clonal Selection Theory
    c. Kitasato & Behring – Discovered tetanus antitoxin
    d. Zinsser – Developed fowl cholera vaccine

43. Pick out the CORRECT pair   [   ]
    a. Griffith – Conjugation in bacteria
    b. Baltimore – Classification of Protozoa
    c. Bassi – Silkworm fungus
    d. Whittaker – Transposons

44. Which of these is the latest strategy of classifying living organisms   [   ]
    a. Five Kingdom System   b. Three Kingdom System
    c. Three Domain System   d. Seven Domain System

45. The alternate name of the ancient disease gonorrhoea was   [   ]
    a. Swarm of gnats   b. Wool sorter's disease
    c. Flow of seed   d. Traveller's diarrhea

46. Which of these is a synonym for *First Life*   [   ]
    a. Protozoa   b. Virus   c. Amoeba   d. Algae

47. The commonest cause of death in children aged 4-10y during 1915-1942 was   [   ]
    a. Tetanus   b. Diphtheria
    c. Congenital syphilis   d. Hepatitis

48. Outbreaks of this disease have always been during the time of wars, killing many
    soldiers off the battle field   [   ]
    a. Shigellosis   b. Plague   c. Diphtheria   d. Anthrax

49. Mary Mellon, a cook from New York is the world famous carrier of   [   ]
    a. AIDS   b. Typhoid   c. Gonorrhoea   d. Hepatitis

50. International Health Organizations came into existence due to the wide spread threat of   [   ]
    a. SARS   b. AIDS   c. Plague   d. Cholera

51. The disease of poverty and insanitation that had been endemic in the Ganges and
    Brahmaputra deltas in Begal between 1817-1923   [   ]
    a. Anthrax   b. Diphtheria c. Cholera   d. Plague

52. The disease that is believed to have killed one quarter of mankind on earth in the 14th century   [   ]
    a. Plague   b. Cholera   c. Tuberculosis   d. Small pox

53. The disease that has been successfully eliminated from the globe   [   ]
    a. Q fever   b. Pinta   c. Small pox   d. Cholera

54. Arrange the following events from the oldest to the latest happenings   [   ]
    A. Pasteur develops anthrax vaccine   B. Domagk develops sulphonamides
    C. Sabin introduces live polio vaccine D. Fleming discovers penicillin

    a. A, D, B, C   b. A, B , C, D c. D, C, B, A   d. A, C, B, D

55. The year 1985 marked the invention of   [   ]
    a. Chick embryo technique
    b. First gene therapy
    c. PCR technique
    d. Use of immunosuppressive agents to perform transplants successfully

Answer Key and Validation

Origin and Scope of Microbiology

1. a	11. c	21. a	31. a	41. a	51. c
2. b	12. c	22. b	32. b	42. d	52. a
3. d	13. d	23. c	33. c	43. c	53. c
4. c	14. b	24. c	34. b	44. c	54. a
5. b	15. c	25. a	35. a	45. c	55. c
6. a	16. a	26. a	36. c	46. a	
7. c	17. b	27. c	37. c	47. b	
8. a	18. d	28. c	38. c	48. a	
9. b	19. c	29. d	39. a	49. b	
10. d	20. d	30. b	40. c	50. d	

42. *Fowl cholera vaccine was developed by Pasteur*

44. *Carl R Woese proposed the Three Domain concept that grouped all living organisms into domains Eukaryotes, Eubacteria and Archaea*

37. **Yersin discovered the plague bacillus Yersinia*

54. ** A. Pasteur develops anthrax vaccine = 1881*
     B. Domagk develops sulphonamides = 1933
     C. Sabin introduces live polio vaccine = 1955
     D. Fleming discovers penicillin = 1929

3

The Cell

Retrace your Subject...
Introduction to cytology, Prokaryotic v/s Eukaryotic cells, cell organelles, Ultrastructures of other microbial cells, signal transduction, Cell cycles ...

- Cell is the structural and functional unit of life. Study of cells is called **Cell Biology** or *Cytology*. All Cells are divided into two main classes: **Prokaryotic** cells lack nuclear envelope and **eukaryotic** cells have a nucleus in which the genetic material is separated from the cytoplasm. The prokaryotic cells are smaller, simpler, lack cytoplasmic organelles and cytoskeleton and their genomes are less complex. However, the same basic molecular mechanisms govern both prokaryotes and eukaryotes.

- The present day prokaryotes include *Bacteria* and are divided into two groups: **Archaebacteria** and **Eubacteria.** The eukaryotic cells are more complex. They are further grouped as plant cells, animal cells, Fungal cells and protists.

- **Experimental models** for cytology include: *E.coli* (bacterial cell), *Saccharomyces* (Yeast cell), *Caenorhabditis elegans* (nematode for understanding multicellular body), *Drosophila melanogaster* (fruit fly, for concepts of Genetics), *Arabidopsis thaliana* (small flowering plant for Mol. Biology of plants), *Xenopus laevis* (Frog for early development of vertebrates) and the *mouse* for studies on mammals.

- **Techniques** employed in cytology include: Microscopy (Electron and Light), cell staining techniques (simple and differential), Subcellular fractionation, centrifugation (ultracentrifugation, differential, equilibrium, density gradient,and other centrifugation principles) and tissue culture techniques.

Feature	Prokaryotic cell	Eukaryotic cell
Nucleus within nuclear membrane	Absent	Present
Size	Upto 1μm	10-100μm
Organelles (mitochondria, chloroplasts, lysosomes, peroxisomes, ER, Golgi complex, etc.)	Absent	Present
Cytoskeleton	Absent, but homologous proteins such as FtsZ, MreB and ParM are present.	Present
DNA base pairs	1×10^6 to 5×10^6	1.5×10^7 to 5×10^9
Chromosomes	Single, circular DNA	Multiple linear DNA molecu

Cell organelles

Organelle	Structure and function
Cell surfaces	
Cell membrane (plasma membrane/plas malemma/prot oplasmic membrane)	A selectively permeable bilipid layer that encloses the cytoplasm in prokaryotic and eukaryotic cells..Structure confirms to *fluid mosaic model* of lipid bilayer with outer hydrophilic ends and inner hydrophobic ends, in which proteins are embedded at places (*extrinsic and intrinsic types.*). The CM has various functions: Control inward and outward passage of certain biomolecules, wastes, and ions, it is the site of respiration and photosynthesis for some cells, the bacterial CM is a trilamellar structure. A central electron translucent layer sandwiched by two electron dense layers.

Table Contd...

Cell walls	A rigid layer external to the cell membrane that occurs in almost all cells except animal cells. It protects the cell from mechanical damage, osmotic imbalances and gives a shape to the cell. Cell walls differ greatly in structure and composition among different groups of organisms. Cell walls of plant cells is made up of cellulose. **Algal cell walls** comprise of microfibrillar polysaccharides cellulose or xylans with matrix polymers such as agar, alginate, carrageenan, and fucoidin. **Bacterial cell walls** comprise of a unique polymer peptidoglycan. In gram positive bacteria, the peptidoglycan or *murein* is thick bed like and contains teichoic acids embedded in it. In gram negative bacteria, the murein layer is very thin, instead an outer membrane consisting of lipopolysaccharide is seen. Some bacterial cell walls contain lipids (*Mycobacterium*). Still others lack a cell wall (*Mycoplasma*). **Fungal cell walls** contain different microfibrillar polysaccharides in different fungi, eg. Cellulose, chitin and other βglucans as inner layer and the outer layers consists of alkali soluble substances such as glycoproteins with mannose/xylose/galactose. In **yeast cell walls** chitin is in very small amounts, instead phosphomannoproteins and βglucans are present.
Capsule	A layer of polysachharide/protein immediately external to the cell-wall of some bacteria. Capsules may be *Microcapsules, Macrocapsules and Slime layers*. Macrocapsules are true capsules, thick and easily demonstrated by negative staining. Microcapsules are only visible under EM, or detected by serological technique like Quellung reaction. Slime layers are more diffused in the medium like a gelatinous bed for a group of cells. Capsules contain homopolysaccharide material (dextrans, cellulose, levan, or colominic acid) or heteropolysaccharide (alginate, colanic acid or hyaluronic acid). Capsules are permeability barriers, prevent desiccation, promote surface adhesion, escape phagocytosis and act as nutrient reserve.

Organelle	Structure and function
Protein Sorting, Transport and Protein synthesis	
Endoplasmic Reticulum (ER)	A protein containing lipid bilayer membrane system that spreads through the cell cytoplasm from nuclear membrane to form internal compartments and chambers. It is made up of tubules and sacs (*cisternae*). At some regions, the outer surface bears many ribosomes. These areas are called *rough ER*, against the *smooth ER* where ribosomes are absent. The lumina of the ER shows early stages of glycosylation of proteins – the addition of mannose and glucosamine residues. The rough ER functions in protein processing while the smooth ER has is involved in lipid metabolism.
Golgi complex	A bunch of flattened membranous vesicles (*cisternae*), oriented such that the *forming end* is nearer to the nucleus and the *distal end* is at the opposite face. Stacks of cisternae are referred to as *dictyosomes*. Functions include formation of lysosomes, formation of secretory vesicles, molecules from the lumen of the ER reach the Golgi apparatus through vesicles, which later fuse with the cell membrane for effective secretion.
Lysosmes	Closed membranous sacs of about 0.5μm containing several enzymes that can degrade a variety of bio molecules. The membranes are made of proton ATPase to maintain acidic (5.0) pH. The lysosomes fuse with food vacuoles and phagosomes to form *secondary lysosomes*. When digestion is complete, the entity remaining is called *residual body*.
Microsomes	Many minute heterogenous membranous vesicles in eukaryotic cells. They are obtained from rough and smooth ER and from cell membranes. They have several protein sorting and transport functions depending on the components within them.
Ribosomes	Large number of ribonucleoprotein organelles that occur on the ER, CM and freely in the cytoplasm. Mediate **protein synthesis**. Each ribosome is made of 2 subunits containing rRNA and Protein. Bacterial ribosomes have a sedimentation coefficient of 70S (30S & 50S subunits), while eukaryotic cells have 80S type ribosomes (60S & 40S). the mitochondrial ribosomes are of 78S (plants), 73S (yeast), 70S (algae) and 60S in mammalian mitochondria. The rRNA are also of various sedimentation coefficients: 16S RNA (in 30S subunit), 5S & 23S (in 50S subunit), rRNA has trole in translation , operon regulatory functions regulate biogenesis of ribosomes.

Organelle	Structure and function
Bioenergetics & Metabolism	
Mitochondrion	Semi autonomous, spherical, ovoid, calyciform or irregularly shaped. 1μm to 10μm in size, made up of two closed membranous sacs and amorphous *matrix that* contains DNA and enzymes of TCA cycle. The inner membrane shows fingerlike projections *cristae* and contains components of ETC, proton ATPases, nucleotides, inorganic phosphate & Calcium The outer membrane contains porins. Respiration and reactions of TCA occur within mitochondria. They are aptly known as the powerhouses of the cell.
Chloroplasts	Semi-autonomous plastids within photosynthetic eukaryotes. They are site for photosynthesis, biosynthesis of chlorophyll, and for certain proteins and lipids. The matrix (*stroma*) contains several flattened sacs (*thylakoids*) which occur in stacks called *grana*. The stroma contains DNA, plastoglobuli, ribosomes, proteins and enzymes of calvin cycle. The photosynthetic apparatus is present in the thylakoids along with chlorophyll, ETC components and ribosomes. Algal chloroplasts are variously shaped – discoid, spiral, oval, circular, etc.
Peroxisomes	Circular membranous microbodies in eukaryotic cells. Electron dense granular matrix if filled with various enzymes (peroxidase, catalase, D- amino oxidase, dehydrogenase and acyl transferase). They perform many functions: gluconeogenesis, photorespiration, oxidation of alcohol and fatty acids, purine metabolism and synthesis of plasmalogens.
Other plastids	All plastids contain the same genome as chloroplasts but differ in structure and function. The other plastids also contain DNA but lack thylakoid membrane system and photosynthetic apparatus. **Chromoplasts:** contain carotenoids, impart yellow, orange and red colours of flowers and fruits. Carotenoids include carotenes like *β carotene* and Xanthophylls like *antheraxanthin, diatoxanthin* and *fucoxanthin*. **Leucoplasts:** Non-pigmented plastids, store energy sources in non-photosynthetic tissues. **Elaioplasts & Amyloplasts:** Leucoplasts that store lipids and starch respectively. **Proplastids:** Small undifferentiated organelles in rapidly dividing cells. All plastids (including chloroplasts) develop from them. **Etioplasts:** when plants are kept in dark, the development of proplastids in leaves is arrested and non-functional etioplasts are formed. Etioplasts devlope into plastids when sunlight is returned to the plant.

Organelle	Structure and function
Nucleus	
Nucleus	A membrane limited body which contains chromosomes and one or more nucleoli. The double unit type nuclear membrane is perforated with many pores. Some cells may contain several nuclei (coenocytic) and many ciliate protozoa contain two nuclei larger macronucleus and smaller micronucleus. Nucleus is a repository for genetic information and acts as the cell's *control centre*. DNA replication, transcription and RNA processing, all takes place within the nucleus, except for gene expression (translation).
Nucleolus	A distinct membranous region within the nucleus in which rRNA are synthsised and assembled into ribonucleoprotein subunits of ribosomes. Normally nuceoli disappear during mitosis. They are absent in micronuclei.
Nucleoskeleton	Intranuclear framework analogous to cytoskeleton. Includes actin, myosin and points of anchorage for complexes in replication and transcription. Protects DNA polymerases from restriction endonucleases. Transport between nucleus and cytoplasm is controlled by actomyosin intranuclearly. The cytoskeleton and nucleoskeleton are connected at the nuclear periphery.
Nucleoid	Term used for various nuclear material like: bacterial chromosome, ribonucleoprotein material in retroviruses, mitochondrial or chloroplast DNA and electron dense core of peroxisomes, any of these may be referred to as nucleoid.

Table *Contd...*

Nuclear membrane	Two concentric membranes around the nucleus. The space between the two membranes is *perinuclear space* and the outer membrane is continuous with ER at certain junctions. The membrane is similar to ER membrane, rough with ribosomes. It provides stuctural framework to the nucleus, separates nucleus from cytoplasm, bears *nuclear pore complexes* which allow regulated exchange of molecules across the nucleus. Also plays critical role in regulation of gene expression.

Organelle	Structure and function
Cytoskeleton, motility and adhesion	
Cytoskeleton	A framework of proteinaceous tubules and fibrils that extend all over the cytoplasm, binding to the cell membrane and various organelles, undergoes continuous modification as and when the need arises. The main components include Microfilaments, Microtubules and Intermediate filaments. *Microtubules* are for cell structure location and orientation of organelles, and motility (axoneme in flagellum). *Microfilaments* are for intracellular movement, amoeboid movement and movement during phagocytosis. *Intermediate filaments* are for strengthening and may even extend from cell to cell in certain tissues.
Flagellum	A thread like appendage that projects from the cell surface. Responsible for motility in bacteria and certain protozoans, may singly (monotrichous), in tufts (lophotrichous), or both ends of the cell (amphitrichous) or all over the cell surface (peritrichous). Bacterial flagellum contains a protein *filament,* which rotates for motility, *hook* which connect the two ends of the flagellum and *basal body* which anchors in cell envelope and bears energy converting apparatus. It consists of the M, S, P and L rings in gram negative bacteria. The P and L ring usually are absent in gram positive bacteria.the rings are essential in the rotatory motion of the flagellum. In archeobacteria, the flagellum subunits (flagellins) are glycosylated and the filament is thinner. In eukaryotes, the flagella are of typical 10+2 pattern and bear minute scales or hairs, etc. and accordingly they are of types *tinsel, whiplash, smooth, hispid.*
Fimbriae	Thin proteinaceous filaments (up to 100nm) that arise from the surface of certain microbial cells, mainly on gram negative bacteria and cyanobacteria, but less common on gram positive bacteria. They are also seen on yeast such as *Candidda and Saccharomyces spp.* Fimbriae are made up of repeating subunits of *fimbrillins (pilins)* arranged helically, each containing non-polar amino acids. All fimbriae are classified into type I,II,III and IV. Functions: Adhesion to surfaces, cell to cell adhesion, adhesion to mucosal linings for pathogenicity, twitching motility of the cell, and many such functions.
Pili	Filamentous, proteinaecious plasmid encoded structures on the surfaces of F^+ gram negative bacteria. Made up of repeating subunits called *pilins* of 70 amino acids each, arranged in ring form. During conjugation, pili (sex pili) aid in formation of the conjugation canal. Some bacteriophages selectively bind to the pili to initiate infection (Eg. MS2, Qβ, f1, fd, and M13.) The exact role of other kinds of pili is not known.
Cilia	Short (7 to 10μm), hair or thread like appendages that extend from members of *Ciliophora.* Like *Paramoecium.* Ultrastructurally similar to eukaryotic flagella. Cilia show rhythmic movements by *effective strokes* followed by recovery strokes. They help in locomotion of the cell and also to drive food towards the cell.

Ultrastuctures of other Microbial cells

Cell Type	Ultrastructure – features
The Yeast & Molds	**Cell wall**: Fungal cell wall lacks cellulose. It contains Chitin, also called *fungal cellulose.* It is different from the insect chitin. **The Protoplast** (cell membrane + Cytoplasm): Lacks chloroplasts, but other organelles are present as in a typical eukaryotic cell. **Cell membrane:** Thin delicate membrane that is closely pressed against the wall. It has a typical *tripartite* structure with two electron dense layers and a less dense central region. Occasional invaginations of the membrane are called *Lomasomes.* **Cytoplasm:** Uniform homogenous matrix comprising of sap filled vacuoles, ER, mitochondria, Golgi apparatus, stored glycogen inclusions, oil droplets and secretory granules. **Flagella:** Although higher fungi lack flagella, lower fungi (*Phytophthora, Saprolegnia, etc.*) spores may bear one or two whip like flagella with typical 9+2 arrangement of microfibrils.

Table *Contd…*

Cell Type	Ultrastructure – features
Algae	**Cell wall:** Algae have cellulosic walls with microfibrillar architecture. Mannan and xylan are found in the walls of many sps. The outer cuticle in most algae is composed of proteins. **Cytoplasm:** the highly granular matrix of an algal cell contains large vacuoles, Golgi apparatus, ER, mitochondria, spherical electron opaque lipid bodies, pyrenoids and chloroplasts. **Choloroplasts:** They are discoid, parietal, reticular or cup shaped. The thylakoids are arranged as stacks of grana with intergrana spaces. Starch is synthesized in chloroplasts or separate amyloplasts. **Pyrenoids** are centers of carbon dioxide fixation within the chloroplasts. They contain high levels of ribulose-1,5-bisphosphate carboxylase/oxygenase. Pyrenoid bodies store starch.
Protozoa	Typical protozoan cell is made up of **cytoplasm** and **nucleus.** The cytoplasm is differentiated into **ectoplasm** (outer rim of cytoplasm along the edge responsible for protective, locomotive and sensory functions) and inner **endoplasm** (inner granular portion responsible for nutrition, excretion and reproduction). Amoebae show temporary protrusions of cytoplasm called **pseudopodia** responsible for locomotion. **Flagella** are long thin thread like filaments seen in Zoomastigophorans. **Cilia** are fine needle like filaments covering the entire surface of the body in ciliates. The **nucleus** consists of network of linin in the nuclear sap. The nucleus contains one or more nucleoli or a central **endosome** or **karyosome.** Extranuclear chromatin material is called **chromodia** or **chromatoid body**, found in certain cysts. In some, a **centrosome** is located just outside the nucleus or in the center of the karyosome whee it is called **centriole.** In ciliates two nuclei are present (**micronucleus and macronucleus)** some trypanosomes contain non nuclear DNA bodies in addition to the nucleus. these are called **kinetoplasts.**

Signal Transduction

- In bacteria, the *two component regulatory system* exists, that regulates gene expression in response to environmental signals such as change in extracellular osmolarity.
- The signal acts through *Signal Transduction Pathway*. The first component of the pathway is the histidine kinase (located in the cell envelope) the second component is the *Regulator protein* they work hand in hand and lead to regulate the right genes thereby controlling the transcription and responding to the signal.
- The signal directly regulates the kinase and kinase is the *sensor* (eg: in Osmoregulation). Some times, there may be a distinct sensor other than the kinase. The signal first acts on the sensor, which then regulates the kinase.
- *Integrins* are receptors that cause attachment of cells to extracellular matrix. They also interact with cytoskeleton and serve as receptors that activate intracellular signalling pathways.
- The components of the *cytoskeleton* also act as receptors and targets in signalling pathways, integrating cell shape and movement in response to other cellular responses.

Cell Cycle

- The sequence of events that occur in a cell right from its formation to the division of the cell to daughter cells is called *cell cycle*. Cell cycles are described both in prokaryotic and eukaryotic cells.
- The major steps include: initiation of DNA replication, replication of chromosomes, segregation of chromosomes, formation of septum, and cell division. In **prokaryotic cell cycle,** involves a growth period followed by division of the cell to give 2 daughter cells in which the same number of chromosomes

Exercises

1. Which of these is an uncommon feature between plant and animal cell   [   ]
   a. Lysosomes
   b. Centriole
   c. Filamentous cytoskeleton
   d. Mitochondria

2. A blastomere is   [   ]
   a. One of the cells formed by the cleavage of a fertilized egg
   b. Constricted region of a mitotic chromosome
   c. Sensory structures in the inner organelles
   d. A cell type that surrounds a developing oocyte.

3. Cytosol   [   ]
   a. Cytoplasmic contents excluding membrane bound organelles
   b. Cytoplasmic contents including the membrane bound organelles
   c. Inner contents of membrane bound organelles
   d. Contents of a cell excluding its cell wall and underlying membrane

4. Cisternae are associated with   [   ]
   a. Golgi complex
   b. Endoplasmic reticulum
   c. Both a & b
   d. Neither a nor b

5. Cilia comprise of a core bundle of microtubules   [   ]
   a. TRUE
   b. FALSE

6. A motor protein that uses ATP to drive movements along actin filaments   [   ]
   a. Myofilaril
   b. Myosin
   c. Myeiln sheath
   d. Microvillus

7. Organelles that use molecular oxygen to oxidize organic molecules   [   ]
   a. Endosomes
   b. Mitogens
   c. Peroxisomes
   d. Sarcomeres

8. What is FALSE about "Plastids"   [   ]
   a. They are often pigmented
   b. They are bounded by double membrane
   c. They carry their own DNA
   d. They are found in plant & animal cells

9. Mast cells are closely related to   [   ]
   a. Germ cells
   b. Blood basophils
   c. Muscle precursor cell
   d. Nerve neurons

10. The cell that is connected to an oocyte & provides macromolecules to the growing oocyte   [   ]
    a. Helper cells
    b. Organizer cell
    c. Nurse cell
    d. Satellite cell

11. The cell that aids in nourishing the developing sperm cell   [   ]
    a. Stem cell
    b. Steroli cell
    c. Spermann`s cell
    d. Promotor cell

12. Schwam cells are responsible for   [   ]
    a. Forming Myelin sheath in the nerve cells
    b. Differentiation of blood leukocytes
    c. Protection from oncogens
    d. Synthesis of membrane bound proteins

13. GTP binding proteins that help relay from cell surface receptors to the nucleous   [   ]
    a. S proteins   b. Src proteins   c. Ras proteins   d. G proteins

14. Chromatin includes   [   ]
    a. DNA material
    b. DNA & histones
    c. DNA & non histone proteins of the nucleus
    d. One copy of chromosome formed by DNA replication and is still joined at the centromere

15. Programmed cell death is referred to as   [   ]
    a. Aster   b. Apoptosis   c. Autocatalysis   d. Zonula adherence

16. Glial cells are associated with   [   ]
    a. nervous system   b. lymphatic system
    c. reproductive system   d. enteric system

17. Green alga used in the studies of actin based cytoplasmic streaming   [   ]
    a *Volvox*   b. *Chondrus*   c. *Nitella*   d. *Spirogyra*

18. Striations in the skeletal and cardiac muscle cells are due to   [   ]
    a. microfilaments   b. stereocilium   c. sarcomeres   d. zippers

19. Lectins may be abundantly derived from   [   ]
    a. Cyanobacteria   b. plant seeds   c. animal muscles   d. all of these

20. The dense covering of microvilli on the surface of epithelial cells in the
    kidney and intestine are referred to as   [   ]
    a. gap junctions   b. brush border   c. zootuft   d. rough coat

21. Histones are associated with DNA in   [   ]
    a. eukaryotic cells   b. prokaryotic cells
    c. both eukaryotic & prokaryotic cells   d. Cyanobacterial cells

22. Ribosomal RNA is transcribed in the   [   ]
    a. Nucleus   b. Nucleolus   c. Mitochondria   d. Cytosol

23. Tracing the ancestry of individual cells in a developing embryo   [   ]
    a. Archeal cytology   b. Evolutionary cytology
    c. Pre-cell analysis   d. Lineage analysis

24. A medium that causes water to move out of a cell due to osmosis   [   ]
    a. Hypertonic   b. Hypotonic
    c. Hyperplasmic   d. Hypoplasmic

25. What is FALSE about hydrophobic molecules.   [   ]
    a. They are non - polar molecules
    b. They do not form bonding interactions with water molecules.
    c. They dissolve readily in water.
    d. a, b, and c are false.

26. Transmembrane proteins involve in the adhesion of cells to the extracellular matrix   [   ]
    a. Integrins   b. Intrinsic proteins   c. Histones   d. Dyneins

27. Adipocytes are   [   ]
    a. defense cells   b. membrane cells   c. fat cells   d. connective cells

28. The core of cilia and flagella in eukaryotic cells constitutes of   [   ]
    a. Cadherins   b. Axonemes   c. Cdk proteins   d. Dyneins

29. Proteins that prevent misfolding of other proteins that may produce inactive states   [   ]
a. Chaperones   b. Chromaffins   c. Connexons   d. Coiled coils

30. The proteins and lipids made in the ER are modified and sorted in   [   ]
a. Nucleus   b. Lysosomes   c. Golgi apparatus   d. Mitochondria

31. Which of these filaments of cytoskeleton are the finest in terms of diameter   [   ]
a. Microtubules   b. Actin filaments
c. Intermediate filaments

32. Vacuoles in a plant cell aid in   [   ]
a. Space filling   b. Intracellular digestion
c. Both a & b   d. Neither a nor b

33. The nuclear contents communicate with the cytosol by means of   [   ]
a. ER   b. Nuclear pores   c. Nucleolus   d. Chromatin

34. Lysosomes are associated with   [   ]
a. Intracellular digestion   b. Generation of hydrogen peroxide
c. Intracellular transport of material   d. Inter cellular transport of material

35. The ER is associated with   [   ]
a. Synthesis and transport of lipids
b. Synthesis and transport of membrane proteins
c. both a & b
d. Neither a nor b

36. Lipid metabolism is a function of   [   ]
a. Golgi apparatus   b. Cytoskeleton   c. Smooth ER   d. Rough ER

37. Chloroplasts in plant cells are thought be a result of symbiosis with   [   ]
a. Cynobacteria
b. Archaeobacteria
c. Purple non sulfur photosynthetic bacteria
d. Purple sulfur photosynthetic bacteria

38. Range of a typical bacterial cell size   [   ]
a. 0.2 - 0.8µm   b. 2 - 8µm   c. 20 - 80µm   d. 200 - 800µm

39. Which of these cells are spiral?   [   ]
a. *Leuconostoc*   b. *Bacillus*   c. *Borrelia*   d. *Trypanosoma*

40. Which of these renders protection from phagocytosis   [   ]
a. Glycocalyx   b. Pilins   c. Flagellins   d. Phospholipids

41. Prokaryotic flagella show   [   ]
a. Rotating movement   b. Whiplash movement
c. Harmonic movement   d. Contractions and relaxations

42. Spirochaetes move by means of   [   ]
a. Pseudopodia   b. Pili   c. Flagella   d. Axial filaments

43. Pick out the TRUE statement in case of eukaryotic cell   [   ]
a. Their flagella consists of multiple microtubules
b. Their plasma membrane consists of sterols and carbohydrates
c. Show presence of cytoskeleton
d. a, b & c are TRUE

44. Sphaeroplasts are formed when lysosome acts on
    a. Cell walls of Gram positive bacteria
    b. Cell walls of Gram negative bacteria
    c. both a & b
    d. Acid fast bacteria

45. Energy is utilized to modify chemicals and transport them across the membrane in
    a. Passive diffusion
    b. Facilitated diffusion
    c. Active transport
    d. Group translocation

46. Dipicolinic acid is characteristically found in
    a. Fungal spores
    b. Bacterial endospore
    c. Gram negative cell walls
    d. all

47. Pick the FALSE statement about Gram positive bacteria
    a. Higher lipid content than Gram negative bacteria
    b. Higher resistance to physical disruption
    c. Higher susceptibility to lysozyme
    d. Higher susceptibility to penicillin & sulfonamide

48. Which of these is/are present in Gram negative cell wall, but absent in Gram positive
    a. Periplasmic space
    b. Outer membrane
    c. Teichoic acids
    d. a & b only

49. The return of endospore to its vegetative state is called
    a. *Encystation*
    b. *Desporulation*
    c. *Germination*
    d. *Reabsorption*

50. Which of these bacteria are square cells
    a. *Leptospira*
    b. *Haloarcula*
    c. *Mycoplasma*
    d. *Rhodobacter*

51. Which of these bacteria are star shaped
    a. *Brevibacterium*
    b. *Dienococcus*
    c. *Neisseria*
    d. *Stella*

52. Metachromatic granules are collectively known as
    a. Volutin granules
    b. Burdan`s granules
    c. Sudan`s granules
    d. None of these

53. In the presence of Iodine, glycogen granules appear
    a. Blue
    b. Reddish brown
    c. Pitch black
    d. Green

54. Sulfur granules in sulfur bacteria serve as
    a. Defense against phagocytosis
    b. Sites for protein synthesis
    c. Energy reserves
    d. All

55. The common lipid storage material in bacterial cells
    a. Fatty acids
    b. Poly β hydroxy butyrate
    c. Poly phosphates
    d. Glycerols

56. Carboxysomes are observed in
    a. Cyanobacteria
    b. Iron bacteria
    c. Archaeo bacteria
    d. Spirochaetes

57. Cells of *Azotobacter* are known for the presence of
    a. Metachromatic granules
    b. Sulfur granules
    c. Lipid inclusions
    d. Caboxysomes

58. Lopotrichous flagellation
    a. Tufts of flagella at both ends of the cell
    b. A tuft of flagella at one pole of the cell
    c. Flagella distributed over the entire cell
    d. Single polar flagellum

59. Outer membrane is present in the cell wall of   [   ]
    a. Gram negative bacteria
    b. Gram positive bacteria
    c. Acid fast bacteria
    d. All

60. Sodium azide resistance is shown by   [   ]
    a. Gram negative bacteria
    b. Gram positive bacteria
    c. Acid fast bacteria
    d. All

61. Endotoxins are primarily produced by   [   ]
    a. Gram negative bacteria
    b. Gram positive bacteria
    c. Acid fast bacteria
    d. All

62. Each phospholipid molecule in the plasma membrane is composed of   [   ]
    a. Polar head & non polar tail
    b. Non polar head & polar tail
    c. Polar head & polar tail
    d. Non polar head & non polar tail

63. Magnetosomes are inclusions of   [   ]
    a. Fe_3O_4
    b. Fe deposits
    c. $FeSO_4$
    d. Ferrous ammonium sulfate

64. Magnetosomes in bacteria   [   ]
    a. Help in destroying other competing bacteria
    b. Protect cell against H_2O_2 accumulation
    c. Act as antigens
    d. Help the cells to attract each other and form aggregates

65. Extracellular signalling molecules   [   ]
    a. Hormones
    b. Growth factors
    c. Neuro transmitters
    d. All

66. What is FALSE about Ras protein   [   ]
    a. Occurs in signal transduction pathways
    b. Ligand binding causes conversion of inactive Ras-GDP to active Ras-GTP
    c. Short stretches of RNA used for DNA synthesis
    d. Active forms of Ras are oncogenes

67. Signal transduction pathways are triggered by   [   ]
    a. Transgenes
    b. Binding of signalling molecules to each other
    c. Binding of signaling molecules to a receptor
    d. All

68. Intracellular signaling molecules   [   ]
    a. G-proteins
    b. Ras proteins
    c. Both
    d. None

69. A short amino acid sequence that directs a protein to a specific location within the cell   [   ]
    a. G-protein
    b. Silencer sequence
    c. Motivator peptide
    d. Signal peptide

70. The Cell Theory was proposed by   [   ]
    a. Robert Hook
    b. Cajal
    c. Schleiden & Schwann
    d. Miller & Urey

71. The internal organelles can be appreciated in the cells by   [   ]
    a. Bright field microscopy
    b. Dark field microscopy
    c. Fluorescence microscopy
    d. Phase contrast microscopy

72. The technique employed to know the site of protein synthesis within a cell [   ]
    a. Cell fractionation
    b. Autoradiography
    c. Scanning Electron Microscopy
    d. Micrometry

73. Sonication and homogenization leads to [   ]
    a. Preparation of an organic soup of disrupted cell organelles
    b. Cell lysis without disrupting structural integrity of cell organelles
    c. Cell fractionation without lysis of the cells
    d. None of the above

74. Cell mass obtained by removing the cell wall [   ]
    a. Meristem
    b. Protoplast
    c. Callus
    d. Mycoplasma

75. Growing complete plants from explant tissue [   ]
    a. Callus culture
    b. Organ culture
    c. Protoplast culture
    d. Meristem culture

76. Regeneration of plant from pith tissue [   ]
    a. Callus cuture
    b. Organ culture
    c. Protoplast culture
    d. Meristem culture

77. Treatment of cells with cellulases and pectinases in solutions containing osmotic stabilizers yields [   ]
    a. Stem cells
    b. Callus
    c. Protoplast
    d. Bergmann's cells

78. Somatic cell hybridization [   ]
    a. Removal of nucleus from a cell
    b. Fusion of two different protoplasts to a give rise to new variety of plant
    c. Aseptic culture of organisms on nutrient medium
    d. Cultivation of protoplasts in liquid media

79. Murashige-Skoog medium [   ]
    a. Culture medium for tissues
    b. Culture medium for fungi
    c. Culture medium for photosynthetic bacteria
    d. Culture medium for algae

80. Match the stains used to detect the respective cell constituents [   ]
    A. Bromophenol blue   i. nucleic acids
    B. Sudan Black B   ii. Lipids
    C. Fuelgen stain   iii. Proteins
    D. Periodic Acid Schiff (PAS)   iv. polysaccharides

    a. A-ii, B-iii, C-i, D-iv
    b. A-iii, B-ii, C-i, D-iv
    c. A-iv, B-i, C-iiii, D-ii
    d. A-i, B-iv, C-ii, D-iii

81. Cold phosphate buffer (pH 8.0), 0.35M NaCl, Mortar, Pestle, Centrifuge, etc are required for the isolation of [   ]
    a. Mitochondria
    b. Inclusion bodies
    c. Chloroplasts
    d. Nucleus

82. Which of these requires highest force of Density Velocity Centrifugation for separation [   ]
    a. Endoplasmic reticulum
    b. Nuclei
    c. Chloroplasts
    d. Mitochondria

83. Tumors originating from ectoderm/endoderm give rise to   [   ]
    a. Sarcomas   b. Carcinomas   c. Benign tumors   d. Malignant tumors

84. What is FALSE about benign tumors   [   ]
    a. They are always locoalised
    b. They never eastablish growth in other parts of the body
    c. They are derived from large undifferentiated aberrant cell
    d. They are generally harmless and can be removed by surgical methods

85. Tumors originating from mesoderm gives rise to   [   ]
    a. Sarcomas   b. Fibromas   c. Carcinoma   d. Metastasis

86. The free radiacal theory and somatic mutation theory are related to   [   ]
    a. Cell ageing   b. Cell permeability
    c. Cell heredity   d. Cell differenciation

87. In a eukaryotic cell cycle, the phase reflecting active RNA and protein synthetic stag is   [   ]
    a. G_1 Phase   b. M phase   c. S phase   d. G_2 phase

Answer Key and Validation

The Cell

1. b	16. a	31. b	46. b	61. a	76. a
2. a	17. c	32. c	47. a	62. a	77. c
3. a	18. c	33. b	48. d	63. a	78. b
4. c	19. b	34. a	49. c	64. b	79. a
5. a	20. b	35. c	50. b	65. d	80. b
6. b	21. a	36. c	51. d	66. c	81. c
7. c	22. b	37. a	52. a	67. c	82. a
8. d	23. d	38. b	53. b	68. c	83. b
9. b	24. a	39. c	54. c	69. d	84. c
10. c	25. c	40. a	55. b	70. c	85. a
11. b	26. a	41. a	56. a	71. d	86. a
12. a	27. c	42. d	57. c	72. b	87. a
13. c	28. b	43. d	58. b	73. b	
14. c	29. a	44. b	59. a	74. b	
15. b	30. c	45. d	60. b	75. d	

7. Peroxisomes contain some enzymes that produce H_2O_2 and some others that degrade H_2O_2.

16. Glial cells are supporting cells of the nervous system. Eg: dendrocytes, astrocytes and schwann cells.

19. Lectins are sugar binding proteins derived from plant seeds and used to detect / purify glycoproteins on cell surface.

24. Hypertonic media contain high conc of solutes that causes water to move out of the cell.

28. Axoneme is bundle of microtubules and associated protein in the core of cilia/flagella and is responsible for their movements.

31. Microtubules = 25nm, Actin filaments = 8nm, intermediate filaments = 10nm in diameter.

40. Glycocalyx is the gelatinous capsule or slime layer surrounding cells of certain bacteria.

44. In the presence of lysozyme, gram positive cell walls are destroyed and the remnant cellular content is called protoplast

47. Lipid and lipoprotein content in Gram negative cell is more than in Gram positive cells

56. Nitrifying bacteria, Cyanobacteria & *Thiobacillus* contain carboxysomes, that contain carboxylase for using CO_2 as sole source of carbon

57. *Bacillus, Mycobacterium, Spirillum* and *Azotobacter* species are known for the presence of lipid inclusions.

62. Each phospholipid molecule contains a hydrophilic polar head of phosphate group and glycerol and a hydrophobic non polar tail of fatty acids.

66. Short stretches of RNA used for DNA synthesis are referred to as RNA primers. Ras protein is a monomeric guanine nucleotide binding protein that occurs in signal transduction pathways.

69. Signal peptides are also called Signal sequences.

84. Benign tumors consist of well differentiated cells similar to the tissue of origin.

87. In a cell cycle, the S is synthetic phase, M is the mitotic phase and G_2 is the active photosynthetic phase.

4

Methods in Microbiology

Retrace your subject...
Sterilization and Disinfection, Stains, Staining Techniques, Principles of Microscopy, Culture Media, Culture Techniques, Chromatographic Techniques, Centrifugation Techniques...

Sterilization and Disinfection

- The sterilization is the process of freeing any article or substance from all microorganisms and other forms of life inclusive of pathogens, non-pathogens and even spores.

- Sterilization can be brought about by use of Physical and Chemical methods. Physical methods include **Sunlight, Drying, Heat, Filtration and Radiation sterilization.**

Method	Principle and Process	Applications
Physical Methods		
Sunlight	Sunlight is bactericidal and plays role in spontaneous sterilization that occurs under natural conditions. The bactericidal activity is due to the combined effect of UV rays and heat rays.	Exposure of dryfruits, fluor, pulses and grains to sunlight is known to extend their shelf-life. However sterilization is not effective as all microbes do not get killed.
Drying	Removal of water from the microbial cells causes death due to dehydration and inactivation of enzymes. Spores remain unaffected by drying.	Natural drying process is widely used to control microbial growth in many foods, metal or plastic articles so as to reduce surface moisture that encourages microbial growth. Mere drying cannot be employed for actual sterilization process.
Moist heat methods		
Boiling	Vegetative cells can be killed by boiling at 90^0C to - 100^0C for 20-30min. Spores remain alive.	Water is subjected to boiling inorder to render it safe for drinking.
Autoclaving	Employs steam under pressure. An autoclave is the most practical and dependable equipment for sterilization. In an autoclave, it is not the pressure that kills the organisms but due to the temperature of the steam. **Principle**: Water boils when its vapor pressure equals to the surrounding atmosphere. The condensed water ensures the moist condition for killing Microbes. Usually all mesophillic and sporeforming bacteria are killed.	Many media solutions, discarded culture and contaminated materials are sterilized in the apparatus Generally the autoclave is operated at pressure of 15 lbs per square inch at 121^0C for 15 minutes.

Table Contd...

Method	Principle and Process	Applications
Tyndallization	Employs steam at room temp to kill microbes. The media is steamed on a boiling water bath for one to two hours followed by incubation at 37^0C. This is repeated for three consecutive days to ensure germination & killing of all spores.	Used for sterilizing high temp sensitive media such as those containing egg or serum.
Dry heat methods		
Incineration	Infected materials are completely burnt into ashes.	Used for rapidly destroying material such as spoiled dressing material, animal carcasses, pathological material etc. The Hindus burn their dead which is a form of incineration.
Hot air oven	It is a thermostatically controlled equipment, That circulates dry air. Killing is due to dehydration and charring.	It is most widely used in the microbiology labs for sterilizing glassware, oils, waxes and also for drying purposes. The temperature is maintained at 160^0C for 2h.
Filtration	It is a method used to free the thermolabile liquids from Microorganisms. Microorganisms are not killed but only removed by filtration. The filters employed are bacteria proof or virus proof depending upon their pore size characters.	Some important biological filters include: Berkefeld or Mandler filter (diatomaceous earth filters), Chamberland-Pasteur filter (porcelian filters) and Seitz filter (asbestos filters)
Radiation Sterilization		
UV	UV rays of $260A^0$ are bactericidal and employed as powerful sterilizing agent. Killing is due to damage of DNA by formation of thymine dimers.	U.V. rays are used for surface disinfection hospitals, food processing units and tables of the veterinarian.
Gamma Radiation	Ionizing radiations with high penetration power. Also called cold sterilization, because gamma radiations directly act on DNA with out much increase in temperature. Large commercial plants use γ radiations for sterilizing syringes, greases, oils etc.	Gram negative bacteria are more susceptible. Yeast and moulds are adequately destroyed. *Dienococcus radiodurans* can resist these rays, as it regains its DNA.
Chemical Sterilization/Disinfection		
Ethanol	75% ethyl alcohol is a very good killing agent and is widely used to disinfect tables, steel equipment and glassware. Death is due to protein coagulation and destruction of lipids. 100% ethanol however cannot penetrate the cells and hence is inefficient in killing.	
Ozone	Ozone is a poisonous unstable gas that is released in ozonation tanks of water to kill all forms of microorganisms including prions. Ozonator units can be easily fitted and find wide application in packaged drinking water plants and in sterilizing air. Ozone being strongly oxidizing agent, first damages the cell membrane structure and then destroys the inner lipoprotein and lipopolysaccharide, changes permeability and causes cytolysis and cell death. In case of viruses, inactivation is due to the oxidation effect that breaks the viral RNA or DNA..	
Bleach	Household bleach consists of 5.25% sodium hypochlorite. It is widely used to disinfect surfaces particularly toilets. Bleach kills microbes almost immediately. However, it is recommended to leave a contact period of 20min for effective sterilization. Bleach decomposes over time when exposed to air, and hence only freshly prepared solutions should be used.	

Table *Contd...*

Chemical Sterilization/Disinfection	
Glutaraldehyde and Formaldehyde	These are liquid sterilizing agents that act on microbes if the immersion time is sufficiently long. Spores take more than 12h of contact period to be killed. Glutaraldehyde and formaldehyde are volatile, and toxic by both skin contact and inhalation. Formaldehyde is also used as a gaseous sterilizing agent. Many vaccines, such as the original Salk polio vaccine, are sterilized with formaldehyde.
Hydrogen Peroxide	Hydrogen peroxide is relatively non-toxic when diluted to low concentrations, such as the 3 % solution. Hydrogen peroxide is strong oxidant due to which it kills several pathogens and it is used to sterilize heat sensitive articles. Peroxide kills in a very short acting time. However, hydrogen peroxide is a skin and eye irritant.
Ethylene oxide	Gas is commonly used to sterilize objects sensitive to temperatures greater than 60 °C such as plastics, optics and electrics. Ethylene oxide treatment is generally carried out between 30 °C and 60 °C with relative humidity above 30% and a gas concentration between 200 and 800 mg/L for at least three hours. Ethylene oxide penetrates well, moving through paper, cloth, and some plastic films and is highly effective. Ethylene oxide sterilizers are used to process sensitive instruments which cannot be adequately sterilized by other methods. EtO can kill all known viruses, bacteria and fungi, including bacterial spores and is satisfactory for most medical materials, even with repeated use. However it is highly flammable, and requires a longer time to sterilize than any heat treatment. The process also requires a period of post-sterilization aeration to remove toxic residues. Ethylene oxide is the most common sterilization method, used for over 70% of total sterilizations, and for 50% of all disposable medical devices. The two most important ethylene oxide sterilization methods are: (1) the gas chamber method and (2) the micro-dose method. To benefit from economies of scale, EtO has traditionally been delivered by flooding a large chamber with a combination of EtO and other gases used as dilutants (usually CFCs or carbon dioxide). This method has drawbacks inherent to the use of large amounts of sterilant being released into a large space, including air contamination produced by CFCs and/or large amounts of EtO residuals, flammability and storage issues calling for special handling and storage, operator exposure risk and training costs Because of these problems a micro-dose sterilization method was developed in the late 1950s, using a specially designed bag to eliminate the need to flood a larger chamber with EtO. This method is also known as gas diffusion sterilization, or bag sterilization. This method minimizes the use of gas.

Stains	
Staining	A technique in which cells or thin sections of tissues are immersed in one or more coloured dyes (stains) to make them more clearly visible through a microscope. Staining enhances the contrast between the the cells and their background in a brightly illuminated microscopic field.
Stains	Stains are aniline type synthetic chemicals that are of acidic or basic in nature. They are with a positive and negative ion.
Acidic Stains	If the colour comes from the negative ion (organic anion), the stain is described as acidic. Acidic stains such as eosin, erythrosine and acid fuchsin have an affinity for cytoplasmic material.
Basic Stains	If the colour comes from the positive ion (organic cation), the stain is described as basic. Basic stains such as Crystal violet, basic fuchsin and haematoxylin have an affinity for nucleic acid material. Use of basic stains is recommended for staining bacterial cells for satisfactory results.

Table *Contd...*

Stains	
Neutral stains	Neutral stains have a coloured cation and a coloured anion; an example is Leishman's stain. Thus it can color the different cell components differently. Cell constituents are described as being acidophilic if they are stained with acidic component, basophilic if receptive to basic component of the stain and neutrophilic if receptive they are to neutral dyes.
Vital Stains	Vital stains are used to stain live cells or their constituents without harming them. Eg. Trypan Blue, highly diluted (1:500 000) crystal violet and saffranin.
Romanowsky stains	The Romanowsky stains are preparations of a combination of eosinate (reduced eosin) and methylene blue. Sometimes the oxidation products of methylene blue azure A and azure B may be employed. Common variants include Wright's stain, Jenner's stain, Leishman stain and Giemsa stain. They are used to examine blood or bone marrow samples.
Counter staining	Counterstaining involves the use of two or more stains in succession, each of which colours different cell type or tissue constituents. For example, in gram staining, crystal violet is the primary stain that stains Gram-positive as well as Gram negative bacteria, but after decolorising, a counterstain such as safranin or basic fuchsin is applied which stains the colorless cells, allowing the identification of Gram-negative bacteria as well.
Electron stains	These are used in the preparation of material for electron microscopy, are described as electron-dense as they interfere with the transmission of electrons. Eg: Lead citrate, phosphotungstic acid (PTA), and uranyl acetate (UA).

Staining Techniques

Technique	Principle
Simple Staining Simple staining is a widely used procedure to study the morphological features and to have a general idea of the overall load of cells in the sample.	When an evenly spread smear of bacterial cells is flooded with a stain such as crystal violet or basic fuchsin for a brief period (1min), the cells take up the stain and can be easily observed under high power and oil immersion objectives of a compound microscope. Since only one stain is used for the purpose, the technique is referred to as simple staining.
Negative staining In this technique, the entire background is made to appear dark but for the cells, they cannot miss the eye.	Usually bacterial cells bear a negative charge on them. They cannot take up acid dyes which are also negatively charged. Hence the dye remains in the background and the cells remain unstained. Nigrosin and India ink serve as negative stains. [Negative staining is a modified version of observing specimen under the light microscope.]
Gram's staining: Gram's staining is one of the most commonly practiced staining procedures in the microbiology laboratory. This technique categorizes bacteria broadly into two groups viz., Gram positive and Gram negative.	a. Both Gram positive as well as Gram negative cells take up the deep violet color of the primary stain crystal violet. b. When treated with a mordant Gram's iodine, a crystal violet-iodine complex is formed resulting in deep purple color in all the cells. Subsequent treatment with the decolorizer alcohol is the differentiating step. The alcohol quickly removes the crystal violet iodine complex from the Gram negative cells with ease. This is attributed to the dissolving of excess lipids present in the outer membrane of Gram negative bacteria. The complex takes a longer time to be removed in Gram positive cells due to thick and impermeable peptidoglycan layer in their cell walls. c. Thus, controlled treatment of alcohol for a limited time period decolorizes all the Gram negative cells while the Gram positive ones retain the purple color. d. A counter stain such as safranine is then used to stain the colorless Gram negative cells red.

Table *Contd…*

Technique	Principle
Flagella staining Certain special techniques are employed to demonstrate flagella under the compound microscope.	Bacterial flagella are so fine that they can be barely seen when stained by ordinary methods of staining. In order to be reasonably visible, the flagella are first thickened several folds by superficial deposition of the stain. Tannic acid treatment allows deposition of stain on the surface of the flagella, and the flagella appear thick and visible.
Capsule staining The extra cellular gelatinous material synthesized by certain bacteria is known as capsule. Capsules are polysaccharides, polypeptides or lipoproteins.	When cells stained with 1% crystal violet are washed for a limited time with 20% $CuSo_4$ solution, the capsule gets decolorized first. The cells and background still remain colored. The cells appear deep blue and capsule appears colorless or light blue. *Note:* Capsules can also be demonstrated by negative staining.
Endospore Staining: Schaeffer and Fulton method	Bacterial endospores are difficult to stain, but once stained; they retain the stain and resist decolorization treatments. The vegetative cells take up the counter stain.
Staining of Metachromatic granules/volutin granules: Albert-Laybourn method	Volutin granules in the cells have an affinity for Toluidine blue in Albert's stain. They stain metachromatically to produce a dark red-purple color. The rest of the cell on the other hand stains light green. Cells that lack volutin granules appear completely light green in color
Lipid Staining: Burdon's method	Lipids in the cell take up Sudan Black stain with ease and resist treatment with xylene. The cytoplasm is counter stained with saffranin
Staining of Spirochete bacteria: Fontana's silver impregnation method	Spirochetes are too thin and cannot be observed by ordinary Gram's staining. The cells are therefore artificially thickened by use of tannic acid and silver impregnation. Heat fixing is avoided as it distorts the delicate cell structure.smears are fixed with formalin fixative.
Acid Fast Staining: Acid fast staining or Ziehl Neelson Carbol Fuchsin (ZNCF) staining is carried out to observe Acid Fast Bacteria (AFB).	Presence of mycolic acids and waxes in their cell walls hinder the intake of normal stains. Treatment with concentrated stain of carbol fuchsin with heat allows staining the cells. Cells once stained do not easily decolorize. *Mycobacterium tuberculosis* resists decolorisation with 20% H_2SO_4, while *Mycobacterium leproae* resists decolorisation with 5% H_2SO_4 (Hence the name acid fast). When a smear of sputum/pus from a nodule is subjected to heating with ZNCF stain, all the cells (acid fast and the non acid fast) are stained red. The mycolic acids and waxes in the cell walls of acid fast bacteria form a complex with the carbol fuchsin that resists decolorisation with acid. A counter stain such as methylene blue is used to stain the non-acid fast cells after decolorisation.

Principles of Microscopy

Technique	Principle & Applications
Light Microscopy	The technique uses ordinary sunlight or electric bulb as the source of illumination. It employs sets of glass lenses through which light is allowed to pass across the specimen. The specimen is magnified several folds up to 1000 times its original size. The major types of modifications of light microscopy include: bright field, dark field, phase contrast and fluorescence microscopic techniques. The magnification and clarity depends upon the numerical aperture and resolving power of the system.
Bright field Microscopy	<ul><li>The background is brightly illuminated and the specimen appears dark.</li><li>To create good contrast, the specimen is stained using appropriate stains.</li><li>Widely used to study morphology of cells.</li><li>The basic caricature comprises of a mirror, condenser, stage, objective and eyepiece</li></ul>

Table Contd...

Technique	Principle & Applications
Dark field Microscopy	• The background is completely dark and the specimen appears brightly illuminated. • A metal dark field stop is fitted inside the condenser such that a hollow cone of light passes through the condensor, across the specimen • Only light passing through the specimen is diffracted into the objective. Hence the background appears totally dark • Is used to observe morphology of cells even without staining. Hence live cells can also be observed comfortably.
Phase contrast Microscopy	• The background is neither totally dark not totally illuminated. • The specimen is variedly illuminated, so that even the inner cell components can be appreciated. • The technique employs creating a ¼ shift in phase of incident light. • The phase contrast is created by fitting an annular diaphragm in the condenser and a phase shifting plate in the eyepiece. • There is no need to stain the specimen and the cells can be seen in living condition.
Fluorescence Microscopy	• The background is totally dark and the specimen glows with fluorescent color. • The source of ilumination is a strong mercury bulb. • The strong white light is allowed to pass through a monochromator so that only UV light passes through it. • Such a blue light is passed through the specimen that is stained with a fluorescent dye such as Acrydine orange. The dye fluresces with a green color under the influence of blue light. • A UV barrier is placed in the eyepiece to protect the eyes from harmful UV light. • Thus, the background appears dark and the cells are fluorescent.

Electron Microscopy (EM)

• Magnifying capacity of the microscopic system is dependent on the wavelength of light. The shorter the wavelength the better is the magnification and resolution.

• In electron microscopy, the source of illumination is an electron beam that has a very short wavelength to enable high magnification with excellent resolution.

• The electron rays are focused by means of electromagnetic lenses placed in a vacuum system. The source of electrons is a fine heated tungsten filament.

• The observation of image is either on the photographic plate or on a computer screen.

• **The cells have to be observed in a killed condition and can never be seen alive under the EM.**

Transmission Electron Microscopy (TEM)	• When the beam of electrons is cast upon the specimen, some of them are transmitted across the specimen, while most are reflected back. • If the image is formed using the transmitted electrons, the technique is said to be TEM. The image is usually formed on a photographic plate. • The specimen requires prior preparation for mounting on the ultra fine grid of the TEM. • Some of the specimen preparation methods include: Ultramicrotomy, Negative staining, Shadow casting and freeze etching. • Since the specimen is sectioned, the inner contents of the cell can be viewed in TEM
Scanning Electron Microscopy (SEM)	• In SEM the surface of the specimen is scanned by means of a scanning coil and the electrons that are scattered from the specimen's surface are used in the formation of the image. • The electrons are passed through scintillation units and amplifiers to obtain corresponding impulses. • The impulses in turn are passed through a picture tube and the image is formed on the TV or computer screen. The image has a 3D appearance and shows minute surface details. • The inner components of a cell however, cannot be observed in SEM as the specimen is neither sectioned nor fractures as in the case of TEM.

Culture Media

Type of Media	Description
Culture Media	A formulation made up of organic and/or inorganic chemicals dissolved in distilled water for the purpose of cultivating microorganisms in the lab is called a culture medium. Culture media are either liquid (Broth Media) or solid (Agar Media). All culture media may be categorized as synthetic or complex media.
Synthetic Medium	A defined medium whose exact composition is known and reproducible is a synthetic medium. These media are usually made up of inorganic components. Eg. Minimal medium, Media for cultivation of Chemolithotrophs and phototrophic bacteria.
Complex medium	A medium comprising of such complex ingredients as peptone, beef extract, blood, serum and other components whose exact makeup is not reproducible is a complex medium. They are used for cultivation of heterotrophic organisms. Eg: Nutrient Agar, MacConkey Agar, etc.
Differential Medium	A type of complex medium that allows growth of different groups of bacteria in a visually different manner. Eg: MacConkey agar differentiates lactose fermenting bacteria (pink colonies) from lactose non-fermenting bacteria (colorless to yellow colonies).
Selective medium	A medium that allows the growth of only the desired group of organism and suppresses or inhibits growth of undesirable organisms from a sample of mixed microbial population is called selective medium. Eg: MacConkey agar is a selective medium as it allows growth of gram negative bacteria and inhibits gram positive bacteria due to the presence of bile salt in it. Saboraud Dextrose agar is also a selective medium as it allows growth of fungi selectively. Bacteria are inhibited due to acidic pH and presence of high conc. of dextrose.
Enriched medium	A medium that contains extra nutritious components such as blood, serum or brain heart infusion to allow growth of certain bacteria that do not grow on simple culture media is called enriched medium. Such organisms that grow only on highly enriched media are called *fastidious organisms*. Eg: Blood agar, Chocolate blood agar, Brain heart infusion agar, etc.
Enrichment Medium	A liquid culture medium that allows cultivation of a particular organism from among a heavy population of other types of organisms even when the desired organism is in very small number, is called enrichment medium. Eg: Selenite F broth serves as an enrichment medium for cultivation of Salmonellae from heavily contaminated samples of food and stool.
Transport media	A medium that allows organisms to remain viable but does not encourage their multiplication is a transport medium. They are employed to transport samples/cultures from site of collection to the processing laboratory without any decrease or increase in cell number. Eg: Stuart's medium for gonococci, Pike's medium for Streptococci and Alkaline Peptone Water for *Vibrio* spp.
Media for Anaerobes	Some media such as Robertson's Cooked meat medium allow the growth of anaerobic bacteria. The cooked meat in the medium serves to create anaerobic environment due to its oxygen reducing capacity. Differential Reinforced Clostridial medium is also a selective medium for anaerobic bacteria.
Semi-solid medium	An agar medium that contains only 1% agar (as against the usual 3%), is called a semi-solid medium or soft agar medium. Such media are employed to study motility patterns of the organisms or to enable easy diffusion of certain ingerients across the entire medium.

Culture techniques

Technique	Applications
Streak Plate Method	Isolation of pure cultures of bacterial cells from a mixed population. Eg. Isolation of bacteria in soil, food and other heavily populated samples.
Spread Plate Method	Isolation as well as enumeration of bacteria in various samples. The counts are expressed as *cfu/ml* (Colony Forming Units). Amount of inoculum is 0.1 ml.
Pour Plate Method	Isolation as well as enumeration of bacteria in various samples. The counts are expressed as *cfu/ml* (Colony Forming Units). Amount of inoculum is 1 ml.
Brewer's Anaerobic Jar Method	Cultivation of anaerobes. After incubation, the plates/tubes are placed in a stainless steel anaerobic jar which is evacuated and refilled with H_2 or CO_2 gas. Alternatively, Gaspak method is employed to create anaerobic environment in the jar.
Anaerobic Glove Box method	This is a closed work area employed to cultivate obligate anaerobes. All the required samples, media and other tools are placed within the sterile box and the box is sealed. It is evacuated and filled with H_2 or CO_2 gas. The manipulations are done only through gloves inserted within the box, by seeing through the glass top.
Agar Deep Shake Tube Method	This is used to differentiate aerobes, anaerobes and facultatively anaerobic bacteria. The culture suspension is inoculated into deep nutrient agar tubes containing only 1% agar. The tube is shaken rigorously to spred the culture deep into the medium. After incubation at 37^0C, 24h, the growth is carefully observed. Aerobes grow at the surface of the tube, anaerobes grow at the bottom. Facultative anaerobes grow through out the tube whereas if the culture is microaerophilic, growth is seen slightly below the upper surface region of the tube.
Single cell isolation	In some cases of advance research, it is possible to select a single cell and inoculate the same in broth culture medium. A micromanipulator device is employed for picking up a single cell while observing under the microscope. The technique is useful for obtaining axenic cultures i.e. cultures that originate from a single parent cell.

Chromatographic Techniques

Term	Description
Chromatography	**Chromatography** is a laboratory technique for the separation of mixture of substances by passing the mixture dissolved in a "mobile phase" through a stationary phase. Chromatography may be preparative or analytical
Preparative Chromatography	Preparative chromatography is for the separation of the components of a mixture for further use. Therefore it is a form of purification.
Analytical Chromatography	Analytical chromatography operates with small amounts of material and aims to measure the relative proportions of analytes in a mixture.
Chromatogram	The actual output of the chromatograph. Eg. different peaks or patterns on the chromatogram correspond to different components of the separated mixture.
Chromatograph	Equipment that enables a sophisticated separation e.g. gas chromatographic or liquid chromatographic separation
Effluent	The mobile phase leaving the column.
An immobilized phase	Stationary phase that is immobilized on the support particles. Eg. Silica layer in thin layer chromatography.

Table *Contd…*

Term	Description
The mobile phase	The phase which moves in a definite direction. It may be liquid or gaseous. The mobile phase consists of the solvent that moves the sample through the column where the sample interacts with the stationary phase and is separated.
The retention time	The time taken by a particular analyte to pass through the system
Column chromatography	A separation technique in which the stationary bed is within a tube called packed column. The particles of the solid stationary phase or the support coated with a liquid stationary phase may fill the inside the column. The mobile phase runs through the column and separation is on the basis of the different retention times of the sample.
Planar Chromatography	A separation technique in which the stationary phase is present on a plane surface. The plane may be a paper, as in paper chromatography or a layer of solid particles spread on a support such as a glass/plastic plate as in thin layer chromatography. Different compounds in the sample travel different distances according to their interaction with the stationary phase as compared to the mobile phase. The specific Retardation factor (R_f) of each chemical is used in the identification of an unknown substance.
Paper Chromatography.	A small dot of sample solution is placed onto a strip/sheet of *chromatography paper*. The paper is placed in a chamber containing a shallow layer of solvent and sealed. As the solvent rises through the paper, it meets the sample mixture which travels along with the solvent. This paper is further subjected to spraying with certain color developing agents and dried to visualize the substances.
Thin layer chromatography	Another widely-employed chromatography technique similar to paper chromatography. However, it involves a stationary phase of a thin layer of adsorbent like silica gel, alumina, or cellulose on a flat, glass/plastic plate. Compared to paper, it has the advantage of faster runs, better separations. High-performance TLC can be used For better resolution and to allow quantization.
Gas Liquid chromatography	A separation technique in which the mobile phase is a gas. Gas chromatography is carried out in a thin coiled capillary or packed column. The solid stationary phase is usually a silicone-based material and the mobile phase is most often Helium gas. It is widely used in analytical chemistry. However, the high temperatures used in GLC make it unsuitable for high molecular weight proteins as heat will denature them.
Liquid chromatography	A separation technique in which the mobile phase is a liquid. The liquid chromatography that utilizes very small column packing particles and a relatively high pressure is referred to as high performance liquid chromatography (HPLC). In the HPLC technique, the sample in the liquid mobile phase is forced through a column that is packed with irregularly or spherically shaped minute particles at high pressure.
Affinity chromatography	This technique is based on selective non-covalent interaction between an analyte and specific molecules. It is often used in biochemistry in the purification of proteins. The proteins are labeled with compounds such as His-tags, biotin or antigens, which bind to the stationary phase specifically. After purification, some of these tags are removed and the pure protein is obtained.
Ion exchange chromatography	This utilizes ion exchange mechanism to separate analytes. It is performed in columns and rarely on planar modes. A charged stationary phase is used to separate charged compounds such as amino acids, peptides, and proteins. The stationary phase is an ion exchange resin that carries charged functional groups which interact with oppositely charged groups of the compound to be retained.

Centrifugation techniques

Technique	Description
Centrifugation	A process of sedimentation or separation of liquid mixtures that uses centrifugal force. The dense and heavier components of the mixture migrate away from the axis of the centrifuge, while less-dense components of the mixture migrate towards the axis. The upper solution after centrifugation is called the supernatant and the settled matter is the pellet. The rate of centrifugation is expressed as revolutions per minute (*rpm*).
Microcentrifuges	These are used to separate small volumes of biological molecules or cells and large cell organelles. Nuclei are often separated by microcentrifugation. Microcentrifuge are run at maximum speeds of 12000-13000*rpm*.
Ultracentrifugation	Small cell components such as ribosomes & proteins, and viruses are sedimented by ultracentrifugation that employs very high centrifugal forces (70000*rpm*). Other applications of ultracentrifugation include membrane fractionation, analytical ultracentrifugation (AUC) for determination of macromolecular properties, amino acid composition of a protein, etc.
Density Gradient Centrifugation	This unique technique enables separation of various components in a mixture simultaneously. Each component in a mixture has a particular density on the basis of which, it can be separated. Eg: It can be used to separate intact organelles from a cell. When cell lysate is added to a tube containing a dense solution such as cesium chloride, sucrose, or glycerol; Centrifugation, creates a gradient in the tube. Heavier organelles migrate to the bottom of the tube and the lighter ones move to the top. Each organelle migrates to its appropriate equilibrium density forming distinct bands at different levels.
Rate Zonal Centrifugation	This is a size and mass dependent sedimentation technique. Eg. Antibodies of various classes are separated based on the mass and not the density. Employed in the separation of cellular organelles such as endosomes or separation of proteins, such as antibodies. For instance, Antibody classes all have very similar densities, but different masses. The path length of the gradient and the time employed for the centrifugation are important factors that account for the success of this technique.

Exercises

1. When a ray of light passes from one medium to another, ___________ occurs   [   ]
   a. Refraction   b. Reflection   c. Interference   d. Repulsion

2. The focal length of a lens is   [   ]
   a. The diameter of the lens
   b. The distance between the centre of the lens and the source of light
   c. The distance between the centre of the lens and the point where the light rays are focused by the lens
   d. The distance between the farthest parallel rays of light those enter the lens

3. Human eye cannot focus on objects nearer than ______   [   ]
   a. 10inches   b. 25inches   c. 6inches   d. 8inches

4. Numerical aperture is ______   [   ]
   a. The light gathering capacity of a microscope
   b. The ability to focus the image with best resolution
   c. The ability to create contrast between the specimen and the background of the microscopic field
   d. The ability to maintain the working distance of a microscope as large as possible.

5. Resolving power is given by   [   ]
   a. ½ (Numerical Aperture) / λ b. 2 (Numerical Aperture) / λ
   c. 2 λ / (Numerical Aperture)   d. ½ λ / (Numerical Aperture)

6. The resolution of a complete microscope is given by ___________   [   ]
   a. λ / N.A._{Objective} $-$ N.A._{Condenser}   b. λ / N.A._{Objective} $+$ N.A._{Condenser}
   c. N.A._{Objective} + N.A._{Condenser} / λ d. N.A._{Objective} - N.A._{Condenser} / λ

7. A hollow cone of light is produced by the condenser in a   [   ]
   a. Bright field microscope   b. Dark field microscope
   c. Phase contrast microscope   d. Fluorescence microscope

8. A mercury lamp is used as a source of illumination in ______   [   ]
   a. Bright field microscope   b. Dark field microscope
   c. Phase contrast microscope   d. Fluorescence microscope

9. Annular diaphragm is used in   [   ]
   a. Bright field microscope   b. Dark field microscope
   c. Phase contrast microscope   d. Fluorescence microscope

10. Bacterial endospores can be best observed by   [   ]
    a. Bright field microscope   b. Dark field microscope
    c. Phase contrast microscope   d. Fluorescence microscope

11. Match the following pairs   [   ]
    A. Bright field   i. *Paramoecium*
    B. Dark field   ii. Staphylococci
    C. Phase contrast   iii. *Treponema*
    D. Fluorescence   iv. *Mycobacterium*

    a. A-iii, B-ii, C-i, D-iv
    b. A-ii, B-iii, C-i, D-iv
    c. A-ii, B-iii, C-iv, D-i
    d. A-iv, B-iii, C-ii, D-i

12. Staining is NOT required for   [   ]
    a. Bright field microscopy   b. Dark field microscopy
    c. Both a & b   d. Fluorescence microscopy

13. The main advantage of phase contrast microscopy is   [   ]
    a. Motility can be observed
    b. Intracellular organelles can be observed
    c. No need of special attachments within the lens systems
    d. No need of a light source

14. The disadvantage of dark field microscopy is   [   ]
    a. Expensive dyes are required   b. A special type of eyepiece is required
    c. Yeast cells cannot be observed d. Artifacts may be mistaken as organisms

15. Phase contrast microscopy is especially useful for detecting _____ in cells   [   ]
    a. Poly β hydroxyl butyrate granules   b. Polymetaphosphate granules
    c. fur granules   d. All

16. The background of the microscopic field appears totally dark in   [   ]
    a. Fluorescence microscopy   b. Dark field microscopy
    c. Both a & b   d. None

17. Atoms on the surface of a solid can be viewed with the help of   [   ]
    a. Scanning tunneling microscopy   b. Scanning electron microscope
    c. Confocal laser microscope   d. Freeze etch microscope

18. The resolution of Tansmission Eectron Mcroscope (TEM) is   [   ]
    a. 100 times better than the light microscope
    b. 1000 times better than the light microscope
    c. 10,000 times better than the light microscope
    d. 1,00,000 times better than the light microscope

19. The most critical requirement of electron microscope is   [   ]
    a. Ultramicroscopy   b. Photomultiplier
    c. High vaccum   d. Palladium spray device

20. In TEM, the electron source is   [   ]
    a. Gold filament   b. Tungsten filament
    c. Uranium filament   d. Any of the above

21. The distinct feature of Scanning Electron Microscope (SEM) is   [   ]
    a. It can achieve magnification upto 100 million times
    b. It can produce image even in absence of vaccum system
    c. It can produce a realistic three dimensional image
    d. It can produce without a photomutiplier

22. Secondary electrons are detected in   [   ]
    a. TEM   b. SEM
    c. EM   d. Atomic force microscope

23. The lens system in EM consists of   [   ]
    a. Quartz lenses   b. Electromagnetic lenses
    c. Magnets   d. Glass lenses

24. The resolution of a microscope   [   ]
    a. Increases with decrease in wavelength
    b. Decreases with decrease in wavelength
    c. Increases with increase in wavelength
    d. Is independent of wavelength

25. In TEM, the denser region of the specimen   [   ]
    a. Appears brighter
    b. Appears darker
    c. Transmit more electrons to the screen
    d. Appear more resolved

26. The final image in a TEM is formed   [   ]
    a. On a photographic film   b. On a fluorescence screen
    c. On a television screen   d. Both a or b

27. What is FALSE about SEM   [   ]
    a. The image is formed from secondary electrons
    b. The dried specimen is coated with a thin layer of metal before viewing
    c. The raised areas appear dark and depressions on the surface of specimen appear bright
    d. Employs a photomultiplier that convets light flashes to electrical current and amplifies

28. The fixative used for specimen to be observed under TEM   [   ]
    a. Ethanol   b. Silver nitrate
    c. Osmium Tetroxide   d. Acetone

29. In a TEM, the prepared specimen is mounted on   [   ]
    a. Fine copper grit   b. Plastic film   c. Uranium plate   d. Thin bed of air

30. Which of these is NOT a method of preparing specimen for observation under TEM   [   ]
    a. Shadow casting   b. Negative staining
    c. Freeze etching   d. Screen plotting

31. In SEM, secondary electrons are   [   ]
    a. Emitted from the staining salt   b. Emitted from the surface of the specimen
    c. Emitted from the tungsten filament   d. Emitted from the mounting surface

32. An axenic culture is   [   ]
    a. A culture belonging to same genus and species
    b. A culture belonging to same genus and species and which has been obtained from
       a single source sample.
    c. A culture belonging to same genus and species and which has been obtained from
       a single parent cell
    d. A culture that has been obtained from an unknown sample source

33. A pure culture contains   [   ]
    a. Organisms belonging to same genus
    b. Organisms belonging to same genus and same species
    c. Organisms belonging to different genera but same species
    d. Organisms belonging to differenrt genera and different species

34. Nutrient agar is a   [   ]
    a. Selective medium   b. Enriched medium
    c. General purpose medium   d. Transport medium

35. Synthetic media are also called   [   ]
    a. Complex media
    b. Simple media
    c. Chemically defined media
    d. Differential media

36. Which of these is an empirical medium?   [   ]
    a. Milk
    b. Peptone water
    c. Tryptone water
    d. Saline

37. Which of these is a pure culture technique?   [   ]
    a. Spread plate method
    b. Streak plate method
    c. Roll tube method
    d. All

38. Serial dilution technique is used in   [   ]
    a. Preservation of microbes
    b. Enumeration of microbes
    c. Disinfection of microbes
    d. Cultivation of microbes

39. One ml of albumin from 10mg/ml solution is added to 90ml of sterile saline. The conc of the new albumin solution is   [   ]
    a. 1mg/ml
    b. 0.1mg/ml
    c. 0.01mg/ml
    d. 0.001mg/ml

40. Micromanipulator technique is used for   [   ]
    a. Enumeration of bacteria
    b. Isolation of a single bacterial cell
    c. Transportation of bacterial cells
    d. Preservation of bacterial cells.

41. Sorbent in chromatography   [   ]
    a. Solid or liquid stationary phase
    b. Solid stationary phase
    c. Liquid stationary phase
    d. Gas mobile phase

42. The mobile phase in chromatography is   [   ]
    a. Always gas
    b. Always liquid
    c. Gas or liquid
    d. Gas, liquid or solid

43. The general process of moving a solute through a chromatographic system is called   [   ]
    a. Effluent
    b. Development
    c. Diffusion
    d. Capillary rise

44. Distribution of a solute between two liquid phases is   [   ]
    a. Partition chromatography
    b. Adsorption chromatography
    c. Gas Liquid Chromatography (GLC)
    d. Column chromatography

45. Which of these techniques uses ion exchange resins that has binding sites for solute molecules   [   ]
    a. Partition chromatography
    b. Adsorption chromatography
    c. Gas Liquid Chromatography (GLC)
    d. Column chromatography

46. Which of these is considered Planar chromatography   [   ]
    a. Paper chromatography
    b. Thin Layer Chromatography (TLC)
    c. a & b
    d. Neither a nor b

47. When a solvent developed plate is sprayed with Conc. H_2SO_4 and heated at 100^0C, all the organic substances spotted will appear as   [   ]
    a. Black spots
    b. Yellow spots
    c. Orange spots
    d. Blue spots

48. In GLC   [   ]
    a. The mobile phase is liquid and stationary phase is gaseous
    b. Both moile and stationary phases are gaseous
    c. The mobile phase is gaseous and stationary phase is liquid

49. Which of these reagents can be used for visualization of lipids   [   ]
    a. Rhodamine B
    b. Ninhydrin
    c. Aniline phthalate
    d. Any of the above

50. TLC comprises of a glass or plastic plate with a thin coating of   [ ]
    a. Silica gel   b. Alumina   c. Cellulose   d. a, b or c

51. The main disadvantage of GLC   [ ]
    a. Slower process compared to TLC   b. Small biomolecules cannot be separated
    c. Very expensive   d. Sample must be heat stable and volatile

52. In GLC the compounds are identified on the basis of   [ ]
    a. R_f value   b. Peak enhancement
    c. Retention time   d. b & c only

53. In a column chromatography the best adsorbing material for separation of sugars is   [ ]
    a. Silica gel   b. Magnesium silicate
    c. Calcium phosphate   d. Hydroxyapatite

54. Gel exclusion chromatography is also called   [ ]
    a. Column chromatography   b. Molecular sieve chromatography
    c. Ion exchange chromatography   d. Planar chromatography

55. In High Pressure Liquid Chromatography (HPLC), high pressure is employed for   [ ]
    a. Improvement of elution rates   b. Exceeding resolution over the other methods
    c. Automation of data analysis   d. Development of improved stationary supports

56. What is FALSE about HPLC   [ ]
    a. The speed of analysis is better than that of GLC
    b. The columns can be reused without repacking
    c. Data analysis is easily automated
    d. The principles of ion exchange and Adsorption chromatography cannot be applicable in HPLC

57. Electrophoresis is a rapid technique for analyzing biomolecules especially   [ ]
    a. Proteins and sugars   b. Proteins and lipids
    c. Proteins and nucleic acids   d. Sugars and nucleic acids

58. In electrophoresis through polyacrylamide gel, the separation is based on   [ ]
    a. Molecular sieving   b. Electrophoretic mobility
    c. Both a & b   d. Neither a nor b

59. The most significant merit of SDS-PAGE over other electrophoretic techniques is in   [ ]
    a. Identification of biomolecules of larger molecular sizes
    b. Excellent resolution
    c. Improvement of speed of separation
    d. Measurement of molecular weights of protein subunits

60. Two different gel layers – Resolving gel and Stacking gel are employed in   [ ]
    a. Pulsed Field Gel Electrophoresis (PFGE)
    b. Agarose gel electrophoresis
    c. Discontinuous gel electrophoresis
    d. SDS-PAGE

61. Which of these is particularly best suited for analysis of large nucleic acid
    fragments (>50,000bp)   [ ]
    a. Pulsed Field Gel Electrophoresis (PFGE)
    b. Agarose gel electrophoresis
    c. Capillary electrophoresis
    d. Two dimensional electrophoresis

62. The technique of detecting DNA-DNA hybridization is called   [   ]
    a. Southern blotting   b. Northern blotting
    c. Western blotting   d. Eastern blotting

63. In a UV spectrophotometer, the source of light is   [   ]
    a. UV lamp   b. High pressure hydrogen lamp
    c. Tungsten halogen lamp   d. Ordinary electric bulb

64. In visible spectrophotometer the source of light is   [   ]
    a. UV lamp   b. High pressure hydrogen lamp
    c. Tungsten halogen lamp   d. Ordinary electric bulb

65. Which of these facilitates pelleting of viruses   [   ]
    a. Clinical centrifuge   b. High speed centrifuge
    c. Ultracentrifuge   d. b & c

66. Which of these can be pelleted by ultracentrifuges but not by high speed centrifuges [   ]
    a. Ribosomes   b. Nuclei   c. Mitochondria   d. Chloroplasts

67. What id TRUE about ultra centrifuges?   [   ]
    a. They have a range of speed 20-80,000rpm
    b. The spin chamber must be refrigerated as intense heat is generated
    c. Must be placed under high vaccum to reduce friction
    d. a, b & c are TRUE

68. In order to sediment cell debris after cell homogenization, the most suitable centrifugation is   [   ]
    a. High speed   b. Ultracentrifuge
    c. Low speed   d. Any of the above

69. Differential centrifugation involves   [   ]
    a. Centrifugation at fluctuating hot-cold temp
    b. Second centrifugation of resuspended pellet of cells
    c. Successive centrifugation at increasing rotor speed
    d. Centrifugation at high speed and low speeds alternately.

70. Solutions of low molecular weight solutes such as sucrose or glycerol are employed in   [   ]
    a. Differential centrifugation   b. Ultracentrifugation
    c. High speed centrifugation   d. Density gradient centrifugation

Answer Key and Validation

Methods in Microbiology

1. a	13. b	25. b	37. d	49. a	61. a
2. c	14. d	26. d	38. b	50. d	62. a
3. a	15. d	27. c	39. b	51. d	63. c
4. a	16. c	28. c	40. b	52. d	64. b
5. d	17. a	29. a	41. a	53. a	65. c
6. b	18. b	30. d	42. c	54. b	66. a
7. b	19. c	31. b	43. b	55. a	67. d
8. d	20. b	32. c	44. a	56. d	68. a
9. c	21. c	33. b	45. b	57. c	69. c
10. c	22. b	34. c	46. c	58. c	70. d
11. b	23. b	35. c	47. a	59. d	
12. b	24. a	36. a	48. c	60. c	

2. *When parallel rays of light strike a convex lens, the rays get focused at a specific point called focal point. The distance between the centre of the lens and the focal point is called its focal length*

15. *Phase contrast microscopy converts differences in cell density into variations in light intensity. Hence dense granules and other particulate matter with the cell can be easily visualized.*

25. *In TEM, the denser region in the specimen scatters more electrons and few electrons are transmitted to the screen. Therefore these regions appear darker. The less dense regions are more electron transparent and so appear brighter.*

26. *In TEM, the final image is formed on a fluorescent screen. When the screen is set aside, there is a photographic film that receives the transmitted electrons.*

27. *The raised areas of the specimen release larger number of secondary electrons and appear bright while the depressions release very few secondary electrons and hence appear darker.*

28. *Gluteraldehyde and osmium tetroxide are used as fixatives to stabilize the cell structure*

36. *Empirical media are naturally occurring substances in which microbes can grow. Eg. Milk, Urine, Blood, etc. these were used as culture media in early days of Microbiology*

47. *H_2SO_4 is a universal solvent, and hence used to produce black spots for all organic substances.*

56. *The greatest merit of HPLC is that all chromatographic modes including partition, adsorption, ion-exchange, chromatofocussing and gel elution are possible*

5

The Microorganisms

Categorization of the living organisms

1. **Two kingdom concept**	- Proposed by **C. Linnaeus** in 1758. - Placed all organisms in *Kingdom Plantae* or *Kingdom Animalia.* - This system is now rejected due to lack of details, on the basis of cell nature, cell type and metabolic diversities.
2. **Three kingdom concept**	- Proposed by **Earnst H Haeckel** in 1866. - Organisms were grouped into *Protista, Plantae and Animalia.* - Failed as the major groups of microbes – algae, fungi, protozoa and bacteria were all placed in one group *Protista*
3. **Four kingdom concept**	- Classified organisms into *Procaryotae, Protista, Plantae* and *Animalia.* - All bacteria were included in *Procaryotae,* Protozoa, algae & fungi in *Protista,* Plants in *Plantae* and animals in *Animalia.*
4. **Five Kingdom Concept**	- This is the most accepted concept of classifying living organisms. Proposed by Robert H Whittaker in 1969. - This concept is comprehensive. Classification is based on three levels of organization. - The scheme is based on how organisms acquire their food. The three modes of nutrition include Photosynthesis, absorption and ingestion. - The Five Kingdoms include: Monera (Eubacteria and Cyanobacteria), Protista (Protozoa & Microalgae), Plantae (Plants), Animalia (Animals), Fungi (Yeast and Molds)
5. **Three Domain System**	- Carl Woese and his co-workers in 1970s proposed a **Three-Domain System.** It is based on new techniques in Molecular Biology and Biochemistry. - In Woese's three domain system, all living organisms are classified into three **domains** or **Superkingdoms** – *The Archaea, The Eubacteria & Protista.* - The *Archaea* includes special group of bacteria that can live under extremely harsh environments. - The *Eubacteria* includes all bacteria except the archaeobacteria. - The *Protista* includes the rest of the beings – plants, animals, all fungi and all protozoa - Initially scientists did not accept the three-domain system as it was placing Archaea as a separate kingdom. - However, studies of DNA revealed that almost two thirds of the DNA of Archaeobacteria was different from that of eubacteria. Hence, finally the three-domain system was accepted.

Characteristics of Major Groups of Microorganisms

Bacteria
- **Bacteria** are unicellular microorganisms. They are only a few micrometres long and have many shapes including spheres, rods, and spirals. The study of bacteria is bacteriology.
- Bacteria are ubiquitous in every habitat. They grow in soil, acidic hot springs, snow layden mountains, radioactive waste, seawater, and even deep in the earth's crust.
- Bacteria are also found as normal flora or pathogens in animals and man. Hans Christian Gram in 1884 developed The Gram stain,which characterised bacteria based on the structural characteristics of their cell walls. The thick layers of peptidoglycan in the "Gram-positive" cell wall stain purple, while the thin "Gram-negative" cell wall appears pink.
- Thus on the basis of morphology and Gram-staining, most bacteria can be classified as belonging to one of four groups (Gram-positive cocci, Gram-positive bacilli, Gram-negative cocci and Gram-negative bacilli).
- Some bacteria however do not take up the Gram's stains easily. They are stained by Ziehl–Neelson carbol fuchsin (ZNCF) stain and are called acid fast bacteria. Eg. Mycobacteria and *Nocardia*.
- Bacteria are prokaryotes. The remarkable difference in the genetic makeup and phenotypic characters has led to division of bacteria into eubacteria (True bacteria) *and* archaeobacteria.
- Identification of bacteria is by morphological, cultural, biochemical, serological and molecular methods.

Algae
- **Algae** include several groups of simple living aquatic organisms.
- Algae are photosynthetic organisms that occur in most habitats. They vary from small, single-celled forms to complex multicellular forms, such as the giant kelps that grow to 65 meters in length
- Algae have conventionally been regarded as simple plants. However, they span more than one domain, including both Eukaryota and Bacteria.
- Algae range from single-cell organisms to multicellular. They all lack leaves, roots, flowers, seeds and other organ structures that characterize higher plants.
- All algae including cyanobacteria have photosynthetic machinery, and so produce oxygen as a byproduct.
- Algae can endure dryness and other adverse conditions in symbiosis with a fungus as lichen.
- The study of algae is called phycology or algology.
- **Microalgae:** A term that includes algae in the phytoplankton. Phytoplankton comprises the base of the food chain in the marine environment. Microalgae can be cultivated under difficult agro-climatic conditions and are able to produce a wide range of commercially interesting byproducts such as fats, oils, sugars and functional bioactive compounds. As a group, they are of particular interest in the development of future renewable energy sources. **Eg:** : *Skeletonema, Thalassiosira, Phaeodactylum, Chaetoceros, Cylindrotheca, Bellerochea, Actinocyclus, Nitzchia, Cyclotella,* Haptophyceae.

Protozoa
- Protozoans (*protos = first, zoan = animal)* are eukaryotic, unicellular microorganisms. They are mainly characterized by their ability to move with the help of some locomotory organelles, and by their lack of cell walls. They are non-photosynthetic (*heterotrophs*).
- Some protozoans live in colonial forms, where the various cells are embedded in a common matrix or joined by fine cytoplasmic threads.
- Study of protozoa is called *Protozoology.*
- Protozoa occur all over the globe, in seawaters, fresh waters, in moist soil, polar regions and also at high altitudes. They are free living, parasitic or symbiotic.
- In unfavorable conditions, protozoa form *cysts* that are more resistant than the vegetative forms.
- *Size wise* protozoa differ widely. Some are very tiny 1 to 4µm (Eg. *Leishmania donovani),* some are up to 600µm (Eg. *Amoeba),* some may be up to 2000µm (eg. Ciliates) and some fossil foraminiferans also measure up to 15cm in diameter !
- *Shape wise* they may assume any shape (Eg. *Amoeba),* leaf like (Eg. *Giardia),* oval (Eg. *Leishmania),* slipper shaped (Eg. *Paramoecium),* Circular (Eg. *Actinosphaerium),* thalloid (Eg. *Ballantidium)* and even siphonous (Eg. *Vortecella).*

	- Locomotory organelles: Flagella, Pseudopodia, Cilia or Myonemes.
	- Modes of asexual reproduction: Binary fission, multiple fission, sporulation, plasmotomy and budding.
	- Modes of sexual reproduction: Syngamy, Conjugation, Autogamyh, Cytogamy, Hemixis, Endomixis.
Fungi	- Fungi are non-photosynthetic, heterotrophic, eukaryotic organisms that occur either unicellularly or in multicellular forms.
	- They live on dead matter (**saprophytes**), in living organisms (**parasites**) or as symbionts.
	- Fungi are found in soil, air, water and foodstuffs. They reproduce by forming characteristic spores of sexual and asexual varieties.
	- Study of fungi is called **Mycology.**
	- Reproduction is by vegetative, asexual and sexual modes.
	- Asexual spores: Chlamydospores, Conidiospores, Sporangiospores, Arthrospores (odia).
	- Sexual Spores: Oospores, Zygospores, Ascospores, Basidiospores.
Viruses	- Refer to Chapter 9: Virology

Special Bacteria

Group & Features	Examples
Archaeobacteria - Principally, archaeaobacteria are similar to eubacteria in most aspects.However, their genetic transcription and translation processes show many typical distinguishing features. For instance, archaean translation uses eukaryotic-like initiation and elongation factors, and their transcription involves TATA-binding proteins and TFIIB as in eukaryotes. Many archaeal tRNA and rRNA genes harbor unique archaeal introns which are neither like eukaryotic introns, nor like bacterial introns. - Like bacteria and eukaryotes, archaea possess glycerol-based phospholipids. However, the archaeal lipids are unique because the stereochemistry of the glycerol is the reverse of that found in bacteria and eukaryotes. - Most bacteria and eukaryotes have membranes composed mainly of glycerol-ester lipids, whereas archaea have membranes composed of glycerol-*ether* lipids. - The archaeal cell walls are also unusual. With the exception of one group of methanogens, archaea lack a peptidoglycan wall. In methanogens, the peptidoglycan is very different from the type found in bacteria. The flagella of Archaeans are also notably different in composition and development from the flagella of bacteria. - Archea are unicellular, prokaryotic, have no nucleus, and have one circular chromosome. - Archaea are usually placed into three groups. These are the halophiles, methanogens, and thermophiles. Halophiles, sometimes known as *Halobacterium* live in extremely saline environments. Methanogens live in anaerobic environments and produce methane. Thermophiles live in places that have high temperatures, such as hot springs.	*Halobacterium, Halococcus, Methanobacterium, Methanococccus, Thermococcus..*
Spirochaetes - Spirochetes are long and slender bacteria, 5 to 250microns long and 0.1microns wide.. They are tightly coiled, and look like miniature springs and hence the name. - They are Gram negative and chemohetrotrophic in nature. - Members of this group bear a unique arrangement of **axial filaments**, similar to bacterial flagella. These filaments run along the outside of the protoplasm within a sheath.t They are responsible for the motility of the organism. - The group of spirochetes includes both aerobic and anaerobic species, in free-living and parasitic forms. - Some spirochetes are well known due to the ability to cause disases such as syphilis and Lyme disease. - Spirochetes are demonstrated in the lab by negative staining or by Fontana's silver impregnation method.	*Leptospira* species, *Borrelia* causes Lyme disease *Borrelia recurrentis*, causes Relapsing fever *Treponema pallidum*, causes syphilis *Leptospiractero haemorrhagiae*, which causes Weil's disease

Table *Contd*...

Group & Features	Examples
Acid fast bacteria - Some bacteria do not take up stains easily. And when they do so, they show marked resistance to decolorization by acids during staining procedures. Such bacteria are referred to as acid fast bacteria (AFB's). - Acid-fast organisms can be stained using heated concentrated dyes such as Ziehl-Neelson Carbol Fuchsin (ZNCF) stain. - Acid-fast bacteria can also be visualized by fluorescence microscopy using specific fluorescent dyes (auramine-rhodamine stain, for example). Some bacteria may also show partial acid-fastedness.	*Mycobacterium tuberculosis, Mycobacterium leproae, Nocardia spp.*
Myxobacteria - The myxobacteria are a family of **gliding bacteria** that produce characteristic fruiting bodies in unfavorable conditions. - They are common in animal dung and organic-rich soils of neutral or alkaline pH. - Some of them grow by utilising cellulose, but many of them secrete antibiotics to kill other bacteria and then produce enzymes to lyse the cells of their prey. - Myxobacteria are aerobic, Gram-negative, elongated rods with either rounded or tapered ends. - In nutrient depletion conditions the cells migrate and form aggregates by chemotaxis. These aggregates then develop into fruiting bodies which typically develop a bright yellow, red or brown pigmentation. The fruiting bodies bear myxospores.	*Archangium Cystobacter Melittangium Stigmatella Myxococcus Angiococcus Chondromyces Nannocystis Polyangium*
Mollicutes - Also called MLO or PPLO. The **Mollicutes** are a special group of bacteria characterised by the absence of a cell wall. They are also referred to as PPLO (pleuropneumonia like organisms). - The class Mollicutes includes phytoplasmas and spiroplasmas. - Mollicutes are commonly called mycoplasmas, and are primarily parasites of various animals and plants, living within the host's cells. - The cells among the smallest procaryotes (0.2-0.3 µm). Most of them have sterols that make the cell membrane somewhat more rigid. Many are able to move about through gliding. - *Mycoplama* is the most representative genus. - *Mycoplasmas* parasitic or saprophytic. Several species are pathogenic in humans causing pneumonia and other respiratory disorders and pelvic inflammatory diseases. - Mollicutes are unaffected by antibiotics that target cell wall synthesis, such as penicillin. - Cholesterol is required for the growth of species of the genus *Mycoplasma* as well as certain other genera of mollicutes. Their optimum growth temperature is often the temperature of their.	*Mycoplasma pneumoniae, M. genitalium, Ureaplasma, Erysipelothrix*
Actinobacteria - Actinobacteria, earlier called Actinomycetes are a special group of Gram-positive bacteria commonly found in the soil. - Many of them form branched spore bearing filaments like those of fungal hyphae, because of which they were once wrongly classified under fungi. - They are usually aerobic and have DNA with a high GC content - Actinobacteria play an important role in formation of humus by degrading cellulose and chitin material from plants. - Some of them may inhabit plants and animals, including a few pathogens. - Most members are aerobic, but a few, such as can grow under anaerobic conditions. Unlike the Firmicutes, the other main group of Gram-positive bacteria, and some Actinomycetes species produce external spores. - Actinobacteria produce hundreds of antibiotics, especially from the genus *Streptomycs*.	*Actinomyces israelii, , Arthrobacter, Corynebacterium, Frankia, Micrococcus, Micromonospora, Mycobacterium, Nocardia, Propionibacterium, Streptomyces*

Table Contd...

Group & Features	Examples
Cyanobacteria - Cyanobacteria (or Blue Green Algae) are one of the oldest, largest and most important groups of bacteria on earth. They are photosynthetic and aquatic prokaryotic organisms. - cyanobacteria lack nucleus and internal organelles. Like all eubacteria, their cell walls contain peptidoglycan. - However, cyanobacteria have a highly organized system of internal membranes which function in photosynthesis. Chlorophyll *a* and several accessory pigments (phycoerythrin and phycocyanin) are embedded in these photosynthetic lamellae. - They are usually unicellular, but often grow in large visible colonies. - Cyanobacteria are able to fix atmospheric nitrogen into ammonia (NH_3), nitrites(NO_2) or nitrates (NO_3), which *can* be absorbed by plants and converted to protein and nucleic acids. Nitrogen fixation is brought about in specialized cells called heterocysts. - BGA provide as nitrogen biofertilizers specially in the cultivation of rice and beans. - The other significant contribution of the cyanobacteria is the origin of plants. The chloroplasts in plant cells are considered to be endosymbiotic cyanobacteria living in association with the plant cell.	*Nostoc, Anabena, Synechococcus, Trichodesmium, Crocosphaera,Lyngbia, Spirulina, Synechococcus, Gleobacter, Dermocarpaß*

Economic importance of Microorganisms

Algae
- Algae serve as primary producers. Phytoplankton are responsible for more than 45% of the Earth's annual primary production.
- Algal blooms may be potentially harmful when they over grow under nutrient-rich conditions.
- Seaweeds are used as fertilisers and even SCP.
- Extracts from the cell walls of brown and red algae provide the polysaccharides agar and carrageen. These are used as thickening agents in food, in surgical dressings and in microbial media
- The exoskeletons of coralline macro-algae often become incorporated into a reef after the alga dies.
- The skeletons of diatoms forms **diatomaceous earth** that is used as abrasives, insecticides, reflective road signs, swimming pool filters.

Fungi
- Fungi are saprophytic and break down dead organic material. They play a significant role in biogeochemical cycles.
- Fungi associate with most vascular plants symbiotically as **mycorrhizae**, that inhabit their roots and supply essential nutrients.
- *Aspergillus niger* is used in the manufacture of citric acid and enzymes.
- *Penicillium notatum & Penicillium chrysogenum* are used in the manufacture of penicillin antibiotic.
- *Penicillium roquefortii* is used in the manufacture of Roquefort cheese
- Yeasts *(Saccharomyces)* are used in alcohol fermentation (wine, beer) and bread making.
- A number of fungi, in particular the yeasts, are important model organisms for studying problems in genetics and molecular biology.
- Fungi also cause several plant and animal diseases: in humans, ringworm, athlete's foot, and several more serious diseases are caused by fungi.
- Plant diseases caused by fungi include rusts, smuts, and leaf, root, and stem rots. These often cause severe damage to crops.

Protozoa
- Protozoa play an important role in biological treatment of **sewage.**
- Bacteria and protozoa are used in **industrial effluent treatment** processes to detoxify some harmful and corrosive material.
- Protozoa occupy the key position as **Primary consumers** in food chains of several ecosystems.
- Some foraminiferans settle at the bottom of the seas and oceans and **form rocks** used as building material.

- Protozoa help locating **oil deposits.**
- **Symbionts** such as *Trychonympha* live in the intestines of termites (white ants) and digest the cellulose eaten by the termites.
- Many protozoans are **harmful**. They cause diseases, produce toxins, destroy wood, cause water pollution and destroy useful soil bacteria.

Viruses
- Refer to Chapter 9: Virology

Major divisions of Protozoa

Class	Locomotory organelle	Example
Flagellata or Mastigophora	Flagella	*Euglena, Trypanosoma, Trichomonas, Synus..*
Rhizopoda or Sarcodina	Pseudopodia	*Amoeba, Euglypha, Globineria, Lithocircus*
Ciliophora	Cilia	*Paramoecium, Balantidium, Votecella,..*
Sporozoa	Absent	*Plasmodium, Monocystis, Sacrcocystis*
Mycetozoa: A new class with multi nucleate forms.		*Didynium*

Major divisions of Fungi

Class	Locomotory organelle	Example
Ascomycota	*Sac and cup fungi, yeasts, mildews*	*Saccharomyces cerevisiae Penicillium chrysogenum Morchella esculentum Neurospora crassa Aspergillus flavusCandida albicans Cryphonectria parasitica*
Basidiomycota	*Club fungi, rusts and smuts*	*Xerula, Amanita phalloides,* rust fungi (Uredinales) and smut fungi (Ustilaginales)
Zygomycota	*Bread molds*	*Mucor, Rhizopus*
Chytridomycota	*Chytrids*	*Allomyces*
Myxomycota	*Slime Molds*	*Physarum polycephalum , Fuligo septica*

Exercises

1. Carl Woese divided prokaryotes into   [   ]
   a. Three distinct Domains
   b. Two distinct Domains
   c. Five distinct Domains
   d. Three distinct Kingdoms

2. Cyanobacteria are considered as ancestors of   [   ]
   a. Chloroplasts   b. Mitochondria   c. Rickettsia   d. Plant cells

3. The endosymbiotic hypothesis explains the origin of   [   ]
   a. Chloroplasts & Mitochondria
   b. Archaeobactereia
   c. Golgi apparatus
   d. Nucleus

4. A population of organisms that arises from a single parent cell is referred to as   [   ]
   a. Species   b. Strain   c. Biovar   d. Serovar

5. The Bergey's Manual contains the currently accepted system of   [   ]
   a. Viral taxonomy
   b. Protozoan taxonomy
   c. Bacterial taxonomy
   d. Microbial taxonomy

6. Phonetic system of classification is based on   [   ]
   a. Phylogenetic analysis
   b. Numerical analysis
   c. Mutual similarity of external morphology
   d. Ecological characteristics

7. Dendrograms are diagrams represented in   [   ]
   a. Phylogenetic classification
   b. Numerical taxonomy
   c. Molecular characterization
   d. Nucleic acid sequencing

8. The similarity between genomes can be compared easily by use of   [   ]
   a. nucleic acid hybridization
   b. Nucleic acid sequencing
   c. Protein sequencing
   d. Conjugation studies

9. Which of these characteristics is used in classification of bacteria?   [   ]
   a. Luminiscence
   b. Secondary metabolites formed
   c. Storage inclusions
   d. All

10. The G + C content of most bacteria is in the range of   [   ]
    a. 20 to 30%   b. 30 to 40%   c. 50 to 60%   d. 70 to 80%

11. The G + C content of fungi is usually   [   ]
    a. Less than the bacterial G + C
    b. More than the bacterial G + C
    c. Same as the bacterial G + C
    d. Same as Algal, Bacterial and Protozoan G + C

12. Organisms in Kingdom Chromista   [   ]
    a. Lack chloroplasts
    b. Have chloroplasts within the lumen of ER
    c. Have chloroplasts well dispersed in the cytoplasmic matrix
    d. Have chloroplasts within the periplasmic space

13. Pick the FALSE statement about the Archea   [   ]
    a. They stain either Gram positive or Gram negative
    b. Some are hyperthermophiles that grow above 100^0C
    c. Some are symbionts in animal digestive systems
    d. Highly susceptible to β Lactum antibiotics

14. Cell walls of Archaeobacteria lack   [   ]
    a. Outer membrane   b. Muramic acid   c. D- amino acids   d. All

15. The most distinctive feature of archaeobacteria from eubacteria & eukaryotes   [   ]
    a. Nature of their membrane lipids   b. Nature of their membrane proteins
    c. Nature of their genome   d. Nature of their membrane sugars

16. Which of these Archaeobacteria is similar to eukaryotes and different from eubacteria   [   ]
    a. T Ψ C arm of tRNA   b. DNA dependent RNA polymerase
    c. Closed DNA circuit   d. mRNA

17. Thermoplasma   [   ]
    a. Sulfate reducing archae   b. Methanogens
    c. Cell wall less archae   d. Extremely thermophilic SO_4 metabolizers

18. The largest group of archaeobacteria   [   ]
    a. Thermoacidophiles   b. Methanogens
    c. Thermoplasms   d. Halobacteria

19. Archaeobic chemoheterotrophs   [   ]
    a. Thermoacidophiles   b. Methanogens
    c. Thermoplasms   d. Halobacteria

20. Strictly anaerobic, autotrophic archaea   [   ]
    a. Thermoacidophiles   b. Methanogens
    c. Thermoplasms   d. Halobacteria

21. Unlike eubacteria the archaeal membrane lipids have   [   ]
    a. Branched hydrocarbons connected to glycerol by ether links
    b. Branched hydrocarbons connected to glycerol by ester links
    c. Unbranched hydrocarbons connected to glycerol by ether links
    d. Unbranched hydrocarbons connected to glycerol by ester links

22. Which of these are anoxygenic, photosynthetic bacteria   [   ]
    a. Green sulfur bacteria & green non sulfur bacteria
    b. Purple sulfur and purple non sulfur bacteria
    c Cyanobacteria
    d. a & b only

23. Filamentous, gliding, thermophilic green sulfur bacterium   [   ]
    a. *Thiothrix*   b. *Chloroflexus*   c. *Beggiatoa*   d. *Cytophaga*

24. Some Cyanobacteria produce dormant structures called   [   ]
    a. Endospores   b. Akinetes   c. Hormogonia   d. Baeocytes

25. Motility of spirochetes is due to   [   ]
    a. Tufts of flagella at both ends
    b. Single polar flagellum
    c. Axial filament underlying the outer sheath
    d. Cilia all over the cell surface

26. Energy parasites that cannot make ATP   [   ]
    a. *Chlamydia*   b. *Archaeobacteria*
    c. *Spirochetes*   d. *Cytophages*

27. Spirochaetes are   [   ]
    a. Gram negative bacteria   b. Gram positive bacteria
    c. Gram variable bacteria   d. Acid fast bacteria

28. Which of these is TRUE in case of Chlamydia　　　　　　　　　　　　　[　]
    a. Gram negative bacteria
    b. Lack peptidoglycon in their walls
    c. Involve elementary and reticulate bodies in their life cycle
    d. All

29. The largest and most diverse group of Eubacteria　　　　　　　　　　　[　]
    a. Actinomycetes　　b. Mollicutes　　　　c. Proteobacteria　　d. Lactic acid bacteria

30. Bacteria that show budding and produce swarmer cell　　　　　　　　　　[　]
    a. *Beggiatoa*　　　b. *Leptothrix*　　　c. *Hyphomicrobium*　d. *Caulobacter*

31. Bacteria with prosthecae and holdfast　　　　　　　　　　　　　　　　[　]
    a. *Caulobacter*　　b. *Sphaerotilus*　　c. *Pseudomonas*　　d. *Leucothrix*

32. Sheathed bacteria with hollow tube like structures around chains of cells　　[　]
    a. *Stigmatella*　　　　　　　　　b. *Campylobacter*
    c. *Sphaerotilus*　　　　　　　　d. *Burkholderia*

33. Bacteria that cause development of plant tumors　　　　　　　　　　　　[　]
    a. *Hyphomicrobium*　b. *Agrobacterium*　　c. *Vampirovibrio*　　d. All

34. Gram negative curved rod shaped bacterium that preys on other gram negative bacteria　　[　]
    a. *Desulfuromonas*　b. *Stigmatella*　　c. *Bdellovibrio*　　d. *Chondromyces*

35. Gram negative curved microaerophilic rods found in the gastric mucosa of humans
    and other animals　　　　　　　　　　　　　　　　　　　　　　　　　[　]
    a. *Helicobacter*　　　　　　　　　b. *Peptostreptococcus*
    c. Citrobacter　　　　　　　　　　d. *Erwinia*

36. Match the columns　　　　　　　　　　　　　　　　　　　　　　　　[　]
    A. *Staphylococcus aureus*　　　　i.　Red
    B. *Pseudomonas aeruginosa*　　　ii. Lemon yellow
    C. *Serratia marscesans*　　　　　iii. Golden yellow
    D. *Micrococcus luteus*　　　　　iv. Green

    a. A- iii, B-iv, C-i, D-ii　　　　　b. A- iii, B-i, C-iv, D-ii
    c. A- iv, B-i, C-iii, D-ii　　　　　d. A- ii, B-iii, C-iv, D-i

37. Pick the FALSE statement about Mycoplasmas　　　　　　　　　　　　　[　]
    a. They cannot synthesize peptidoglycan precursors
    b. Many species require sterols for growth
    c. They grow on agar to give colonies of *bisected pearl appearance*
    d. They are gram negative cell wall less bacteria

38. Endospores are used by bacteria　　　　　　　　　　　　　　　　　　　[　]
    a. To reproduce
    b. To survive unfavorable conditions
    c. For easy Dissemination
    d. To change from gram positive to gram negative state or vice versa

39. Which of these does NOT show homolactic fermentation　　　　　　　　　[　]
    a. *Streptococcus*　　b. *Leuconostoc*　　c. *Enterococcus*　　d. *Lactococcus*

40. Which of these produce asexual spores　　　　　　　　　　　　　　　　[　]
    a. *Actinomycetes*　　b. *Enterobacter*　　c. *Shigella*　　　d. *Rickettsia*

41. Which of these is an actinomycete   [   ]
    a. *Arthrobacter*   b. *Actinomyces*   c. *Dermatophilus*   d. All

42. *Propionibacterium* is important in   [   ]
    a. Cheese manufacture   b. Acne vulgaris (skin infection)
    c. Both a & b   d. None

43. *Bifidobacterium* is known for   [   ]
    a. High lipid content and mycolic acids
    b. Colonization of intestinal tract of nursing babies
    c. Its unusual rod-coccus growth cycle
    d. ability to produce aerial mycelia with spores

44. Filamentous hyphae surrounded by cell walls that contain chitin   [   ]
    a. Fungi   b. Algae   c. Actinomycetes   d. Acid fast bacteria

45. What is FALSE about Archaeobacteria   [   ]
    a. They multiply either by binary fission, budding or fragmentation
    b. They are aerobic, facultatively anaerobic or strictly anaerobic
    c. They may be chemolithotrops or organotrophs
    d. They are susceptible to lysozyme and β lactum antibiotics

46. Membrane lipids of Archaeobacteria are characterized by   [   ]
    a. Lack of glycerol groups   b. Glycerol fatty acid ether linkages
    c. Glycerol fatty acid ester linkages   d. Glycerol fatty acid tetra esters

47. Pseudomurein is seen in the walls of   [   ]
    a. Methanogens   b. Purple sulfur bacteria
    c. Actinomycetes   d. Mycobacteria

48. Actinomycetes are characterized by   [   ]
    a. High G + C content   b. Low G + C content
    c. Highly variable G + C content   d. Uniform G + C content

49. The most outstanding function of Actinomycetes in soil is   [   ]
    a. To control protozoan population in soil
    b. Mineralization of organic matter in soil
    c. Production of organic acids in soil
    d. Neutralization of alkaline soils

50. The genus *Nocardia* belongs to   [   ]
    a. Actinomycetes   b. Archaeobacteria
    c. Spirochaetes   d. Mycoplasmas

51. Mollicutes are commonly known as   [   ]
    a. Actinomycetes   b. Archaeobacteria
    c. Spirochaetes   d. Mycoplasmas

52. Asexual reproduction by formation of hormogonia is exhibited by   [   ]
    a. Actinomycetes   b. Cyanobacteria   c. Archaeobacteria   d. Mycoplasmas

53. The pleasant smelling volatile substance *Geosmin* is produced by   [   ]
    a. *Ureaplasma*   b. *Treponema*   c. *Streptomyces*   d. *Halobacterium*

54. Resting spores called akinetes are produced by   [   ]
    a. Cyanobacteria   b. Fungi   c. Algae   d. Actinobacteria

55. *Frankia, Arthrobacter* and *Corynebacterium*   [   ]
    a. All three are actinomycetes   b. All three are Spirochaetes
    c. All three are Archaeobacteria   d. All three are Mycoplasmas

56. Pleuro Pneumonia Like Organisms (PPLO) were later called   [   ]
    a. Mycoplasmas   b. Archaeobacteria
    c. Actinobacteria   d. Euryarchaeota

57. Which of these is regarded as ancient bacterium   [   ]
    a. *Leuconostoc*   b. *Yersinia*   c. *Dienococcus*   d. All

58. Firmicutes   [   ]
    a. Low G+C gram positive bacteria   b. High G+C gram positive bacteria
    c. Archaeobacteria   d. Actinobacteria

59. Holocarpic Fungi   [   ]
    a. Ingest the whole food in a live condition
    b. Saprophytic fungi that become parasitic occasionally
    c. The entire fungal body becomes a reproductive unit that produces asexual or sexual cells
    d. Can absorb solid food directly through cell membrane.

60. Balanced fungal parasites   [   ]
    a. The host plants are infected without any visible symptoms
    b. Cause severe disease of plants that destroy the entire field crops.
    c. Infect the plant hosts over a long period so that the host continues to live and supplies the needs of the parasites.
    d. Can grow only upon specific living host plants.

61. Lower Fungi   [   ]
    a. Basidiomycota   b. Ascomycota   c. Chytridomycota   d. Zygomycota

62. Mushrooms belong to   [   ]
    a. Basidiomycota   b. Ascomycota   c. Chytridomycota   d. Zygomycota

63. Which of these hyphae are coenocytic   [   ]
    a. aseptate multinucleate   b. septate multinucleate
    c. septate uninucleate   d. both a & b

64. Chitin   [   ]
    a. Fungal cellulose   b. Algal cellulose
    c. Fungal & Algal cellulose   d. Insect & Fungal cellulose

65. Invaginations of fungal cellmembrane along the hyphae are called   [   ]
    a. mesosomes   b. lamosomes
    c. parenthesomes   d. dolipores

66. Match the columns   [   ]
    A. Basidiomycota   i) *Aspergillus*
    B. Ascomycota   ii) *Plasmodiophora*
    C. Zygomycota   iii) *Agaricus*
    D. Chytridomycota   iv) *Mucor*
    a. A-iii, B- ii, C- iv, D- i
    b. A-iv, B- iii, C- i, D- ii
    c. A-ii, B- i, C- iii, D- iv
    d. A-iii, B- i, C- iv, D- ii

67. What is not seen in fungal hyphae? [ ]
    a. starch granules   b. glycogen   c. vacuoles   d. all

68. Fungi are broadly classified into [ ]
    a. Eumycetes and Deuteromycetes   b. Eumycetes and Myxomycetes
    c. Eumycetes and Phycomycetes   d. Eumycetes and Basidomycetes

69. Lack well defined sexual cycle of reproduction [ ]
    a. Deuteromycetes   b. Myxomycetes   c. Phycomycetes   d. all

70. Father of Indian Mycology [ ]
    a. Lt. Col. K.R.Kiritikar   b. Sir. E.J.Butler
    c. S.N. Das Gupta   d. J. Sen Gupta

71. The largest group of fungi [ ]
    a. Sac fungi   b. Mushrooms   c. Slime molds   d. Bread molds

72. Slime molds are also known as [ ]
    a. Chitrids   b. Fungi imperfectii
    c. Rust fungi   d. Smut fungi

73. Aquatic fungus that colonizes decaying tissues and dead insects under water [ ]
    a. *Pilobolus*   b. *Neurospora*   c. *Saprolegnia*   d. *Sphaerotheca*

74. The number of basidiospores on each basidium [ ]
    a. 4   b. 6   c. 8   d. 16

75. *Sphaerothece* is known for [ ]
    a. Causing powdery mildew disease   b. Causing ergot disease
    c. Causing spoilage of meat   d. Being used as edible fungus

76. The fungus that has become a very good material for genetic studies [ ]
    a. *Claviceps*   b. *Erysiphe*   c. *Neurospora*   d. *Ustilago*

77. Dimorphic fungi [ ]
    a. Nonsporulating mycelial thallus
    b. Change from mold in the animal body to yeast in the external environment
    c. Change from yeast in the animal body to mold in the external environment
    d. Obligate parasitic fungi of plants and animals

78. Neustonic algae are found [ ]
    a. in soil   b. in water-atmosphere interface
    c. suspended in water   d. deep in the water

79. The alga *Acetabularia* is commonly called [ ]
    a. *Laughing old man*   b. *Mermaid's wine goblet*
    c. *Kaleidoscope*   d. *Fish killer*

80. Dense proteinaecious area associated with synthesis and storage of starch in algae [ ]
    a. Karyolymph   b. Pyrenoid   c. Eye spot   d. Epitheca

81. Match the columns   [   ]
    A. Brown algae
    B. Dinoflagellates
    C. Golden algae
    D. Diatoms
    i) Pyrrophyta
    ii) Chrysophyta
    iii) Phaeophyta
    iv) Bacillariophyta
    a. A-i, B- iv, C- iii, D- ii
    b. A-iv, B- iii, C- i, D- ii
    c. A-iii, B- i, C- ii, D- iv
    d. A-iii, B- iv, C- ii, D- iii

82. Which of these is not an alga   [   ]
    a. Elphidium   b. Gonium   c. Monostroma   d. Chrysocapsa

83. Which of these is an alga   [   ]
    a. Cyclotella   b. Physarum   c. Phoma   d. Lycoperdon

84. What is common among Chlorella, Volvox and Ulva   [   ]
    a. All are brown algae
    b. All are stonewarts
    c. All are dinoflagellates
    d. All are green algae.

85. The two piece silica wall of diatoms is called   [   ]
    a. Frustule   b. Epitheca   c. Laminarin   d. Pellicle

86. The correct sequence of evolutionary advancement   [   ]
    a. Entamoeba, Fungi, slime molds, plants & Animals
    b. Fungi, slime molds, Entamoeba, plants & Animals
    c. Fungi, Entamoeba, slime molds, plants & Animals
    d. Fungi, Entamoeba, plants & Animals, slime molds

87. Majority of Protozoa are   [   ]
    a. parasitic and pathogenic
    b. free living organisms in aquatic habitat
    c. soil-borne
    d. symbiotic & commensals

88. Specific vector for Trypanosomes   [   ]
    a. Anapheles mosquito
    b. Tsetse fly
    c. Ades mosquito
    d. both b & c

89. Pick out the FALSE statement   [   ]
    a. In some protozoa, mitochondria are absent
    b. In some protozoa, mitochondria and cristae are absent
    c. In some protozoa, mitochondria have discoid cristae
    d. In some protozoa, mitochondria are extensions of ER

90. Protozoa harbour   [   ]
    a. contractile vacuoles
    b. food vacuoles
    c. secretary vacuoles
    d. all

91. Trypanosomes are characterized by the presence of   [   ]
    a. undulating membrane
    b. cilia
    c. silicified skeleton
    d. Axostyle

92. Whittaker placed protozoa under kingdom   [   ]
    a. Animalia   b. Monera   c. Protista   d. Plantae

93. Non flagellate amoeba are placed in   [   ]
    a. Mastigophora   b. Sarcodina   c. Ciliphora   d. Myxozoa

94. Which of these is NOT a protozoa   [   ]
    a. *Nosemia*   b. *Babesia*   c. *Nitella*   d. *Opalina*

95. Which of these is a protozoa?   [   ]
    a. *Radiolaria*   b. *Trentepohlia*   c. *Bryopsis*   d. *Caulerpa*

96. The yeast of commerce   [   ]
    a. *Saccharomyces cerevisiae*   b. *Phoma*
    c. *Candida*   d. *Rhodotorula*

97. Filamentous yeast is   [   ]
    a. *Candida albicans*   b. *Ashbi gosypii*
    c. *Torulopsis*   d. *Hansenula*

⚷ Answer Key and Validation

The Microorganisms

1. b	18. b	35. a	52. b	69. a	86. c
2. a	19. d	36. a	53. c	70. b	87. b
3. a	20. b	37. c	54. a	71. a	88. b
4. b	21. a	38. b	55. a	72. a	89. d
5. c	22. d	39. b	56. a	73. c	90. d
6. c	23. b	40. a	57. c	74. a	91. a
7. b	24. b	41. d	58. a	75. a	92. c
8. a	25. c	42. c	59. c	76. c	93. b
9. d	26. a	43. b	60. c	77. c	94. c
10. c	27. a	44. a	61. c	78. b	95. a
11. a	28. d	45. d	62. a	79. b	96. a
12. b	29. c	46. b	63. d	80. b	97. b
13. d	30. c	47. a	64. a	81. c	
14. d	31. a	48. a	65. b	82. a	
15. a	32. c	49. b	66. d	83. a	
16. b	33. b	50. a	67. a	84. d	
17. c	34. c	51. d	68. b	85. a	

1. *Carl Woese's concept categorises all living organisms into three domains and prokaryotes into two domains i.e. Eubacteria and Archaea. The Third domain is the Eukarya that comprises of all eukaryotic organisms.*

13. *All Archaeobacteria are resistant to attack by lysoszyme and to β Lactams.*

14. *c & d are similar to eukaryotes. a is different from both eukaryotes and eubacteria.*

37. *Mycoplasmas produce fried egg colonies.*

39. *Leuconostoc carries out heterolactic fermentation through Phospho ketolase pathway*

45. *Archaeobacteria resist attack by lysozyme and β lactum antibiotics since they do not have the muramic acid and D amino acids as in eubacterial peptidoglycon. See also question no. 13.*

57. *Dienococcus represents a group of gram positive cocci in pairs or tetrads, characterized by lack of teichoic acids and presence of an outer membrane in the cell wall like that of gram negative bacteria.*

83. *Cyclotella is a representative member of the algal group Chrysophyta. The other options b, c & d are fungi.*

94. *Nitella is an alga*

95. *b, c & d are algae.*

6

Principles of Bacterial Classification

Retrace your subject...
Two Kingdom Classification, Three Kingdom Classification, Four Kingdom   Classification, Five Kingdom Classification, Three Domain Classification, Bergeys Manual of Systematic Bacteriology.

The arrangement of organisms/substances in a systematic order is called classification. The classification of living forms is referred to as **Taxonomy**. Taxonomy aims at identifying living organisms in terms of their relationship between one group and another.

The current system of classification began in 1758 with the famous publication *Systema Naturae* by Carolus Linnaeus based on the characterisitics they have in common. The Linnaean System assigns two names to every organism. This is said to be **binomial nomenclature.**

System	Features
Two Kingdom Classification	• Linnaeus recognized two Kingdoms: Kingdom Plantae and Kingdom Animalia • Kingdom Plantae: Bacteria, Algae, Fungi, Bryophytes, Pteridophytes, Gymnosperms and Angiosperms • Kingdom Animalia: Protozoa, Invertebrates and vertebrates.
Three Kingdom Classification	• Prposed by Ernst Haeckel in 1886 • The three Kingdoms include Plantae, Animalia and Protista. • Kingdom Protista included Algae, Protozoa, Fungi and organisms lacking nucleus (bacteria).
Four Kingdom Classification	• Proposed by Copeland in 1959 a. Lower Protists (Prokaryotes) b. Higher Protists: (Eucaryotes) c. Plantae d. Animalia
Five Kingdom Classification	• Proposed by Robert H Whittaker in 1969 • Kingdoms Monera, Protista, Fungi, Plantae and Animalia • Considered the nutritional strategies of organisms – Photosynthesis, absorption & ingestion. • Bacteria, Cyanobacteria and Mycoplasmas were placed in Monera. • Molds, Slime molds, Protozoa and Algae were placed in Protista • Yeast, molds and mushrooms were in Kingdom Fungi
Three Domain Classification	• Proposed by Carl Woese in 1978 • Employs data from advanced scientific techniques of molecular biology and Biochemistry. • Woese and his co-workers discovered three types of ribosomes and accordingly three cell types. • Thus, Three Domains were poposed. Eubacteria, Eucarya and Archaea.

Domain Archaea	Domain Eubacteria	Domain Eukarya
Prokaryotic Cell Structure	Prokaryotic Cell Structure	Eucaryotic Cell Structure
Cell wall contains no peptidoglycan. Instead a Pseudomurein (false peptidoglycan) is present	Primarily characterized by presence of peptidoglycan in their cell walls	Cell walls are basically made from carbohydrates. Peptidoglycan is absent.
Lipids in the membrane consist of ether linkages between branched carbon chains and glycerol	Lipids in the membrane consist of ester linkages between unbranched carbon chains and glycerol	Lipids in the membrane consist of ester linkages between unbranched carbon chains and glycerol
Usually show resistance to antibiotics	Usually show susceptibility to antibiotics	Usually show resistance to antibiotics
Genetic material comprises of circular chromosomes + histone like proteins + Plasmids	Genetic material comprises of circular chromosomes + Plasmids. Histones are absent	Genetic material comprises of nucleus with more than one linear chromosome. Histones are present.
Methionine is the start signal for protein synthesis	Formylmethionine is the start signal for protein synthesis	Methionine is the start signal for protein synthesis

Bergeys Manual of Systematic Bacteriology

A comprehensive directory, first published in 1923 by David Hendricks Bergey. The Manual is describes all known bacteria by classifying them based on their structural and functional characteristics. Based on a variety of parameters, the First Edition arranged bacteria- into families, orders and tribes. This has been modified subsequently and a more empirical approach followed for classification based on the advanced biological facts that were discovered one after the other. Bergey's manual provides an effective system of explanation for determining the genetic position of an unknown organism. The entire manual is presented in definite volumes (currently 5) and the contents are updated regularly in subsequent editions. Several publications are made available and some are currently in print too, each trying to make the arrangement of bacteria as user friendly as possible.

Exercises

1. Gametogenesis and Zygote formation is NOT seen in   [   ]
   a. Bacteria   b. Algae   c. Green Plants   d. Animals

2. All bacteria are fundamentally categorized as   [   ]
   a. Gram positive and Gram negative
   b. Eubacteria & Archaeobacteria
   c. Cell wall less bacteria and bacteria with cell walls
   d. Capsulated and Non-capsulated bacteria

3. Kingdom Protista was first proposed by   [   ]
   a. Carl von Nageli   b. Roger Stanier   c. Ernst Haeckel   d. R G E Murray

4. Which of these statements is thought to be TRUE   [   ]
   a. Eucaryotic cells evolved from Procaryotic cells
   b. Procaryotic cells evolved from Eucaryotic cells

5. Robert H Whittaker is known for   [   ]
   a. Three Kingdom System   b. Five Kingdom System
   c. Two kingdom System   d. Three Domain System

6. Carl Woese is known for   [   ]
   a. Three Kingdom System   b. Five Kingdom System
   c. Two kingdom System   d. Three Domain System

7. which of these is TRUE in case of Five Kingdom System   [   ]
   a. All prokaryotes are included in Kingdom Monera
   b. Fungi and plants are included in a common Kingdom
   c. Protozoa and animals are included in Kingdom Protista
   d. Archaeobacteria are placed in a distinct Kingdom

8. Which of these did NOT find any place in Five Kingdom Systema.   [   ]
   a. Cyanobacteria   b. Mycoplasmas   c. Viruses   d. Archaeobacteria

9. The Three Domain system categorises organisms as   [   ]
   a. Eucaryotes, Eubacteria and Archaeobacteria
   b. Eucaryotes , Procaryotes and Viruses
   c. Protista, Monera and Eucaryota
   d. Plants, animals and Microorganisms

10. Which of these cannot be considered eubacteria   [   ]
    a. Methanogens   b. Coliforms
    c. Capsulated bacteria   d. Sporulating bacteria

11. Which is the odd man out among: Micrococcaceae, Pseudomonadaceae and
    Enterobacteriaceae   [   ]
    a. Micrococcaceae   b. Pseudomonadaceae
    c. Enterobacteriaceae

12. What is common between the groups Archaeobacteria, Lactobacilli & Bacillaceae   [   ]
    a. All three contain bacteria with thick Gram positive cell walls
    b. All three contain bacteria with thin Gram negative cell walls
    c. They contain bacteria that lack cell walls
    d. They are all bacteria with unusual cell wall structures

13. The main taxonomic difference between *Pseudomonas* and *Klebsiella* is [ ]
    a. Sucrose fermentation   b. Lactose fermentation
    c. Catalase fermentation   d. Bacitracin sensitivity

14. The start signal for protein synthesis is formylmethionine in [ ]
    a. Eubacteria
    b. Eubacteria and Archaeobacteria
    c. Eucaryotes
    d. Eucaryotes ans Eubacteria

15. In the field of taxonomy Aristotle is known for [ ]
    a. Two Kingdom system
    b. Three Kingdom system
    c. Five kingdom system
    d. Three Domain system

16. In the Three Kingdom system, the Kingdom Protista included [ ]
    a. All bacteria
    b. Bacteria & Fungi
    c. Bacteria, Fungi and Protozoa
    d. Algae, Bacteria, Fungi and Protozoa

17. The idea of dividing Protista into lower protists (Procaryotes) and Higher protists
    (Eucaryotes) was given by [ ]
    a. W F Whittingham   b. Charles Darwin   c. Roger Stanier   d. R H Whittaker

18. Whittaker's system of classification considered [ ]
    a. Cellular organization   b. Mode of nutrition
    c. Both a and b   d. Mode of reproduction

19. Procaryotes were further classified into Eucaryotes and Archaea by [ ]
    a. Louis Pasteur   b. C R Woese   c. R H Whittaker   d. Linneus

20. The Archaeae comprise of [ ]
    a. Methanogens
    b. Methanogens + Extreme halophiles
    c. Methanogens + Extreme halophiles + Thermoacidophiles
    d. None of these

21. In the Bergey's Manual which of these Genus is classified under *Facultatively anaerobic,*
    Gram negative rods [ ]
    a. *Desulfotobacter*   b. *Bacteroides*   c. *Hafnia*   d. *Thermomicrobium*

22. Which of these comes under appendaged bacteria [ ]
    a. *Pediococcus*   b. *Caulobacter*   c. *Leptothrix*   d. *Listeria*

23. Long slender branched filamentous bacteria with external spores [ ]
    a. Acid fast bacteria b. Water molds   c. Actinobacteria   d. Spirochaetes

24. Which of these are placed in Vol 3 of Bergey's Manual of Systematic Bacteriology
    under the title low G+C Gram Positives [ ]
    a. Spirochaetes   b. Mollicutes   c. Fusobacteria   d. Actinobacteria

25. Which of these are placed in Vol 4 of Bergey's Manual of Systematic
    Bacteriology under the title high G+C Gram Positives [ ]
    a. Spirochaetes   b. Mollicutes   c. Fusobacteria   d. Actinobacteria

26. The general taxonomic similarity between two organisms can be best assessed by   [   ]
    a. Phage typing
    b. Serological reactions
    c. Association co-efficient in numerical taxonomy
    d. % G+C value

27. The Three Domain System largely considers   [   ]
    a. % G+C value
    b. rRNA sequences of the organisms
    c. Presence of locomotory organelles
    d. Toxin producing capability

28. One of the following structure is found particularly in bacteria   [   ]
    a. Mitochondria
    b. Membrane bound vacuoles
    c. Ribosomes
    d. Mesosomes

29. Sterols are usually present in the cell membranes of   [   ]
    a. Eukaryotes
    b. Procayotes
    c. a & b
    d. Viruses

30. Which of these are considered unicellular eukaryotic microorganisms   [   ]
    a. Bacteria
    b. Viruses
    c. Yeast
    d. Algae

Answer Key and Validation

Principles of Bacterial Classification

1. a	6. d	11. a	16. d	21. c	26. c
2. b	7. a	12. a	17. c	22. b	27. b
3. c	8. c	13. b*	18. c	23. c	28. d
4. a	9. a	14. a	19. b	24. b	29. a
5. b	10. a	15. a	20. c	25. d	30. c

10. *Methanogens are a group of Archaeobacteria that are significantly different from eubacteria*

13. *Klebsiella is lactose fermentor while Pseudomonas is lactose non-fermenting bacterium.*

7

Microbial Metabolism and Physiology

Retrace your Subject

Terminology, Anaerobic metabolism, Aerobic metabolism, ED Pethway, Gluconeogenesis, HMP Shunt, Fat metabolism, Protein metabolism, Photoautotrophy, Photoheterotrophy, Chemoautotrophy, Chemoheterotrophy...

Terminology

All chemical processes that occur in living organisms are collectively termed as metabolism. The energy requiring processes constitute *anabolism* while the processes that release energy form *catabolism*.

Anabolism is needed for growth, reproduction and repair of cellular structures. **Catabolism** is required for movement, transport and synthesis of complex molecules.

Catabolic reactions involve transfer of electrons, which allows energy to be captured in high energy bonds (ATP). Electron transport is related to sequential oxidation and reduction.

Enzymes: Anabolic and catabolic reactions are absolutely dependent on specialized proteins called enzymes. Enzymes lower the activation energy to start a reaction. They provide the surface on which reaction takes place. Enzymes bear a specific shape, size and active site to bind to a particular substrate. Enzymes are synthesized within the cell and either act inside the cell (*endoenzymes*) or in the immediate environment of the cell (*exoenzymes*)

Coenzyme and Cofactor: The protein portion of the enzyme is the apoenzyme. A coenzyme is the non-protein organic molecule associated (or bound) to the enzyme. Coenzymes are mainly synthesized from vitamins. Eg. CoA from Pantothenic acid and NAD from Niacin. Cofactors are inorganic ions such as magnesium, zinc or manganese. Cofactors improve the formation of enzyme-substrate complex.

Carrier Molecules: Certain molecules such as cytochromes and coenzymes carry hydrogen atoms or electrons in oxidative reactions.

Enzyme inhibitors: Some molecules which are similar to substrate can bind at the active site of the enzyme and inhibit the actual substrate to bind to the enzyme. These are called *competitive inhibitors*. Eg. Sulfa drugs are competitive inhibitors in folic acid synthesis, malonate for succinate dehydrogenase. Some other substances attached to the enzyme at an allosteric site that is a site other than the active site. They cause distortion in the tertiary protein structure of the enzyme and the substrate cannot bind to the active site due to change in shape. This is called *non-competitive inhibition*. Eg. Hydrogen ions for the enzyme chymotrypsin. In some other cases, the end product binds to the enzyme and inactivates it. This is *feedback inhibition*. Lead, mercury and other heavy metals also cause irreversible damage to some enzymes. Enzyme inhibition reactions help in regulating the rates of microbial growth and production of certain products.

Anaerobic metabolism: Glycolysis and fermentation reactions constitute anaerobic metabolic processes. Glycolysis (EMP) is almost universal pathway seen in aerobic and anaerobic autotrophs and heterotrophs. It is a 10-step process of breakdown of glucose to pyruvate. Glycolysis is followed by TCA in aerobes and specific fermentation reactions in anaerobes.

Fermentation: Metabolism of pyruvate in the absence of oxygen is usually referred to as fermentation by biochemists. Fermentation is aimed at passing electrons from reduced NAD to other molecules, thus recycling the reducing power to keep glycolysis going. Two common pathways are the homolactic acid fermentation and alcoholic fermentation. Other fermentations include mixed acid fermentation, propionic acid fermentation, butanediol fermentation and butyric-butylic fermentation. Definition and details of fermentation from industrial point of view are included in Chapter 14).

Aerobic fermentation: Several aerobes and facultative anaerobes indulge in more productive processes to metabolize pyruvate. In the presence of oxygen, pyruvate is converted to acetylcoA, which enters the TCA cycle (also called Kreb's cycle), citric acid being the key intermediate. The Kreb's cycle is significant in terms of energy yield, generation of reducing power and synthesis of a variety of intermediate molecules that find utility in anabolic pathways.

Aerobic Metabolism

Electron Transport Chain (ETC): The ultimate process in aerobic respiration constitutes transfer of energy rich electrons from the reduced substrate to oxygen through a series of seven carriers. ATP is generated through oxidative phosphorylation whenever the amount of energy liberated is sufficient for phosphorylation.

During aerobic metabolism of glucose molecule, 30 ATPs are produced from 10 pairs of electrons from NAD + 4 ATPs are produced from 2 pairs of electrons from FAD + 2 ATPs from glycolysis + 2 GTPs from Kreb's cycle; making a total yeild of 38 ATPs per glucose molecule.

ED Pathway

- A metabolic pathway for degradation of glucose in a wide range of bacteria such as: *Pseudomonas, Rhizobium, Serratia, Xanthobacter, Xanthomonas* and *Zymomonas*.
- These orgasnisms lack phosphofructokinase and hence cannot follow the usual EMP. Instead, the key enzyme in ED pathway is KDPG aldolase.
- Bacteria such as *E.coli* that follow the EMP (glycolysis) pathway, metabolise gluconate and other aldonates via ED pathway.
- The pathway produces pyruvate and glyceraldehyde-3-phosphate via the key intermediate KDPG (2-keto, 3-deoxy, phospho gluconate).
- Only one ATP is produced (net) in ED pathway. However, two molecules of reducing power are generated as in EMP.

Gluconeogenesis

Synthesis of glucose and (other hexoses) from non-carbohydrate sources such as acetate, glycerol, pyruvate, succinate, etc. it is achieved by first converting the source molecule to an intermediate on glycolysis and then carrying out the reversal of reactions of EMP to get glucose. The three irreversible reactions are bypassed by other enzymes. Eg. Glusose is synthesised by gluconeogenesis in *E.coli* from pyruvate, oxaloacetate and malate.

HMP Shunt

This is variously known as Hexose monophosphate pathway, pentose phosphate pathway/cycle, phosphogluconate pathway and Warburg-Dicken's pathway.

- Present in several Procaryotes, eukaryotes, plants and animals.
- It is useful to the cell in providing NADPH, providing precursors for various biosynthetic pathways, providing energy even in the absence of TCA cycle and metabolism of those pentoses which can be converted to intermediates of this pathway.
- It involves oxidative decarboxylation of glucose 6 phosphate to ribulose 5 phosphate followed by series of interconversions in which hexose and triose phosphates are formed from pentose phosphates.

Fat Metabolism

Fats are hydrolyzed to glycerol + 3 fatty acids by lipase enzyme. Glycerol is metabolized by glycolysis to form pyruvate. The fatty acids with even number of carbon atoms undergo a series of β oxidation. In each step the length of the fatty acid is reduced by two carbon atoms and acetylCoA is liberated. The acetylCoA molecules enter the TCA.

Protein metabolism

Proteins are first hydrolyzed into individual amino acids by various proteases. The amino acids are deaminated (removal of amino groups) and the deaminated molecules enter glycolysis or fementation or Kreb's cycle.

Photoautotrophy

The photoautotrophs capture energy from light and manufacture carbohydrates from CO_2. Green and purple bacteria, cyanobacteria, algae and higher plants carryout photosynthesis. In green plants, algae and cyanobacteria, oxygen is evolved during photosynthesis (oxygenic photosynthesis). They show two sets of reactions called *light reactions* and *dark reactions*. In the light reactions, light energy is converted to chemical energy – ATP through photophosphorylation. In the dark reactions, the chemical energy is used to manufacture carbohydrates through the Calvin Benson cycle.

In case of green and purple bacteria, the photosynthesis does not evolve oxygen (anoxygenic). Bacteriocholorophyll absorbs light of shorter wavelengths than that by chlorophyll *a*. Photosynthetic bacteria split H_2S instead of H_2O and hence produce elemental sulfur or strong sulphuric acid in place of oxygen.

Photoheterotrophy

Non sulfur purple and green bacteria derive energy from light but use organic substances such as alcohols, fatty acids or carbohydrates as carbon source.

Chemoautotrophy

Chemoautotrophs or chemolithotrophic bacteria oxidize inorganic substances for energy and CO_2 as carbon source. Nitrifying bacteria, non photosynthetic sulfur bacteria, iron bacteria and hydrogen bacteria are chemoautotrophs.

Chemoheterotrophy: Bacteria that are entirely dependent on organic substances for energy as well as carbon are termed chemoheterotrophs. These occur as parasites or saprophytes in aerobic or anaerobic conditions. Cultivation of these organisms in the lab requires complex media with organic nutrients such as peptone and beef extract.

Sr	Group	Representative genera
1	Photoautotrophs a. Green sulfur bacteria b. Purple sulfur bacteria	*Chlorobium* *Chromatium*
2	Photoheterotrophs a. Non sulfur purple bacteria b. Non sulfur green bacteria Chemoautotrophs (chemolithotrophs)	*Rhodomicrobium, Rhodospirillum* *Chloroflexus,* *Chloronema*
3	a. Nitrifying bacteria b. Nonphotosynthetic sulfur bacteria c. Iron bacteria d. Hydrogen bacteria	*Nitrobacter,* *Nitrosomonas* *Thiothrix* *Thiobacillus* *Siderocapsa* *Alkaligenes*
4	Chemoheterotrophs	*E.coli* and other commonly occurring bacteria

Sr	Group	Source of energy	Source of carbon	End product(s)
1	Photoautotrophs	Light	CO_2	Sulfur or sulphuric acid
2	Photoheterotrophs	Light	Organic substances	Generally $CO_2 + H_2O$
3	Chemoautotrophs i.e. Chemolithotrophs	HNO_2, NH_3, H_2S, S, Fe^{2+}, H_2	CO_2	HNO_3, $HNO_2 + H_2O$, $H_2O + S$, H_2SO_4, $Fe^{3+} + OH$, H_2
4	Chemoheterotrophs	Organic substances	Organic substances	$CO_2 + H_2O$, Organic acids, alcohols.

Exercises

1. Glycolysis is   [   ]
   a. Anaerobic metabolic process   b. Aerobic metabolic process
   c. Semi aerobic process   d. Amphibolic process

2. Net harvest of ED Pathway   [   ]
   a. 2ATP + 2molecules of reducing power
   b. 1ATP + 2molecules of reducing power
   c. 2ATP + 1molecules of reducing power
   d. 1ATP + 1molecules of reducing power

3. Which of these is part of aerobic respiration   [   ]
   a. Butyrate Production   b. Ethanol Production
   c. Lactic Production   d. Citrate Production

4. Which of these exhibits ED pathway   [   ]
   a. *Pseudomonas*   b. *Xanthomonas*   c. *Rhizobium*   d. All

5. Which of these exhibits ED pathway   [   ]
   a. *Serratia*   b. *Salmonella*   c. *Saccharomyces*   d. All

6. ATP generated in ETC is as a result of   [   ]
   a. Oxidative phosphorylation   b. Substrate level phosphorylation
   c. Anaerobic phosphorylation   d. Photophosphorylation

7. Apoenzyme is the protein portion of the enzyme   [   ]
   a. TRUE   b. FALSE

8. *Beggiatoa* utilizes H_2S as energy source. Hence it is a   [   ]
   a. Chemotroph   b. Autotroph   c. Phototroph   d. Organotroph

9. The main link between anabolism and catabolism   [   ]
   a. Pyrvate   b. ATP   c. AcetylcoA   d. Pi

10. Hydrolysis of sucrose to glucose and fructose is catalysed by   [   ]
    a. Glucosidase   b. Phosphoglucomutase
    c. Fructuranosidase   d. Glycogenphosphorylase

11. The first enzyme in the glycolytic pathway   [   ]
    a. Hexokinase   b. Glucosidase
    c. Phosphoglucoisomerase   d. Enolase

12. Which of these enzymes facilitates TCA cycle   [   ]
    a. Malate synthase   b. Isocitrate lyase
    c. Succinate dehydrogenase   d. All of these

13. Which of these may be called as *Anaplerotic pathway*   [   ]
    a. Glyoxylate Pathway   b. TCA Cycle
    c. Cyclic Photophosphorylation   d. ETC

14. The reaction of formation of AcetylcoA from Pyruvate involves release of   [   ]
    a. NADH   b. CO_2 c. Both   d. None

15. Pentose Phosphate Pathway (PPP) is also referred to as   [   ]
    a. Hexose Monophosphate Shunt   b. Phosphogluconate Pathway
    c. Both   d. None

16. The PPP is particularly useful for generation of   [   ]
    a. Reducing power   b. ATP   c. Pentoses   d. None of these

17. *Rhodomicrobium*   [   ]
    a. Purple non-sulfur bacterium   b. Green sulfur bacterium
    c. Non-photosynthetic sulfur bacterium   d. Hydrogen bacterium

18. The source of Carbon for chemolithotrophs   [   ]
    a. Alcohols   b. CO_2 c. Organic acids   d. CO

19. Which of these are iron bacteria   [   ]
    a. *Siderocapsa*   b. *Thiobacillus*   c. *Helicobacter*   d. *Leuconostoc*

20. Which of these is involved with iron metabolism in bacteria   [   ]
    a. Siderophores   b. Porins   c. CoA   d. Lipid A

21. Penicillin causes   [   ]
    a. Lyses of endospore bearers
    b. Lysis of actively growing bacterial cells.
    c. Lysis of cells in their stationary phase
    d. All.

22. The enzyme superoxide dismutase help bacteria to survive in   [   ]
    a. Oxygen rich environment   b. Oxygen depleted environment
    c. CO_2 rich environment   d. High pressure environment

23. Beta oxidation pathway followed during   [   ]
    a. Fatty acid degradation   b. Nucleic acid degradation
    c. Amino acid degradation   c. Lactic acid degradation

24. Gluconeogenesis is   [   ]
    a. Modified glycolytic pathway for higher energy yields
    b. Pathway for generation of pentoses from polysaccharides
    c. Supply of glucose from aminoacids, lactic acid, glycerol, etc.
    d. Pathway for polysaccharides from glucose and other sugars.

25. Photosynthetic bacteria are characterized by   [   ]
    a. Presence of chloroplasts   b. Non-cyclic photophosphorylation
    c. Cyclic photophosphorylation   d. Glyoxylate pathway

26. The unique polysaccharide seen only in bacteria   [   ]
    a. Chitin   b. Xyloglucan   c. Carrageenan   d. Murein

27. The polysaccharide in bacterial slime layer   [   ]
    a. Inulin   b. Dextran   c. Hyaluronic acid   d. Amylose

28. Which of these bacteria are sterol dependent   [   ]
    a. Rickettsias   b. Chlamydias   c. Mycoplasmas   d. Actinobacteria

29. All these are aromatic amino acids except   [   ]
    a. Cystein   b. Phenylalanine   c. Tyrosine   d. Tryptophan

30. Which of these enzymes catalyses ATP producing reaction in glycolysis   [   ]
    a. Phosphoglucoisomerase   b. Phosphoglyceromutase
    c. Enolase   d. Hexokinase

Answer Key and Validation

Microbial Metabolism and Physiology

1. a	6. a	11. a	16. c	21. b	26. d
2. b	7. a	12. c	17. a	22. a	27. b
3. d	8. a	13. a	18. b	23. a	28. c
4. d	9. b	14. c	19. a	24. c	29. a
5. a	10. c	15. c	20. a	25. c	30. b

12. *a and b facilitate the Glyoxylate cycle*

16. *PPP is a multipurpose pathway for generation of ATP, reducing power, and pentoses. However, ATP and reducing power may be obtained from other conventional catabolic pathways, but PPP is the primary source of pentoses.*

8

Microbial Genetics and Genetic Engineering

Retrace your Subject

Significant contributions, The genetic code, The gene, Mutations, DNA repair, Plasmids, Bacterial Recombination....

Significant contributions

- Fred Griffith's transformation experiment (1928) on *Streptococcus pneumoniae* lead to the understanding the some component of cells can transfer virulence to non-pathogenic strains from pathogenic ones.
- Oswald T Avery, C M MacLeod & M J McCarty (1944) proved for the first time that DNA can transform many characters into a cell, and that *DNA carried genetic information.*
- Alfred D Hershey & Martha Chase (1952) proved that *DNA was the genetic material in T2 bacteriophage.* They mixed *E. coli* cells with T2 viruses whose DNA was labelled with ^{32}P (or protein coat labelled with ^{35}S) and showed that only the DNA entered the bacterial cell leaving back protein coat. Yet complete bacteriophages developed within the cell which means DNA carries the genetic material.
- Marshall Nirenberg, Heinrich Matthaei, Philip Leder & Har Gobind Khorana (early 1960s) discovered *codon* a code word that involves a nucleotide triplet for a particular amino acid.

The genetic Code

- Robert W Holley (1968) won Nobel Prize for being the first person to sequence a nucleic acid (phenyalanyl-tRNA) along with Nirenberg & Khorana.
- The genetic code shows degeneracy. Up to six different codons may code for the same amino acid. Of the total 64 types of codons, 61 code for some or the other amino acid and are called *sense codons.* The remaining three are *non-sense* or *stop codons* (UGA, UAG and UAA) which are involved in the termination of translation.
- Although there are 61 sense codons, 61 corresponding tRNAs do not exist. Generally if the second and third anticodon positions complement with the first two bases on the mRNA codon, proper amino acid binding occurs. Thus, the 5′ nucleotide in the anticodon can vary. This is known as *wobble.* Wobbling avoids synthesis of large types of tRNAs and also minimises ill effects of DNA mutations. [Eg: Glycine mRNA codons: GGU, GGC, GGA & GGG. Glycine tRNA anticodons: ICC, CCC].
- The code is *non-overlapping* and there is a single starting point with one reading frame along one direction (some viruses and bacterial genes however do show overlapping genes rarely).

The Gene

- The gene is defined as a *linear sequence of nucleotides or codons with a fixed start point and end point, that codes for a polypeptide or rRNA or tRNA.*
- Earlier *one gene - one enzyme* hypothesis was postulated, but eventually modified to *one gene – one polypeptide* because some enzymes consist of 2 or more polypeptides coded by different genes. The segment that codes for a single polypeptide is referred to as *cistron.*

- All genes are not involved in protein synthesis. Some of them code for tRNA and rRNA.
- Prokaryotic and viral genes differ from eukaryotic genes. In almost all the cases (with very rare exceptions), the coding information in the cistron is continuous in bacteria and viruses. However, in eukaryotic genome, many genes contain the actual coding information (*exons*) and the non-coding sequences (*introns*) which interrupt the continuity of the coding.
- The beginning of a gene is at the 3′ end of the *sense strand* of DNA. An RNA polymerase recognition and regulatory site known as the *promoter* is located at the start of the gene.
- In *E.coli,* the promoter has two functions related to two specific segments of the promoter. i.e. the *RNA polymerase recognition site* (a consensus sequence of 5′ TTGACA$^{3′}$, located about 35bp before the transcriptional start point) and the *RNA polymerase binding site* (a consensus sequence of 5′ TATAAT$^{3′}$ located at the   minus 10 region. It is also called the *pribnow box).*
- The initially transcribed portion of the gene is not always the sense material. It is a *leader sequence* which is not actually translated to protein, but is imp in initiation of translation and in regulation of transcription. In procaryotes, the leader is the *Shine-Dalgarno sequence* i.e. 5′AGGA3′).
- The actual coding region of every prokaryotic gene begins with 3′TAC5′ which produces 5′AUG3′ the *initiation codon* which codes for *N-formylmethionine* a modified form of methionine.
- Transcription stops at a *terminator sequence* which lies after a non-translated *trailer sequence.*
- Besides the promotor, leader, coding region, trailer and terminator sequences, the genes also have regulatory sites such as the *operator* and *CAP binding sites,* where specific proteins bind to stimulate or inhibit the protein synthesis.

Mutations

- Mutation [Latin = to change] is a *stable heritable change in the nucleotide sequence of DNA.* Many of these changes are largely harmful, affecting the phenotypic characteristics of the organisms some times even causing death of the organisms (lethalmutations). However, certain mutations generate newer strains of organisms with desirable traits. Many microbial mutations are deliberately executed, and put to uses in industries, agriculture, pollution control and research laboratories.
- **Conditional mutations** are those that are expressed only under certain environmental conditions. Eg. Death of a mutant at temp slightly above the optimum growth temp.
- **Directed or Adaptive mutations:** The cells undergo mutations after 'choosing' the best change to adapt to the surroundings for survival. Eg: Strains of *E.coli,* which lost the ability to use lactose, suddenly become lactose fermentors if they are in a medium where lactose is the only source of carbon.
- **Spontaneous mutations:** Changes that develop in the absence of any added agent are spontaneous mutations. These occur occasionally in nature and not necessarily helpful.
- **Induced mutations:** These are a result of deliberate or accidental exposure of the organism to some physical or chemical mutagen. They are of great value in generating potentially beneficial strains.
- **Hypermutation:** Some starving bacteria might rapidly generate multiple mutations through activation of special mutator genes. In such random process, the rate of production of favorable mutants would increase, many unfavorable mutants would die.
- **Transition mutations:** Sometimes, bases can take imino or enol form instead of the keto form (tautomeric shifts). This results in allowing purine for purine or pyrimidine for pyrimidine substitution that lead to stable alteration of the sequence. These are called transition mutations and occur relatively more frequently.
- **Transversion mutations:** A purine is substituted for a pyrimidine, or vice versa. These mutations occur very rarely due to steric problems.

- **Frameshift mutations:** Usually caused by addition or deletion of a nucleotide base or DNA segments resulting in an altered codon reading frame. (Transition, transversion and frameshift mutations are regarded as spontaneous mutations).

- **Mutations at protein level :** *Silent mutations* (triplet codes for same amino acid AGG to CGG both code for Arg), *Neutral mutations* (codes for different, but functionally equivalent amino acid AAA to AGA Lys to Arg), *Missense mutations* (codes for structurally and functionally different amino acid), *Nonsense mutations* (codes for chain temination CAG (Gln) to UAG (nonsense codon).

- **Reverse mutations:** *True reversion* AAA (wild type Lys) to GAA (mutant Glu) again to AAA (reversion - Lys wild type). Equivalent reversion (UCC (ser- wild type) to UGC (Cys mutant) again to AGC (Ser - phenotypically wild type).

- **Mutagens:** Four modes of mutagen action are: Base analog incorporation, specific mispairing, intercalation and bypass of replication.

Induced Mutagen Principle	Mode	Example
Base analog	Substances similar to nitrogenous bases and get incorporated in the growing polynucleotide chain during replication. They show different base pairing properties and cause stable mutations.	**5-Bromouracil** (5-BU) an analog to thymine. It undergoes enol form shift and causes pairing with guanine in place of adenine.
Specific mispairing	Mutagen changes the base's structure and alters its pairing characteristics. Many of these mutagens are selective and cause defined damage to the DNA.	**Methylnitrosoguanidine** an alkylating agent that adds methyl groups to guanine causing mispair with thymine. (Results in GC-AT transition). **Ethylmehylsufonate** and **hydroxylamine** are other examples.
Intercalating agent	Distort DNA to induce single nucleotide pair insertions and deletions. This results in mutations due to formation of a *loop* in the DNA.	Acridines such as **proflavin** and **acridine orange.**
Carcinogens	Directly damage the bases severely, base pairing is impaired and DNA cannot serve as template. The damage to DNA could be lethal, but generally but repair mechanism is triggered that restores the damaged DNA material with minimum errors.	**UV radiation** generates cyclobutane type dimers (thymine dimers) between adjacent pyrimidines. Other examples of carcinogenic mutagens include ionizing radiations, Aflatoxin B1 and other benzopyrene derivatives.

DNA repair

- The processes by which a cell corrects the damages done to its DNA is called DNA repair

- Exposure to UV light or radiations may cause mild to serious lesions in the DNA molecule that can lead to mutations. Hence, DNA repair mechanisms are constantly active within the cell and reverts any damages caused to the cells. The common types of damages include oxidation, alkylation or hydrolysis of bases.

- The repair mechanisms are dependent on type of the cell, age of the cell, nature of environment and the extent of damage caused.

- Damage beyond repair leads to cell dormancy (senescence), cell suicide (apoptosis) or unregulated cell division (cancer).

- Major Types of repair mechanisms:
  - Direct reversal: Enzymes such as photolyases and methyl guanine methyl transferases are involved.
  - Base excision repair (BER): The damaged base is removed by a DNA glycosylase, resynthesized by a DNA polymerase, and a DNA ligase performs the final nick-sealing.

- Nucleotide excision repair (NER): Employs several NER enzymes at the time of transcription to check major helix distorting lesions.
- Mismatch repair(MMR): Employs factors that repair mispairing at the time of replication or recombination

Plasmids	
Definition	Extra-chromosomal DNA molecule distinct from the chromosomal DNA capable of replicating independently of the chromosomal DNA. The term *plasmid* was first introduced by the American molecular biologist Joshua Lederberg in 1952.
Size, Structure and Number	Most commonly plasmids are circular and double-stranded, ranging from 1 to over 200 kilobase pairs (kbp). The number of identical plasmids within a single cell can range anywhere from one to even thousands under some circumstances.
Occurrence	Occur naturally in bacteria, but are sometimes found in eukaryotic organisms such as *Saccharomyces cerevisiae*.
Transfer	Plasmids are transferred directly by conjugation. They may also be intentionally transferred through transformation and specialized transduction.
Application	<ul><li>Plasmids serve as excellent means of horizontal gene transfer among bacteria.</li><li>Some of them carry genes that provide resistance to naturally occurring antibiotics</li><li>The proteins encoded by certain plasmids serve as toxins</li><li>Plasmids may also can transform non-nitrogen fixers to nitrogen fixing bacteria under conditions of nutrient deprivation.</li></ul>
Types	<ul><li>F-plasmids (Fertility): Contain tra-genes that bring about conjugation.They are capable of conjugation (transfer of genetic material between bacteria which are touching).</li><li>R Plasmids (Resistance): Contain genes that can induce resistance against antibiotics. They were known to man even before the official discovery of plasmids, then called R-factors.</li><li>Degradative plasmids: Impart the ability to degrade certain unusual substances, e.g., toluene or salicylic acid</li><li>V_i Plasmids (Virulence): Resposible for the pathogenicity of a bacterium. Eg. V_i Plasmids in *Salmonella, Yersinia* and *Shigella*.</li><li>Col-plasmids: Contain genes responsible for production of bacteriocins, proteins that can kill other bacteria.</li></ul>

Bacterial Recombination	
Conjugation	<ul><li>Transfer of genetic material between bacteria through direct cell-to-cell contact.</li><li>Discovered in 1946 by Joshua Lederberg and Edward Tatum</li><li>Sometimes regarded as sexual reproduction of bacteria, but no gametes are formed and no new progeny of cell is obtained.</li><li>The process involves a donor call and a recipient cell. The donor cell contains a conjugative (sex) plasmid and bears sex pili on the cell surface.</li><li>The conjugative plasmid is the F plasmid and hence the donor cells are also called F^+ cells while the recipient cells are F^- cells.</li><li>F-plasmid carries a *tra* and a *trb* locus that consist of about 40 genes. The *tra* locus includes the *pilin* gene and regulatory genes, for the formation of pili on the cell surface. Some proteins coded in the *trb* loci also seem to be responsible for the actual exchange of DNA.</li><li>After conjugation the recipient cell gains certain new properties by which they may be identified. These include: antibiotic resistance, xenobiotic tolerance, ability to utilize a new metabolite.</li><li>*Hfr* Strain: An F^+ cell, in which the conjugative plasmid is integrated into its genome is called *Hfr* cell (high frequency recombination cells). When an *Hfr* cell conjugates with an F^- cell, it attempts to transfer their *entire* DNA through the mating bridge. Along with the plasmid DNA, the bacterial genome is also dragged into the recipient</li></ul>

	cell. Such mating is especially useful in gene mapping and study of genetic linkage. The experiments are called interrupted mating experiments. • F prime (F′) cell: Sometimes the integrated plasmid in an *Hfr* cell dissociates from the actual genome. During dissociation, the plasmid may erroneously bring out some bacterial genes with it. Such cells, bearing the F plasmid with bacterial genes are called F prime cells and the plasmid is the F prime plasmid. Such cells are instrumental in transfer of bacterial genes from cell to cell. The process is called sexduction.
Transformation	• The process by which bacterial cells take up naked DNA molecules • Transformation was first demonstrated by **Fred Griffith** in 1928in *Streptococcus pneumoniae* (Earlier *Pneumonococcus* or *Diplococcus*). In 1944, **Avery, McCarty** and **MacLeod** proved that DNA was the transforming principle. • The process of transformation is similar in both gram negative and gram positive bacteria. The DNA fragments first binds to the exterior of the cell. The fragment is reaches the periplasm in gram negative bacteria. Finally, one DNA strand is taken up into the cytoplasm while the other is degraded. Recombination is quick in gram negative bacteria than in gram positive bacteria. Gram positive bacteria require a DNA-binding protein on the surface of the cell. • Natural transformation: Some cells are competent enough to take up the DNA matter from the medium naturally due to genetically programmed physiological state. • Artificial Transformation: In this, the cells are made competent to receive DNA by using $CaCl_2$ and heat shock treatment. The competent cells are made to take up the DNA by techniques such as electroporation, protoplast formation, and microprojectiles. • Transformation of eukaryotic cells in tissue culture is usually called transfection • Bacterial transformation is commonly demonstrated by using the genes for drug resistance. • Transformation can be used to assess gene linkage and the possible distance between two genes. Natural transformation gifts the recipient bacteria with a pool of extra nucleotides.
Transduction	• The process by which DNA is transferred from one bacterium to another by a virus Bacteriophage). This is commonly used by molecular biologists to introduce a foreign gene into a host cell's genome • Transduction in Lysogenic cycle: In a typical lysogenic cycle, the phage chromosome is integrated into the bacterial chromosome and remain as prophage so for several generations. Sometimes, the prophage may undergo induction (usually by UV light) and the phage genome dissociates from the bacterial chromosome and initiates the lytic cycle. The lytic cycle leads to the production of new phage particles which are released by lysis of the host. • However, due to faulty dissociation, the phage genome may be excised along with some bacterial genes which in turn get packed into phage heads during replication. The phages thus produced infect fresh hosts and cause transfer of bacterial genes from cell to cell. • Transduction in Lytic cycle: In lytic cyle, a piece of bacterial genome may get packed into the phage head. The new virus capsule now loaded with bacterial DNA continues to infect another bacterial cell. This bacterial material may become recombined into another bacterium upon infection. • Generalised Transduction: The random and natural process of insertion of any part of the bacterial DNA into phage head and its subsequent transfer to other cells. • Restricted transduction: Generally carried out under controlled conditions. The genes packed into the viral head are intentional and known so that selectively the desired genes can be transferred from cell to cell via viral vectors. • Transduction was discovered by Norton Zinder and Joshua Lederberg in 1951

Exercises

1. The work of Fred Griffith on the transfer of virulence was with   [   ]
   - a. *Staphylococcus aureus*
   - b. *Streptococccus pneumoniae*
   - c. *Klebsiella pneumoniae*
   - d. *E.coli*

2. DNA is the genetic material in the T_2 bacteriophage   [   ]
   - a. F W Redi
   - b. Avery & MacLeod
   - c. Har Gobind Khorana
   - d. Hershey & Chase

3. Griffith's transforming principle is DNA and therefore DNA is the genetic material   [   ]
   - a. F W Redi
   - b. Avery, MacLeod & MacCarty
   - c. Har Gobind Khorana
   - d. Hershey & Chase

4. There are up to six different codons for a given amino acid. This is referred to as   [   ]
   - a. Code wobbling
   - b. Code degeneracy
   - c. Code specificity
   - d. Code universality

5. Beadle and Tatum are known for   [   ]
   - a. Discovery of generalized transduction
   - b. Description of bacterial conjugation
   - c. One-gene-one enzyme hypothesis
   - d. DNA carries information during transformation

6. Match the discoverers with the discoveries   [   ]
   - A. Lederberg & Tatum
   - B. Griffth
   - C. Zinder & Lederberg
   - D. Jerne & Burnet
   - i. Bacterial transformation
   - ii. Generalized transduction
   - iii. Bacterial conjugation
   - iv. Clonal selection theory

   - a. A-ii, B-iii, C- i, D- iv
   - b. A-ii, B-i, C- iv, D- iii
   - c. A-iii, B-i, C- ii, D- iv
   - d. A-i, B-iii, C- iv, D- ii

7. Lysogenic bacteriophages were introduced by   [   ]
   - a. Ruska
   - b. Chatton
   - c. Lwoff
   - d. Twort

8. Pick out the WRONG pair   [   ]
   - a. Jacob & Wollman – F factor is a plasmid
   - b. Jacob & Monod - Elucidation of genetic code
   - c. Arber & Smith – Restriction endonucleases
   - d. Kohler & Milstein – Monoclonal Abs

9. The Ames test is   [   ]
   - a. Bacterial assay for detection of mutagens and carcinogens
   - b. Detection of R plasmids in pathogenic bacteria
   - c. Assay of immune tolerance levels
   - d. Assay of endonuclease activities

10. The Polymerase Chain Reaction [PCR] was developed by   [   ]
    - a. Baltimore
    - b. Chase
    - c. Khorana
    - d. Mullis

11. In the conventional understanding of the genetic code, the correct number of codons   [   ]
    a. 61 sense codons + 3 Nonsense codons = 64
    b. 60 sense codons + 3 Nonsense codons = 63
    c. 57 sense codons + 3 Nonsense codons = 60
    d. 62 sense codons + 4 Nonsense codons = 65

12. The segment that codes for a single polypeptide is referred to as   [   ]
    a. Cistron   b. Reading frame   c. Sense segment   d. Trailor sequence

13. One-gene-one hypothesis is more appropriately modified as   [   ]
    a. One gene one RNA   b. One gene one protein
    c. One gene one polypeptide   d. One gene one function

14. The *Shine Dalgarno* sequence in prokaryotes   [   ]
    a. 5'AGG A3'   b. 3' TAC 5'   c. 5' AUG 3'   d. 5' TTGACA3'

15. The *Consensus sequence* of the – 10 region Pribnow Box   [   ]
    a. 5' AGGA3'   b. 5' TATAAT3'   c. 5' TTCACA3'   d. 5' AUe,3'

16. Autotrophic mutants are result of   [   ]
    a. Conditional mutations   b. Biochemical mutations
    c. Adaptive mutations   d. Hypermutations

17. A mutation in which purine is substituted for a pyrimidine is   [   ]
    a. Reverse mutation   b. Transversion
    c. Transition   d. Frame shift mutation

18. Mutations that affect only one base pair in a given location are   [   ]
    a. Forward mutation   b. Intercalating mutation
    c. Point mutations   d. Missense mutations

19. A second mutation that changes a mutant to a wild type organism is   [   ]
    a. Back mutation   b. Reversion mutation
    c. both a & b   d. Wild type mutation

20. A second mutation that overcomes the effect of the first mutation is   [   ]
    a. Superior mutation   b. Secondary mutation
    c. Forward mutation   d. Null mutation

21. AAA (Lys) → GAA (glu) → AAA (Lys)   [   ]
    a. True reversion   b. Intragenic suppression
    c. Equivalent Reversion   d. Physiological suppression mutation.

22. UCC (Ser) → UGC (Cys) → AGC (Ser)   [   ]
    a. True reversion   b. Intragenic suppression
    c. Equivalent Reversion   d. Physiological suppression mutation

23. UAG → UAU mutation is   [   ]
    a. Nonsense Suppression   b. Transition
    c. Nonsense mutation   d. Neutral mutation

24. CAG → UAG mutation is   [   ]
    a. Nonsense Suppression   b. Transition
    c. Nonsense mutation   d. Neutral mutation

25. Replica plating finds use in detecting & isolating   [   ]
    a. Auxotrophs   b. Autotrophs
    c. Improved strains   d. Plasmids

26. Repair of thymine dimers is brought about by [ ]
    a. Excision repair   b. Recombination repair
    c. Oligodynamic reactions   d. Photoreactivation

27. SOS repair is brought about using [ ]
    a. Lex A repressor   b. RecA protein
    c. UV rays   d. Visible light exposure.

28. A DNA molecule or sequence that has a replication origin and is capable of
    being replicated is called [ ]
    a.. Plasmid   b. Episome   c. Cistron   d. Replicon

29. Plasmids can be eliminated from host cells in a process known as [ ]
    a. Transposition   b. Deplasmidation   c. Curing   d. Lysogeny

30. F' x F⁻ conjugation leads to [ ]
    a. Transposition   b. Transduction   c. Sexduction   d. Transformation

31. The low frequency transduction lysates can be corrected to high frequency transduction lysates
    if there is a normal λ phage in the same host cell [ ]
    a. TRUE   b. FALSE

32. Reverse transcriptases can be used to construct [ ]
    a. c DNA   b. Viral envelopes   c. mRNA   d. All

33. Which of these is associated with detection of specific DNA fragments [ ]
    a. Nothern blotting   b. Southern blotting
    c. Western blotting   d. Eastern blotting

34. Taq polymerase obtained from the bacterium the *Thermus aquaticus* finds application in [ ]
    a. PCR Technique   b. DNA electrophoresis
    c. Western blotting   d. All

35. Plasmids that contain λ phage cos sites and can be packaged into phage capsids are [ ]
    a. Somatostatin vectors   b. Cosmids
    c. λ plasmids   d. Chimera

36. Nitrogen base molecules that contain single ring structure [ ]
    a. Pyrimidines   b. Purines
    c. Adenine & Cytosine   d. Guanine & Thymine

37. Which of these in NOT a component of RNA [ ]
    a. Phosphoric acid   b. Adenine   c. uracil   d. Thymine

38. Number of hydrogen bonds between cytosine & guanine [ ]
    a. 2   b. 3   c. 4   d. none

39. Structurally, uracil closely resembles [ ]
    a. Thymine   b. Cytosine   c. Adenine   d. Guanine

40. The sugar-phosphate strand of a DNA grows from [ ]
    a. 5' to 3' end   b. 3' to 5' end   c. either a or b   d. both ends.

41. Passing genetic information from one nucleic acid to another [ ]
    a. Replication   b. Transcription   c. Translation   d. all three

42. Introns are absent in [ ]
    a. Procaryotes   b. Eucaryotes   c. Archaea only   d Protozoa only

43. Direct Linkage of nucleic acid fragments [ ]
    a. Insertion   b. Restriction   c. Ligation   d. Promotion

44. What is TRUE about gyrases   [   ]
    a. Type II topoisomerases
    b. Convert relaxed DNA to super helical form
    c. ATP dependent
    d. a, b & c are TRUE

45. Transcription of RNA nucleotide sequence into cDNA requires   [   ]
    a. Beta Replicase   b. Rec A Protein   c. Capping enzyme   d. Reverse transcriptase

46. This was largely used in the Human genome project   [   ]
    a. Shotgun method
    b. cDNA method
    c. Fragmentation by Mechanical shearing.
    d. Automated Polynucleotide synthesizer.

47. Nucleic acid probe is used to detect the cloned gene in the recombinent cell   [   ]
    a. TRUE   b. FALSE

48. Friedrich Miescher (1844-1895) described the material found in cell nuclei as __________.   [   ]
    a. Nuclein   b. Nucleic acid   c. Nucleoprotein   d. Protamine

49. Histone   [   ]
    a. Protein in fish sperm
    b. Protein part of the nucleoprotein
    c. A short *Arg* polypeptite chain in the nucleotide
    d. Nucleic acid portion of nucleoprotein

50. A nucleotide consists of   [   ]
    a. A sugar + N containing base
    b. A sugar + PO_4 group
    c. A sugar + PO_4 group + N containing base
    d. PO_4 group + N containing base

51. The deoxyribose sugar lacks   [   ]
    a. Oxygen atom in the 2nd C atom
    b. Oxygen atom in the 3rd C atom
    c. Oxygen atom between the 1st and 4th C atoms
    d. OH groups

52. In a DNA/RNA, the PO_4 group is attached   [   ]
    a. To the 1st C of pentose sugar
    b. To the 5th C of pentose sugar
    c. To the N containing base
    d. To the oxygen between 1□and 5□C of the pentose sugar

53. In case of a nucleotide, which of the statement is CORRECT?   [   ]
    a. The N containing base is always attached to the sugar at its number one C position
    b. The purine bases are attached to the sugar at its number one and pyrimidines at its number three C position
    c. The purine bases are attached to the sugar at its number three and pyrimidines at its number three C position
    d. The N containing base is attached to the sugar and PO_4 group on either side.

54. DNA and RNA are termed 'acids' because   [   ]
    a. Their sugars contain COOH group
    b. Purines / Pyrimidines contains COOH group
    c. Nucleotides are esters of phosphoric acid
    d. All of these

55. The two main pyrimidines found in RNA are [   ]
    a. C & T b. C & G c. T & U d. C & U

56. The difference between purines and pyrimidines [   ]
    a. Pyrimidines contain one C N ring, while purine contain 2 C N rings
    b. Amino group present in purines, absent in Pyrimidines
    c. Keto group occurs in Pyrimidines, absent in purines
    d. All

57. 5 Bromouracil is an analog to [   ]
    a. Uracil b. Adenine c. Thymine d. Cytosine

58. F' Cells [   ]
    a. Contains no F plasmid
    b. Are the same as *Hfr* cells
    c. Contain F palsmid with a small segment of bacterial chromosome
    d. Cells that have a high potential to become *Hfr* cells

59. Transfer of bacterial chromosome form one cell to the other through F palsmid [   ]
    a. Retrotransfer b. Restricted transduction
    c. Sexduction d. High resolution mapping

60. Holliday model explains [   ]
    a. Role of surface proteins in conjugation
    b. Genetic map of F palsmid
    c. Low frequency transduction
    d. General recombination.

61. The change of non virulent bacteria into virulent bacteria was demonstrated via [   ]
    a. Transduction b. Transformation
    c. Conjugation d. rDNA technology

62. DNA is the genetic material [   ]
    a. D Hershey and Martha Chase b. O T Avery, C M MacLoed and M J MacCartey
    c. Fred Griffth d. Oswald T Avery

63. The Amino acid that can be coded by 6 different codons [   ]
    a. *Arg* b. *Gly* c. *Val* d. *Pro*

64. *The genetic code is degenerate* implies that [   ]
    a. There may be different codons for a single amino acid
    b. There may be same codon for more than one amino acid
    c. There may be more than one nonsense codons
    d. Every codon is always a triplet

65. Wobble hypothesis [   ]
    a. Anticodons are not exclusive to a particular codon
    b. Proper amino acid will bind, even if only the second and third positions of the anticodon are complementary to the codon.
    c. Proper amino acid will bind, even if only the central nucleotide of the anticodon is complementary to the codon.
    d. Proper amino acid will bind, even if only the first and second positions of the anticodon are complementary to the codon.

66. Cistron　　　　　　　　　　　　　　　　　　　　　　　　　　　　　　[　]
    a. The cluster of DNA segment that code for a complete enzyme made of many polypeptides
    b. The cluster of genes that operate collectively to regulate a particular metabolic function
    c. The sequence of DNA that promotes protein synthesis
    d. The segment of DNA that codes for a single polypeptide

67. Which of these is most acceptable　　　　　　　　　　　　　　　　　　　[　]
    a. One gene-One enzyme hypothesis　　　b. One gene-One polypeptide hypothesis
    c. One gene-One promotor hypothesis　　d. One gene-multiple protein hypothesis

68. In a typical bacterial Structural Gene, translation of mRNA begins with the codon　[　]
    a. AUG　　　　　b. AUA　　　　　c. AGG　　　　　d. AGU

69. Most genes contain coding information (*exons*) interrupeted by noncoding sequences
    (*Introns*) – This is TRUE in case of　　　　　　　　　　　　　　　　　[　]
    a. Bacteria　　　b. Viruses　　　c. Eukaryotes　　　d. Archae

70. 5′TTGACA 3′　　　　　　　　　　　　　　　　　　　　　　　　　　　[　]
    a. RNA polymerase recognition site　　b. RNA polymerase binding site
    c. The – 10 region　　　　　　　　　　d. The -35 region

71. *Pribnow Box*
    a. 5′TTGACA 3′
    b. The start point of polypeptide synthesis on the mRNA　　　　　　　　[　]
    c. The region where RNA polymerase begins to unwind the DNA for transcription
    d. The region where DNA transfer begins in conjugation

72. The first part of mRNA which is not usually translated, but useful in initiation of translation　[　]
    a. Leader sequence　　　　　　　　　b. Regulator sequence
    c. Trailer sequence　　　　　　　　　d. Nonsense sequence

73. In prokaryotes, mRNA begins with a special sequence of *Shine Dalgarno* which is　[　]
    a. 3'AAGA 5'　　　b. 5'AAGA 3'　　　c. 3'AUGU 5'　　　d. 5'AUGU 3'

74. A letheal mutation in *E.coli,* which is expressed in high temp but not seen in low temp
    incubation is an example of　　　　　　　　　　　　　　　　　　　　　[　]
    a. Auxotrophy
    b. Hypermutation
    c. Conditional mutation
    d. Adaptive mutation

75. Microbial strains that can grow on minimal medium are best described　　　　[　]
    a. Auxotrophs　　　b. Prototrophs　　　c. Transvertants　　　d. Phototrophs

76. DNA was first proved as the genetic material based on the experiments on　　　[　]
    a. *E.coli*　　　　b. *Pneumococci*　　　c. *Neurospora*　　　d. T$_2$ Bacteriophage

Answer Key and Validation

Microbial Genetics and Genetic Engineering

1. b	12. a	25. a	38. b	51. a	64. a
2. d	13. c	26. d	39. a	52. b	65. b
3. b	14. a	27. b	40. a	53. a	66. d
4. c	15. b	28. d	41. b	54. c	67. b
5. c	16. b	29. c	42. a	55. d	68. a
6. c	17. b	30. c	43. c	56. a	69. c
7. c	18. c	31. a	44. d	57. c	70. a
8. b	19. b	32. a	45. d	58. c	71. c
9. a	20. a	33. b	46. a	59. c	72. a
10. d	21. a	34. a	47. a	60. d	73. b
9. a	22. c	35. b	48. a	61. b	74. c
10. d	23. a	36. a	49. b	62. a	75. b
11. a	24. c	37. d	50. c	63. a	76. d

8   *The genetic code was elucidated by Nirenberg, Khorana & others. Jacob & Monod are known for he operon model of gene regulation*

23   *When nonsense triplet codes for an a acid, chain termination is suppressed. Hence it is called a ense suppression*

24.   *When triplet codes for chain termination, it is a nonsense mutation.*

27.   *In some mutations, the DNA damage leads to large. The Rec A Protein binds to such gaps & initiate strand exchange this is referred to a SUS Repair*

42.   *Introns are non coding sequence of DNA that are enzymatic ally removed from the transcribed RNA, before translation*

57.   *5 Bromoaracil is an analog to theymine and causes pairing with guanine in place of adenine*

62.   *Fortunately Hershey and Chase happened to work on T_2 bacteriophage which is a DNA virus. If they had done the experiment with any RNA virus, it would have created much confusion about the universal genetic material.*

63.   *Arg is coded by the codons CGU, CGC, CGA, CGG, AGA and AGG*

73.   *Shine Dalgarno sequence in prokaryotes is the Leader Sequence. The beginning of mRNA leader with 16S rRNA orients the mRNA on the ribosome properly.*

9
Virology

Retrace your Subject...

Discovery of viruses, General characteristics, Structure and Composition of viruses, viral groups, Important viral Profiles, Viral pathology, Viroids, prions, Emerging viruses

Discovery of viruses

Scientist	Observation/Discovery
Dimitri Ivanowsky (1892) – Russia	Reported pathogenic agents smaller than bacteria. The causative agent of *tobacco mosaic disease* could be filtered through bacteria proof porcelain filters.
Martinus Beijerinck (1898) – Holland	Made the same observation and established that *Some pathogens are smaller than bacteria and can pass through all bacteria proof filters.* He called these agents as *contagium vivium fluidum* (contageous poisonous fluid). Later the term was simplified to **Virus** (Latin meaning *poison*).
Friedrich Loeffler & Paul Frosch (1898) – Germany	The causative agent of *Foot and Mouth disease* of cattle was also filterable through bacteria proof filters. *These pathogens are smaller than any known microorganisms and they cannot multiply ouside the host body.*
Viklhelm Ellerman & Olaf Bang (1908)	Observed viral leukemias in chicken.
Peyton Rous (1911)	Rous Sarcoma Virus (RSV) causes solid tumors in chickens.
Fredrick Twort (1915)	Discovered viruses that infect bacteria.
Felix d'Hérelle (1917)	Named bacterial viruses as **bacteriophages** (bacteria eaters). Twort & d'Hérelle observed *glassy phenomenon* of viral plaques on agar media contaminated with bacteria. This laid a foundation for the field of Molecular Biology.

General characteristics, structure and composition of viruses

- Viruses are non-cellular microbes, consisting of nucleic acid within a protein coat.
- They can replicate only within specific animal, plant or microbial cells.
- The viral genome (DNA or RNA but never both) encodes structural proteins.
- The virus has no intrinsic metabolism. It depends entirely on the host cell for energy, nucleotides and amino acids.
- An individual virus particle (*virion)* may be isometric, rod shaped, filamentous or even pleomorphic.
- Most commonly, viruses are made up of genome enclosed in a shell like protein capsid, the **nucleocapsid** structure. The capsids (protein coats) are further made of repeating subunits called **capsomeres** Some virions may also have an external lipoprotein **envelope.**
- Size wise, viruses range from 20-25nm (Eg. Parvoviridae, picornaviridae) to 200-450nm (Eg: Poxviridae) and sometimes even as large as 2000nm (Eg: Closterovidae).
- Viral **replication cylce** includes *attachment or adsorption* to host cell by receptors, *Penetration* of the virion or its genome into the cell, *uncoating* of viral genome, *gene expression.* The *gene replication*

requires virus encoded enzymes and host cell components. *Virus assembly* occurs in host cytoplasm or nucleus, followed by *maturation and Release* by cell lysis, exocytosis or budding.

- Viral **Classification** is based on host cell specificity, structure, composition, genome and replication strategy. Some viruses are classified into families (ending with *viridae*), subfamilies (ending with *virinae*), genera (ending with *virus*) and even sometimes into species. Strains of viral species are differentiated by antigenic determinants.

- Viral **cultivation** includes different techniques: Animal viruses: *Chick embryo technique, tissue culture technique, cultivation in continuous cell lines and cultivation in live animals.* Plant viruses: *tissue culture technique, sap application to healthy plants.* Bacteriophages: Isolation of *plaques* on agar media seeded with specific bacterial host cells.

- Viral **assay** is done by enumeration of *pocks* in chick embryo, CPE areas (cytopathic effect areas) in tissue culture media, local lesion on host leaves and *pfu* (plaque forming units) for bacteriophages. Some laboratories count viruses by electron microscopy but this does not differentiate the live and killed virions.

- Viral **Identification** is based on certain diagnostic tests as: ELISA, complement Fixation Tests, haemadsorption inhibition test, haemagglutination inhibition test, immunofluorescence, immunosorbant electron microscopy tests and neutralisation tests.

- Virus **inactivation:** Viruses are inactivated usually by heat $50\text{-}80^0\text{C}$, UV light, ionic radiations, dessication, antiseptics and disinfectants. Enveloped viruses are inactivated by ether & chloroform and by detergents. Specific antiviral agents are also employed for therapeutic purposes.

Viral Groups

Viral Group	Examples
dsDNA Animal viruses	
Herpes viridae	Herpes simplex types I & II, Epstein Barr (EB) virus
Poxviridae	Vaccinia virus, variola major (small pox)
Adenoviridae	Human adenovirus
Papovaviridae	Polyoma virus, SV40
Others: Baculoviridae, Hepadna viridae, Iridoviridae	
ssDNA Animal viruses	
Parvoviridae	Adeno Associated Virus (AAV), mouse minute virus (MMV)
Porcine circo virus	
dsRNA Animal viruses	
Reoviridae	Infantile gastroenteritis virus
Birnaviridae	
ssRNA Animal viruses	
Coronaviridae	Human common cold, GI disease viruses
Retroviridae	Sarcoma, leukemia, RVS Mouse viruses.
Togaviridae (Arboviruses)	TypeA (alpha) encephalitis viruses,
Flaviviridae	TypeB (Flavi) encephalitis viruses.
Picornaviridae	Polio virus, Coxackie, Rhinovirus, Foot & mouth disease virus
Paramyxoviridae	Mumps virus, Measles virus, Sendai virus
Rhabdoviridae	Rabies virus, Vesicular stomatitis virus (VSV),
Arenaviridae	Lassa viruses (genralised infections)
Bunya viridae	Bunyawera virus
Orthomyxoviridae	Influenza virus
Others: Toroviridae, Caliliciviridae, Nodaviridae.	

Viral Group	Examples
dsDNA Plant viruses Caulimovirus	Cauliflower Mosaic Virus, Dahlia Mosaic Virus
ssDNA Plant viruses	Geminivirus
dsRNA Plant viruses Reoviridae	Plant reoviruses: Wound tumor Virus (WTV), Rice Dwarf Virus (RDV), Fiji Disease Virus (FDV), Maize Rough Dwarf Virus (MRDV), rice ragged stunt virus.
ssRNA Plant viruses Tobamoviridae Potexviridae Carlaviridae Potyviridae Tobra virus Tombusviridae Tymoviridae Comoviridae Nepoviridae	 TMV, Cucumber green mottle mosaic virus, Tomato Mosaic virus Potato virus X, Papaya mosaic virus, Cassava common mosaic virus Carnation latent virus, Chicory blotch virus, Hop latent virus Potato virusY , Beet mosaic virus, pea mosaic virus, pepper mottle & soy bean mosaic viruses Tobacco Rattle virus (TRV), Pea early browning virus Tomato bushy stunt virus (TBSV), Turnip Crinkle Virus. Turnip yello mosaic virus (TYMV), cocoa yellow mosaic virus. Cowpea mosaic virus, Bean Pod mottle virus, Radish mosaic virus, Ring spot viruses of tobacco, tomato, raspberry & strawberry
Bacteriophage	
ssDNA Phages	Icosahedral:ΦX174, ΦR, α3, St 1, 6SR, BR2, U3 and G series (G4, G6, G13, G14). Helical: Ff group (fd, f1, M13, Zj/2, Ec9, AE2, HR and δA). If group (If1, If2).
dsDNA Phages	T even phages of *E.coli,* T2, T4, T6. T odd phages of *E.coli,* T1, T3, T5, T7. *E.coli* phages μ1, λ and Φ80. The hemophilous phage HP1, P1 phage of *E.coli,* and *Salmonella,* the *Salmonella phages* P22 and X, *Bacillus subtilis phages* PBSX, PBS1, PBS2, SPO1 & 2, *Pseudomonas* phage PM2, N1 phage of cyanobacteria.
dsRNA Phages	Φ6
ssRNA Phages	f2, MS2, R17, M12, fr and FH5 and QB.

Important viral Profiles

Virus	Structure, composition and Multiplication
TMV	Rod shaped (300nm x 15-18nm diameter), helical, ssRNA virus. The virion consists of 2130 identical protein subunits (each of mol. Wght. 17,500 daltons & 158 amino acid residues) arranged around a central orifice of 4nm. The subunits are made of 158 amino acid residues each. The protein capsid encloses a stranded RNA helix (Mol. Wght. 2.06 x 10^6 daltons), with 49 nucleotides and a pitch of 23^0. Each protein subunit is associated with 3 nucleotides of RNA i.e. 49 subunits associated with three turns of RNA helix. **Multiplication:** Entry of virus into healthy cell of tobacco leaf, dissociation of capsid within, formation of *replicative RNA* in the host nucleus, formation of several viral RNA strands in the cytoplasm, synthesis of viral subunits with host ribosomes and host tRNA, assembly, release of TMV without host cell lysis.
Adenovirus	Different types of Adenoviruses (*adeno = gland*) cause respiratory diseases in birds and mammals and are also responsible for production of tumors. They are icosahedral dsDNA viruses of 80nm, with 252 spherical subunits of which 240 are *hexons* and 12 are *pentons*. From the 12 vertex penton points of the capsid, 12 *spikes* are seen emerging out.The dsDNA inside has 100-140 nucleotides, 35,000 base pairs long, with terminal redundant sequences. The circular form of DNA is much more infectious than the linear form.

Table *Contd...*

Virus	Structure, composition and Multiplication
HIV	Spherical virus consisting of two molecules of ssRNA in an icosahedral capsid surrounded by membrane envelope .**Envelope:**100nm in diameter,made of lipid bilayer and glycoproteins. The *lipid bilayer* is from the host cell membrane, and the *glycoproteins* (gp41 & gp120) are virus coded. **Protein capsid:** Icosahedral, consists of a central mass and surrounding core subunits P7, P8, P17, P18, P24, etc. **Genome:** There are two identical macromolecules of ssRNA each of 9749 nucleotides, attached to the enzyme *reverse transcriptase* essential for copying RNA genome into a DNA. The RNA codes for *gag, pol* and *env* genes, which code for capsid proteins, polymerases and envelope proteins respectively. **Replication:** Recognition of CD_4 markers on T_4 lymphocytes, fusion of envelope and entry of nucleocapsid, release of RNA genome, formation of *complementary DNA*, formation of dsDNA, interation into host DNA , *Provirus* stage, activation of viral genes, expression, and assembly of nucleocapsids, release of viruses by process of budding.
Lytic phages (T_4)	The *coliphage* (multiplies within coliform bacteria), most commonly studied representative bacteriophage among all T even phages. **Structure:** Hexagonal *icosahedral head* ($900A^0$) containing a highly coiled *dsDNA* molecule ($52,000\ A^0$). The phage DNA contains *hydroxymethyl cytosine* in place of usual cytosine.A *cylindrical tail* surrounded by protein coat is attached to the head. The tail is connected to the head by means of a *collor*. At the end of the tail is the hexagonal *base plate* with six *tail fibres* each emerging from each corner. **Multiplication:** Adsorption, penetration of viral DNA into host cell, eclipse period for formation of viral components, assembly and maturation, lysis and release of phages.
Lambda phages	dsDNA *temperate phages* that exhibit lysogenic cycle. **Structure:** The λ. phage head is *icosahedral* (55nm) made up of 300-600 capsomeres, which are, arranged in clusters of pentamers or hexamers. There is a tail bears collor but has only *one thin tail fibre* (25nm) The tail is non-contractile and bears no sheath. **Multiplication:** Adsorption, penetration of viral DNA into host cell, integration of viral genome into host cell genome (lysogeny), triggering lytic cycle by drugs or UV light, formation of viral components, assembly and maturation, lysis and release of phages.

Viral pathology

Disease	Pathogen	Treatment
Viral diseases of the skin		
Warts	*Papilloma virus*	Acid/laser therapy, cryotherapy.
Variola (small pox)	*Variola virus*	No specific treatment
Varicella (Chickenpox)	*Varicella-Zoster virus*	Acyclovir
Shingles (Herpes Zoster)	*Varicella-Zoster virus*	Acyclovir
Herpes simplex	*Herpes simplex type 1 virus*	Acyclovir
Rubeola (Measles)	*Measles virus*	No specific treatment
Rubella (German Measles)	*Rubella virus*	No specific treatment
Viral diseases of the skin		
Herpetic keratitis	*Herpes simplex type 1 virus*	Trifluridine
Viral diseases of the NS		
Poliomyelitis	*Polio virus*	No specific treatment once the virus reaches the Nervous System,
Rabies	*Rabies virus*	Post exposure prophylaxis with series of vaccination.
Encephalitis (Mosquito borne) [EEE = Eastern Equine Encephalitis, WEE = Western EE, SLE = St. Luis Encephalitis, CE = California Encephalitis.	*Arboviruses through Culex & Aedes mosquitoes.*	No specific treatment once the virus reaches the NS, Local control of mosquitoes is the preventive measure.

Table *Contd...*

Viral diseases of the Lymphatic and cardiovascular system		
Burkitt's Lymphoma Infectious mononucleosis	*Epstein Barr Virus (EBV)* *EBV*	Tumorous infection, no specific treatment.
Cytomegaloirus disease	*Cytomegalovirus*	Infection with lobed lymphocytes with no specific treatment.
Yellow fever	*Yellow fever virus (Arbovirus)*	Sexually transmitted, no treatment and no vaccine available.
Dengue fever	*Dengue fever virus (Arbovirus through Aedes mosquito)*	No specific treatment

Disease	**Pathogen**	**Treatment**
Viral diseases of the Respiratory system		
Common cold	*Corona viruses & Rhinoviruses*	Cough suppressants and antihistamines (Cetrizine).
Respiratory syncytial viral (RSV) disease.	*RSV (serious disease of infants)*	Ribavirin
Influenza	*Several serotypes of influenza viruses.*	Amantadine & Rimantadine, Zanamivir & oseltamivir.
Viral diseases of the Digestive system		
Mumps	*Mumps virus*	Parotid gland infection, No specific treatment, MMR vaccine is preventive.
Hepatitis A, B, C, D, E	*HAV,HBV,HCV,HDV,HEV*	No specific treatment, administration of interferons helps in some cases.
Viral gastroenteritis	*Rotavirus, Calcivirus (or Norwalk virus)*	Oral/intravenous rehydration
Viral diseases of the Reproductive system Genital herpes	*Herpes simplex virus type 1, HSV type 2*	Suppressive chemotherapy, no cure.
Genital warts	*Papilloma virus*	Electrodessication, acid/laser therapy, cryotherapy.

Viroids

- Viroids are short pieces of naked RNA of 300-400 nucleotides without any protein coat. The nucleotides are internally paired so that the viroid gets a three dimensional structure, protected from cellular enzymes.
- Like *introns,* the Viroid RNA does not code for any proteins. So far, viroids have been found to be pathogenic to plants only. (Eg. Potato spindle tuber viroid).

Prions

- Stanley Pruisner (1982) proposed that srapie disease of sheep is caused by an *infectious protein*. He coined the term **prion** for these protenaecious infectious agents.

- Later, other prions were discovered. Prions cause encephalopathies in man and animals leading to development of large vacuoles in the brain. Egs: Mad cow disease of cattle, kuru, Creutzfeldt-Jacob disease, Gerstmann-Straussler-Sheinker syndrome and fatal familial insomnia.

- These diseases are caused by converting a normal glycoprotein (PrP^C) in the host cell into abnormal protein (PrP^{Sc}). The abnormal prion proteins enter other healthy cells and convert the normal PrPs therein.

- Soon the abnormal prion proteins accumulate in the brain and form plaques, causing tissue damage.

Emerging viruses

New viral strain	Family	Remarks / causes of emergence
Influenza virus	*Orthomyxoviridae*	New strains have emerged due to integrated pig, duck agricultural practices
Dengue virus	*Flaviviridae*	Mosquito breeding in open water storages and urban population density cause new strains.
Sin Nombre virus	*Bunyaviridae*	Increased human-rodent contacts.
Rift valley fever virus	*Bunyaviridae*	Increased dams and irrigation practices
Hantaan virus	*Bunyaviridae*	Increased human-rodent contacts
Machupo virus	*Arenaviridae*	Increased human-rodent contacts
Junin virus	*Arenaviridae*	Increased human-rodent contacts
Ebola virus	*Filoviridae*	Monkeys in Europe and US.
Marburg virus	*Filoviridae*	Monkeys in Europe
Human immnodeficieny virus	*Retroviridae*	Abnormal sex practices, transfusions, drug abuse and needle transfer.
Human Tcell leukemia virus	*Retroviridae*	Blood transfusions, drug abuse and needle transfer, social factors.
Norwalk virus	*Caliciviridae*	Newer diagnostic procedures

Exercises

1. Match the columns [ ]

   i.   Ivanowsky   A. Bacteriophages
   ii. Loeffler   B. *Contagium vivum fluidum*
   iii. Twort & d'Herelle   C. TMV
   iv. Beijerinck   D. Foot & mouth disease

   a. i – C, ii – D, iii – B, iv - A
   b. i – A, ii – B, iii – C, iv - D
   c. i – C, ii – D, iii – A, iv - B
   d. i – D, ii – C, iii – A, iv – B

2. Crystals of tobacco mosaic virus were first obtained by [ ]
   a. Wendell Stanley   b. Felix d'Herelle
   c. Dimitri Ivanowsky   d. Paul Frosch

3. First viral pandemic disease documented in humans. [ ]
   a. Influenza   b. AIDS
   c. Hepatitis   d. Yellow fever

4. Discovery of electron microscopy revealed that viruses are of two types [ ]
   a. Plant & animal viruses   b. Helical & spherical
   c. Circular & tadpole shaped   d. Simple & complex

5. Adolf Mayer is known for [ ]
   a. Transmission of mosaic lesions in plants
   b. Cultivation of bacteriophages
   c. Explanation of lysogenic cycle
   d. Transduction.

6. Hershy & Chain experiment clarified that [ ]
   a. Viral nucleic acid can be alone infectious
   b. Viruses cannot be cultivated in ordinary culture media
   c. Viruses can cause human diseases
   d. Phages have icosahedral heads

7. Whom of these is associated with viral classification? [ ]
   a. Lwoff   b. P. Tournier   c. R.W.Horne   d. All of these

8. The common method of cultivation of most viruses currently is [ ]
   a. Cell culture   b. Live animal inoculation
   c. Chick embryo inoculation   d. Dead animal inoculation

9. The cytopathic effect *pyknosis* is associated with [ ]
   a. Large vacuoles in cytoplasm   b. Nuclear shrinking
   c. Rounding up of tissues cells   d. Breaking of chromosomes

10. The cytopathic effect *syncytia* is associated with [ ]
    a. Large vacuoles in cytoplasm   b. Nuclear shrinking
    c. Fusion of cells in the tissue culture   d. Breaking of chromosomes

11. Counts of bacteriophages are expressed as [ ]
    a. *pfu per ml*   b. *cfu per ml*
    c. *ifb per ml*   d. *cpe per ml*

12. What is *False* about cell culture? [   ]
    a. Epithelioid and fibroblastic cells attach to plastic and form a monolayer.
    b. Primary cell cultures are those prepared directly from animal tissues.
    c. Secondary cell cultures are obtained from secondary lymphoid organs.
    d. Continuous cell lines consist of a single cell type that can be propagated indefinitely in culture.

3. Baltimore system of viral classification is based on [   ]
    a. Presence and absence of envelope
    b. Host of the virus
    c. The genetic system of virus
    d. Symmetry of the capsid

14. What is *True* about cultivation of viruses in tissues [   ]
    a. Many viruses multiply in tissue cells without causing obvious cytopathic effect.
    b. Mouse embryos cannot be used for developing primary cell cultures.
    c. Negri bodies are clumps of ribosomes in the host cell due to viral attack.
    d. a, b, and c are *True*.

15. Herpes simplex virus grows best in [   ]
    a. Chorioallantoic membrane of chick embryo
    b. Yolk sac of chick embryo
    c. Both a & b
    d. None of these

16. Formation of RBC lattice due to the binding of viral proteins to erythrocytes is called [   ]
    a. Immunoblotting   b. Hemoprecipitation
    c. Pock   d. Hemagglutination

17. The best method of carrying out virus neutralization test is by the use of [   ]
    a. Type specific antigens   b. RBCs
    c. Monoclonal Antibodies   d. Tissue cultures

18. Northern blot hybridization analysis is used to detect [   ]
    a. DNA   b. RNA   c. Proteins   d. All

19. In RNA (+) stranded viruses [   ]
    a. The RNA can be reverse transcribed to DNA
    b. The RNA can be directly used as mRNA
    c. The RNA is double stranded
    d. The RNA is always found with polymerases

20. The viral particle may be called as [   ]
    a. Nucleocapsid   b. Protomer   c. Capsid   d. Viroid

21. The smallest biochemical unit of a virus [   ]
    a. Capsid   b. Capsomere   c. Pentamere   d. Protomer

22. Capsid is defined as [   ]
    a. Protein shell surrounding the nucleic acid genome
    b. Lipid bilayer surrounding the virus
    c. genome enclosed within the protein coat of the virus
    d. Surface structures projecting from the viral particle.

23. The key regions on the virus that determine the surface receptors on host cells [   ]
    a. Capsomeres   b. Core   c. Glycoproteins   d. Monomers

24. The most widely used method in the examination of viruses.   [   ]
    a. Nuclear magnetic resonance method
    b. Electron microscopy
    c. X-ray diffraction
    d. Cryoelectron microscopy

25. Rod shaped viruses are actually   [   ]
    a. Helical   b. Discoid   c. Icosahedral   d. Polymeric

26. The icosahedron symmetry represents a solid with   [   ]
    a. 12 triangular faces and 20 vertices
    b. 20 triangular faces and 12 vertices
    c. 20 circular faces and 12 triangular inserts
    d. 12 circular faces and 20 triangular inserts.

27. Pick out the structurally odd virus   [   ]
    a. TMV   b. Sendai virus
    c. Vesicular stomatitis virus   d. Herpes virus

28. A complex virus   [   ]
    a. Has a helical body
    b. Has a circular body
    c. Has both helical as well as circular structures in its body
    d. Has lipids in its capsid

29. Adenoviruses are known for their   [   ]
    a. Large icosahedral structure   b. Bullet shape
    c. Complex glycoproteins   d. Lack of protein capsid

30. What is *False* about the viral envelope?   [   ]
    a. Basic structure is the lipid membrane acquired from host cell membrane.
    b. Embedded in the membrane are viral glycoproteins
    c. Glycoproteins carry covalently linked oligosaccharides
    d. The glycoproteins interfere with the attachment and entry of the virus into the host cell

31. RNA dependent DNA polymerases are seen in   [   ]
    a. Retroviruses.   b. Rhabdoviruses   c. Poxviruses   d. Orthomyxoviruses

32. Match the colums   [   ]
    A. Zoonosis   i. Herpes simplex
    B. Respiratory transmission   ii. Prions
    C. Sexual contact   iii. Hanta viruses
    D. Genetic   iv. Paramyxo viruses

    a. A - ii, B - i, C - iv, D – iii
    b. A - iii, B -ii, C - iv, D – i
    c. A - ii, B - i, C - iii D – iv
    d. A - iii, B - iv, C - i, D – ii

33. Which of these is NOT transmitted sexually   [   ]
    a. Human papilloma virus   b. Adenovirus
    c. Molluscum contagiosum   d. HTLV III

34. Prodromal symptoms   [   ]
    a. Earliest specific symptoms
    b. Non-specific early symptoms
    c. Symptoms seen during convalescence
    d. Symptoms of secondary infections

35. An abortive infection is said to occur when   [   ]
    a. the virus fails to replicate after entry into the body
    b. infection leads to death of the cells
    c. infection occurs with out cell death
    d. the infection is asymptamatic

36. Which of these inhibits cellular protein synthesis   [   ]
    a. Togavirus   b. Herpes simplex virus
    c. Polio virus   d. All

37. Cowdry type A inclusion bodies are seen in   [   ]
    a. Rabies   b. Polio
    c. Subacute sclerosing panencephalitis   d. Chicken pox

38. Viruses that causes transformation or immortalization   [   ]
    a. Reoviruses   b. Retroviruses
    c. Oncogenic viruses   d. Papilloma viruses

39. Slow viruses that have long incubation periods   [   ]
    a. Lentiviruses   b. Adenoviruses   c. Arboviruses   d. Rhinoviruses

40. Intranuclear 'Owl's eye" inclusion bodies are produced by   [   ]
    a. Reoviruses   b. Rabies viruses
    c. Cytomegaloviruses   d. Paramyxoviruses

41. Break bone fever   [   ]
    a. Japanese encephalitis   b. Dengue fever
    c. West Nile fever   d. Sand fly fever

42. The first case of AIDS was reported in   [   ]
    a. USA   b. Japan   c. Chandipura   d. Kyasanur Forest

43. Causative agent of chikungunya   [   ]
    a. Nirovirus   b. Phlebovirus   c. Flavivirus   d. Alphavirus

44. *Aedes aegypti* mosquito is associated with   [   ]
    a. SARS   b. Bird flue
    c. Dengue fever   d. Venezuelan equine encephalitis

45. SARS – A new syndrome emerged in   [   ]
    a. Southern China   b. Western India   c. South Korea   d. Japan

46. SARS is caused by   [   ]
    a. Karla viruses   b. Nepo viruses   c. Corona viruses   d. Tobra viruses

47. Intracytoplasmic inclusion bodies called *Negri Bodies* are diagnostic of   [   ]
    a. Kuru   b. Rabies   c. German measles   d. Small pox

48. The knobs and stalks of glycoproteins in the envelope of HIV   [   ]
    a. gp81 and gp120   b. gp120 and gp41   c. gp24 and gp17   d. gp17 and gp24

49. Dr. Stanley was awarded the Nobel Prize for   [   ]
    a. Crystallizing TMV
    b. Elucidating the structure of TMV
    c. Designing a virus proof filter
    d. All

50. Pick out the WRONG statement about TMV   [   ]
    a. Though it infects the tobacco plants, there is a slight increase in the nicotine content post infection.
    b. It maily damages the Solanaecous plants.
    c. The virus is a helical array of 2130 protein subunits
    d. It is a ssRNA virus

51. The most common mode of infection of TMV is   [   ]
    a. Insect pest vector
    b. Through seed infection
    c. through stem/leaf wounds duding cultivation operations
    d. Direct plant to plant contact

52. Which of these is a viral disease   [   ]
    a. tomato leaf curl
    b. Rattoon stunt of sugarcane
    c. Little leaf of cotton
    d. All

53. Which of these is NOT viral disease   [   ]
    a. Fire blight of apple
    b. Broom corn stunt
    c. Bunchy top of banana
    d. Citrus canker

54. Infectious mononucleosis is caused by   [   ]
    a. Varicella zoster virus
    b. Echo virus
    c. Coxackie virus
    d. Epstein Barr virus

55. Rous sarcoma virus is associated with   [   ]
    a. Citrus canker
    b. Solid tumors in chicken
    c. Sarcoma in mice
    d. Stunt of brinjal

56. What is TRUE about Adenoviruses   [   ]
    a. They can cause respiratory disese in birds
    b. They are dsDNA viruses
    c. Their capsids are icosahedral with 252 subunits
    d. a, b abd c are TRUE

57. Short pieces of Naked RNA without protein coat   [   ]
    a. viroids
    b. prions
    c. virions
    d. nucleotides

58. Pruisner is known for   [   ]
    a. Discovery of Sin nombre virus
    b. Elucidation of lysogeny
    c. Discovery of prions
    d. Discovery of rubeola virus

59. The most recent system of viral classification   [   ]
    a. LHT system
    b. Casjens and King system
    c. Baltimore system
    d. ICTV system

60. The host cell for the bacteriophage ΦX174 is   [   ]
    a. *E.coli*
    b. *Bacillus subtilis*
    c. *Pseudomonas*
    d. *Klebsiella*

Answer Key and Validation

Virology

1. c	11. a	21. d	31. a	41. b	51. c
2. a	12. c	22. a	32. d	42. a	52. d
3. a	13. c	23. c	33. b	43. d	53. c
4. b	14. a	24. b	34. b	44. c	54. d
5. a	15. d	25. a	35. a	45. a	55. b
6. a	16. d	26. b	36. d	46. c	56. d
7. d	17. c	27. d	37. c	47. b	57. a
8. a	18. b	28. c	38. c	48. b	58. c
9. b	19. d	29. a	39. a	49. a	59. d
10. c	20. a	30. d	40. c	50. a	60. a

3   *a Pandemic is a disease that occurs worldwide during the same period. Pandemic of Influenza was documented over 100years ago.*

27   .* Herpes virus has an irregular shape while the others are helical*

39   *AIDS viruses are examples of Lentiviruses (lenti = slow)*

50   *The TMV infection reduces nicotine content by 20-30%*

53   * Diseases in options a, b & d are caused by bacteria*

10

Medical Microbiology

Retrace your subject…
Normal flora of the human body, type study of some major groups of bacterial pathogens, fungal, helminthes and viral parasites…

Clinical Microbiology is concerned with the study of pathogens (disease carrying organisms). Each pathogen is studied for its morphological, cultural and biochemical characteristics, antigencity, pathogencity, lab diagnosis and treatment. Host defense mechanisms, host parasite interactions and normal microflora of human body also form a part of the clinical microbiology.

Normal Microbial Flora of the Human Body

Normal flora indicates all the population of microorganisms that inhabit a healthy human body. Those organisms that are found more or less on a permanent basis for a given age and sex are called *resident flora*. The nonpathogenic or potentially pathogenic microorganisms that inhabit the body for hours, days, or weeks due to the influence of the environment, but can never establish a permanent population in the body are called *transient flora*.

Body part	Normal flora	Remarks
Skin	Staphylococcus aureus, Staphylococcus epidermidis, Micrococcus species, Nonhaemolytic streptococci Diphtheroids, Propionibacteria Peptostreptococcus, Few candida *Acinetobacter, etc..*	Low pH, fatty acids in sebaceous secreation and presence of lysosomes control the flora. Improper cleaning, profuse sweating, use of unclean garments tend to cause qualitative and quantitative alterations in the flora
Nasophraynx Mouth and Respitatory tract	Alpha haemolytic streptococci, nonhaemolytic streptococci, pneumococci, *Staphylococcus epidermidis, Staphylococcus aureus,* Gram negative rods, Anaerobic cocci, *Fusobacterium Neisseria meningitidis, Moraxella* Yeast *(Candida)*	Few bacteria found in the bronchi and alveoli are usually sterile.mouth and nasophraynx is sterile at birth. Lysozyme in saliva, sticky mucous linening,hair, Excessive blood capillaries in the alveoli and other factors help control the number of flora in the respiratory tract.
GI tract and rectum	*E.Coli, Enterobacter, Klebsiella, Citrobacter, Lactobacillus lactis, Streptococci, Staphylococci, Anaerobic bacteroids, Bifidobacteria, Clostridium, Candida, Pseudomonas, Proteus, Peptostreptococci etc.*	Stomach lacks resident flora due to extreme acidity. Colon harbours bile tolerant flora. 96 to 99% of adult colon consist of anaerobic and facultative anaerobes, with 10^8 to 10^{10} Bacteria per gram. The mouth is sterile at birth. In breast fed children the intestine contain streptococcus lactis and lactobacilli. In bottle fed children the bowel contains mixed flora.
Genitourinary tract flora	*Staphylococcus, Acinetobacter, Candida albicans, Mycobacterium smegmatis, Bifidobacterium spp, Corynebacterium spp, Gardnerella, Hemophilus, Lactobacillus, Mobilincus, Enterococcus, Actinomyces, Ureaplasma, etc.*	Urethra consists of a variety of resident and transient flora including enteric bacteria. Urinary bladder is normally sterile. Vaginal flora is influenced by hormones. It harbours lactic acid bacteria at birth and puberty. *Candida* is a common inhabitant of vagina. Kidneys, ureters and testes are free from microflora in healthy individuals.

Features of some of the major groups of pathogens are enlisted in the following tables

Pathogen	Morphology and cultural characteristics	Pathogenicity/Strains	Diseases/ symptoms/ syndrome caused.	Lab diagnosis
E. coli	Gram negative motile short rods to coccobacilli, pink colonies on MacConkey agar, metallic sheen on EMB agar.	• Entero toxigenic *E.coli* (ETEC) – invasion • Enteroinvasive *E.coli* (EIEC) - invasion • Enteropathogrenic *E.coli* (EPEC) – adherence - Enterohemorrhagic *E.coli* (EHEC) - cytotoxin • Enteroaggregative *E.coli* (EaggEC) • –aggregative adherence	• Traveller's diarrhoea • Dysentery with bloody stools • Infant diarrhoea, non bloody stools • hemorrhagic colitis • - Infant diarrhea, vomiting and dehydration	Lactose fermented with acid and gas, IMViC = + + - - Strains are identified by serotyping, phage typing, colicin typing, MUG test and antibiotic sensitivity patterns.
Klebsiella pneumoniae	Gram negative rods, capsulated, non motile, lactose fermentor, produce highly mucoidal pink colonies on MacConkey agar	*K.pneumoniae* is the most common pathogen with few strain variations. Capsule plays very imp role in pathogenesis.	Strains cause pneumonia and other respiratory infections, UTI, septicemia, nosocomial infections bovine mastitis and equine diseases,	Acid with/without gas from lactose and glucose, growth seen in KCN media, H_2S negative, Urease positive, MR positive VP negative, Indole neg.
Enterobacter	Gram negative motile, short rods that ferment lactose. Grow well on NA and MacConkey agar.	*E.aerogenes* is most common. Usually normal flora of the colon, but acts as opportunistic pathogen when it comes in contact with other systems.	Involved in UTI, meningitis, pneumonia and septicemia.	IMViC = --++, H_2S neg, Acid and gas from glucose and lactose. Sorbitol positive, lysine decarboxylase positive. Spp variations occur.

Pathogen Morphology and cultural characteristics	Pathogenicity/Strains	Diseases/ symptoms/ syndrome caused.	Lab diagnosis
Proteus Gram negative short rods, active motility, swarming (spread out) colonies on culture media. Selective growth is on CLED (Cystine Lactose Electrolyte Deficient) medium.	*Proteus mirabilis* *P. vulgaris,* *P.rettgeri,* *P.shigelloides* *Morganella morganii*	Opportunistic pathogen, UTI, Nosocomial infections, pneumonia and septicaemia.	Urease positive, swarming colonies on agar medium (Dienes Phenomenon), H_2S positive, gelatin liquefaction positive, lipase positive, mannose negative,
Pseudomonas Gram negative, aerobic, chemoorganotrophic to chemolithotrophic, motile mono flagellate bacteria. Pigmented growth (blue-green) occurs on NA and cetrimide agar.	*P.aeruginosa* is most imp pathogen, *P.mallei,* *P.psuedomallei,* *P.maltophil* and several others are also pathogenic.	UTI, skin infections, burn abscesses, stitch infections and nosocomial infections are common.	Catalase positive, oxidase positive, twiching motility, urease positive, growth at 42°C, pigment production help in identification.

Table *Contd...*

Pathogen Morphology and cultural characteristics	Pathogenicity/Strains	Diseases/ symptoms/ syndrome caused.	Lab diagnosis
Shigella Gram negative rods, non motile, lactose non fermenter, Small colonies appear on NA, MacConkey agar, EMB agar and Salmonella Shigella agar.	*Shigella dysenteriae* *S.boybii* *S.flexneri* *S.sonnei* (late lactose positive)	Bacterial dysentery (shigellosis) M cells of Peyer's patches take up the pathogens where they kill macrophages. Some strains produce Shiga tyoxin	IMViC = -+ --, H_2S neg, Lysine decarboxylase Neg, sugar fermentation without gas, no growth on KCN media.
Salmonella Gram negative motile rods, shows good growth on NA, MacConkey agar, DCA, SS agar. Selenite F broth and tetrathionate broth serve for enrichment.	More than a 1000 serotypes exist. *S.typhi*, *S.paratyphi A*, *S.paratyphi B* cause enteric fervers, *S.typhimurium* and *S.hadar* cause food poisoning, *S.pullorum*, *S.dublin*, *S.cholerasuis* cause diseases in animals. (Kauffmann-White classification scheme presents all serotypes of salmonellae)	Typhoid, paratyphoid, gastroenteritis, fowl typhoid, cattle typhoid, etc.	IMViC = -+-+, urease neg, H_2S positive, Acid and gas from glucose, lysine decarboxylase positive. Variations among spp and strains are common.
Vibrio Gram negative bacteria, curved rods, monoflagellate, motile, halophilic, show good growth on TCBS agar shows green colonies of *V.parahaemolyticus* and yellow colonies of *V.cholerae*.	*V.cholerae and , V.parahaemolyticus* are imp pathogens. Production of cholera toxin and toxin regulated pili are factors for *V.cholerae*. O1 strains show 2 biotypes *Eltor* and *Classical* varieties. Ogawa, Inaba and Hikojima are the major O1 strain varients.	*V.cholerae* causes cholera, *V.parahaemolyticus* causes gastroenteritis, particularly through improperly cooked sea food consumption	Oxidase positive, lactose neg, glucose fermented to acid without gas. Growth in 2-4% NaCl, pH 8.0 Sucrose is fermented by *V.cholerae* but not by *V.parahaemolyticus*. Kanagawa test (hemolysis) positive for *V.parahaemolyticus*.
Bacillus anthracis Gram positive, non motile, aerobic to fac. anaerobic rods in chains with square ends. Grows well on NA and blood agar.	The bacterium produces the anthrax toxin which has 3 components – oedema factor (EF), Protective Ag (PA) and Lethal factor (LF). The components together render pathogenicity in various ways.	Anthrax may be infection of skin (Hide Porter's disease/ cutaneous anthrax), lung infection (pulmonary / woolsorter's disease) or the gut infection (intestinal anthrax). It is a primary disease of animals sheep, horse and cattle.	Catalase positive spore bearers. Weak lecithinase positive, non hemolytic on sheep blood agar, shows string of pearl effect - rounded cells when grown in presence of benzylpenicillin.

Table *Contd...*

Pathogen Morphology and cultural characteristics	Pathogenicity/Strains	Diseases/ symptoms/ syndrome caused.	Lab diagnosis
Corynebacterium diphtheriae Gram positive, aerobic to facultatively anaerobic, non motile, asporogenous curved, coryneform rods arranges in Chinese letter configuration. Selectively grown on tellurite blood agar and Loeffler's serum slants.	Occurs in 3 varieties: *mitis, intemedius* and *gravis*. Lysogenic strains produce a heat labile exogenic diphtheria toxin. It inhibits protein synthesis by ADP ribosyl transferase activity. The ribosylation is inactivated for EF2.	Diphtheria spreads through droplet inhalation, food and milk. Fever, sore throat, headache, formation of grey membrane (diphther) in the nasopharynx/ nose/ larynx/tonsils. Leads to respiratory obstruction.	Urease negative, acid without gas from glucose, maltose & mannose. Typical Daisy head colonies (*gravis*), several metachromatic granules (*mitis*), help in identification. Elek's gel ppt test identifies the toxigenic strains.
Clostridium Gram positive, spore forming, obligate anaerobes to aerotolerant bacteria. Show good growth in Robertson's cooked meat medium, McClung Toabe Egg yolk agar, reinforced clostridial medium and blood agar, all incubated under anaerobic conditions.	*C.botulinum* – causes botulism a toxin mediated serious food poisoning with headache, nausea, diplopia, disphagia, asphyxia and even death (flaccid paralysis). *C.barati & C.butyricum* also cause botulism. *C.perfringens* – cause gas gangrene, food poisoning and enteritis due to production of a spectrum of toxins. Other gangrene causing clostridia: *C.histolyticum, C.noyvi, C.septicum* and *C.sordelli*. *C.tetani* – Causes tetanus, an anaerobic necrotic wound infection that leads to painful convulsive spasms, asphyxia and death. *C.difficle* – Antibiotic associated diarrhoea .		Demonstration of sporebearers in appropriate clinical samples, cultural characters, stormy fermentative reactions and more than 10^5 cells in food/feces are diagnostic.
Mycobacterium leproae Acid fast bacteria (AFB), slow growing, non motile, asporogenous, aerobic to facultatively anaerobic. Growth does not occur in cell free lab media, are cultured in footpads of mice.	Mycolic acids, chord factor and Wax D are characteristic features.	Causal agent of leprosy (Hansen's disease). Affects peripheral nervous system. Depending on the T cell response, the disease is lepromatous leprosy, borderline leprosy or tuberculine leprosy.	Acid fast staining of skin lesions, nasal secretions. Foamy Virchow cells filled with AFB are characterisitic. Lepromin test: A skin test that shows *Medina, Fernandez or Mitsuda reaction,* to classify the type of leprosy.
Mycobacterium tuberculosis Acid fast rods, , non motile, asporogenous, aerobic to facultatively anaerobic. Growth observed on prolonged incubation on Lowenstein Jenson medium.	Some bacteria survive phagocytosis and kill macrophages.	Causal agent of tuberculosis. A chronic infectious disease of the lungs and URT.	AFB in sputum, VSF, urine sediments and biopsy specimens. Catalase negative, no growth at 42°C, nitrate reduction negative. Mantoux skin test is indicative of sensitisation and possible susceptibility.

Table *Contd...*

Pathogen Morphology and cultural characteristics	Pathogenicity/Strains	Diseases/ symptoms/ syndrome caused.	Lab diagnosis
Nocardia Gram positive to acid fast bacteria, (actinomycete), non motile, aerobic, chemoorganotroph, occur as branched mycelia, short rods, coccoidal forms. Growth on enriched media for actinomycetes.	*N.asteriodes* causes nocardiosis. The disease is symptomatic with lesions in lungs, brain or kidneys. Infections via wounds may cause suppurative subcutaneous lesions like in maduromycosis. Cell walls contain nocerdomycolic acids. Granulomatous, suppurative tumors (mycetomas) appear on the feet or hands. *Actinomadura madurae* and *Madurella spp.* Also cause this madura foot disease.		
Staphylococcus aureus Gram positive cocci in bunches, aerobic to fac. anaerobic, non motile, produce golden yellow pigment, grow on NA, Blood agar, salt mannitol agar and SM110 medium.	Pathogenicity is due to production of coagulase, hyaluronidase, Lipid A, hemolysins, DNAse, lecithinase, heat stable enterotoxin, etc.	Strains cause various symptoms: Pimples, boils, bronchitis, arthritis, carbuncles, furuncles, mumble foot, endocarditis, cystitis, food poisoning, impetigo, joint-ill, kearatoconjunctivitis, meningitis, osteomyelitis, pneumonia, scalded skin syndrome and toxic shock syndrome	Coagulase lecithinase, DNAse and catalase positive. Mannitol is fermented with acid and no gas.
Streptococcus pyogens Gram positive cocci in chains, asporogenous, grow well on NA and blood agar.	Classified as per Lancefield typing.. *S.pyogens* (group A) - Pharyngitis, impetigo, rheumatic fever, glomerulonephritis. *S.bovis* (group D) - Endocarditis, colon cancer *S.dysgalactiae* (group C,G) - Pharyngitis, pyogenic infections *S.anginosus* (group F) - pyogenic infections, brain abscesses Viridans Streptococci (Eg. *S.acidominimus*) – Dental carries (*S.mutans*) and endocarditis abscesses. *S.agalactiae* (group B) - neonatal sepsis meningitis	Catalase negative, oxidase negative.
Neisseria Gram negative cocci in pairs with adjacent sides flat or concave. Aerobic bacteria that are fastidious and grow on Theyer Martin Agar, Muller Hinton agar and chocolate blood agar.	*N. gonorrhoeae* and *N.meningitis* are imp pathogens. Fimbriae, IGA1 proteases and antigenic variations help in pathogenicity.	Gonococci cause STD gonorrhoea and ophthalmia neonaturum of the eye and meningococci cause bacterial meningitis.	Diplococci in phagocytes of blood or pus specimen, CNS, etc. Oxidase positive, lactose non fermenters, ferment glucose without gas production.

Pathogen	Disease/syndrome caused	Pathogen	Disease/syndrome caused
Spirochaetes		**Rickettsia**	
Treponema pallidum	Syphilis (venereal)	*Rickettsia rickettsii*	Rocky mountain spotted fever (tick borne)
T. pertenue	Yaws	*R. tsutsugamushi*	Scrub typhus (mite borne)
T. carateum	Pinta	*R. prowazeki*	Epidemic typhus (louse borne)
T. endemicum	Bejel	*R. quintana*	Trench fever (louse borne)
Borrelia recurrentis	Epidemic relapsing fever	*R.typhi*	Murine typhus (flea borne)
Borrelia spp (many)	Endemic relapsing fever	*Coxiella burnetii*	Q fever
Leptospira interrogans	Leptospirosis		
Mycoplasmataceace		**Chlamydia**	
Mycoplasma hominis	Pyelonephritis	*C. trachomatis*	Lymphogranuloma venereum
M. pneumoniae and others	Respiratory disorders	*C. pneumoniae*	Bronchitis, pneumonia,
M. genitalium	Non gonococcal urethritis	*C. psittaci*	Psittacosis (bird mediated pneumonia)
Ureaplasma urealyticum	Non gonococcal urethritis		

Pathogen	Disease/syndrome caused
Fungi	
Aspergillus flavus	Aflatoxicosis
Claviceps purpurea	St. Anthony's fire
Superficial mycoses	Pityriasis vesicolor
Malassezia furfur	Tinea nigra (skin)
Exophiala werneckii	Black piedra (hair)
Piedraia hortae	White piedra (hair)
Trichosporon beigelli	
Cutaneous mycosis	
Microsporum spp	Nails and skin infection
Trycophyton spp	Hair and skin infection
Epidermophyton spp	Athlete's foot
Subcutaneous mycoses	
Sporothrix schenkii	Lymphocutaneous sporotrichosis
Cladosporium and others	Chromoblastomycosis
Systemic mycosis	
Histoplasma capsulatum	Reticuloendothelial cytomycosis (Darling's disease/ Spelunkler's disease)
Blastomyces dematitidis	Respiratory blastomycosis (Chicago disease)
Coccidioides immitis	Coccidiomycosis fever
Cryptococcus neoformans	Buss-Buschke disease (pulmonary infection through birds)
Opportunistic mycosis	Candidiasis (intertrigo, thrush, UTI, meningitis, RTI, fungemia, etc.)
Candida albicans, C.tropicalis, C.viswanathii, etc.	
Pneumocystis carinii	Pneumonia and extra pulmonary infections in debilitated patients (eg. AIDS patients).

Pathogen	Disease/syndrome caused
Viral Pathogens	
Rhinovirus, Coronavirus, Adenovirus, Influenza virus, Epstein Barr virus, Respiratory syncytial virus.	Common cold, pharyngitis, tosillitis, Bronchiolitis, pneumonia.
Measles virus	Rubeola
Rubella virus	German measles
Epstein Barr virus	Infectious mononucleosus
Varicella Zooster virus	Chicken pox/Shingles
Hepatitis viruses, Yellow fever virus & Epstein Barr virus	Liver infections
Enterovirus, Echo virus, Coxsackie virus	Meningitis
Enterovirus, , Coxsackie virus & Polio virus	Paralysis
HIV	AIDS
Parasites	
Protozoa	
Plasmondium	Malaria
Giardia lamblia	Diarrhoea
Entamoeba histolytica	Amoebic dysentery
Trypanosoma cruzi	Chaga's disease
Leishmania donovani	African sleeping sickness
Toxoplasma gondii	Toxoplasmosis (lymphadenopathy)
Helminthic Worms	
Ascaris lubmricoides	Round worm infection
Taenia solium (pork)	Tape worm infection
Taenia saginata (beef)	Tape worm infection
Necator americanus	Hook worm infection
Ancylostoma duodenale	Hook worm infection
Enterobius vermicularis	Pin worm infection
Whuchereria bancrofti	Filariasis
Echinococcus granulosis	Hydatid worm infection
Strongyloides stercoralis	Thread worm infection
Schistosoma haematobium	Blood fluke

Exercises

1. Which of these is known to produce heat stable enterotoxin   [   ]
   a. *Staphylococcus aureus*   b. *Clostridium tetani*
   c. *Vibrio cholerae*   d. All of these

2. *Shigella dysentriae* is characteristically known to be   [   ]
   a. Citrate positive   b. Oxidase positive   c. H_2S negative   d. Manitol positive

3. Apart from respiratory infections, *Klebsiella pneumoniae* is also known to cause   [   ]
   a. Septicemia   b. Meningitis   c. Peptic ulcers   d. Urinary tract infections

4. The levels of IgE are known to be high during the illness by   [   ]
   a. *Psuedomonas*   b. *Ascaris*   c. *Entamoeba*   d. Pnuemococci

5. The lodgement and multiplication of an organism in or on the host's body is   [   ]
   a. Infection   b. Immunity   c. Disease   d. Inflamation

6. The ability of an organism to cause a disease   [   ]
   a. Pathogenesis   b. Prognosis   c. Pathogenicity   d. Toxigenicity

7. A syndrome is   [   ]
   a. The final stage of an infectious disease
   b. A collection of pathological conditions due a particular infection
   c. Simultaneous infections by different pathogens in the same host
   d. A viral infection

8. Preparation of vaccines is difficult when   [   ]
   a. The pathogen undergoes frequent strain variations
   b. The pathogen is non-toxigenic
   c. The pathogen is capsulated
   d. The causative agent is opportunistic

9. Chocolate blood agar is different from blood agar as,   [   ]
   a. Some chocolate powder is added as additional ingredient
   b. The amount of agar added is only 1% in chocolate agar
   c. The blood used in chocolate agar is devoid of leucocytes
   d. The blood used in chocolate agar is lysed

10. An infection that is caused due to a contaminated catether   [   ]
    a. Nosocomial infection   b. Secondary infection
    c. Herd infection   d. Zoonotic infection

11. Which of these is manifested by a toxin   [   ]
    a. Cholera   b. Amoebiasis   c. Poliomyelitis   d. Typhoid

12. Capsules are protective against   [   ]
    a. Antibodies   b. Antitoxins   c. Phagocytosis   d. All

13. An infection that is confined to an individual only is regarded   [   ]
    a. Sporadic   b. Endemic   c. Epidemic   d. Pandemic

14. *S. aurerus* is commonly found in   [   ]
    a. Water with Fecal contamination
    b. Soil
    c. Air
    d. Mucosal linings of the nasal cavity of mammals.

15. *Vibrio parahaemolyticus* is a pathogen of   [   ]
    a. The central nervous system
    b. The urinary tract
    c. The Reproductive system
    d. None of these

16. Pathogenic protozoa in stool sample usually occur in the form of   [   ]
    a. Trophozoites   b. Spores   c. Cysts   d. Zygotes

17. Latent period of an infection means   [   ]
    a. The period during which no symptoms are seen even though the pathogen is present
    b. A stage of the disease in which no treatment is possible
    c. The stage of infection after treatment has been stopped
    d. The period during which lab diagnosis is not possible

18. Coagulase test for *S.aureus* gives most reliable results if the plasma is from   [   ]
    a. Horse blood   b. Elephant blood   c. Chimp blood   d. Human blood

19. Toxic Shock Syndrome (TSS) spreads through   [   ]
    a. Use of contaminated tampons
    b. Use of contaminated catethers
    c. Use of contaminated condoms
    d. None of these

20. *Bacillus anthracis* causes   [   ]
    a. Hide Porter's disease
    b. Wool sorter's disease
    c. Both
    d. None of these

21. Colonies of *Bacillus anthracis* produce   [   ]
    a. Metallic sheen
    b. Medusa head morphology
    c. Pitch black pigment
    d. Seminal drop morphology

22. Diphtheria toxin   [   ]
    a. Causes inflammation of alveolar walls
    b. Inhibits protein synthesis by inactivating Elongation Factor 2 (EF$_2$)
    c. Inhibits formation of mucosal lining within the respiratory tract
    d. All of these

23. Cells of *Corynebacterium diphtheriae* is known to show   [   ]
    a. Safety pin morphology
    b. Chinese letter configuration
    c. Budding yeast morphology
    d. Queen's necklace morphology

24. Mycetoma is caused by   [   ]
    a. *Mycobacterium smegmatis*
    b. *Nocardiopsis*
    c. *Eikenella*
    d. *Vibrio harveyi*

25. Waterhouse Friderichsen Syndrome   [   ]
    a. Gonococci   b. Pneumococci   c. Meningococci   d. None of these

26. Traveller's diarrhoea   [   ]
    a. *Entero Pathogenic E.coli (EPEC)*
    b. *Entero Toxigenic E.coli (ETEC)*
    c. *Entero haemorrhagic E.coli (EHEC)*
    d. *Entero Invasive E.coli (EIEC)*

27. Which of these pathogens does not come under enterobacteriaceae   [   ]
    a. *Vibrio spp.*   b. *Citrobacter spp.*   c. *Proteus spp.*   d. *Yersinia spp.*

28. Pick out the WRONG pair   [   ]
    a. *Helicobacter pylori* - Peptic ulcers
    b. *Campylobacter jejuni – Guillian-Barre Syndrome*
    c. *Pseudomonas aeruginosa* – Swimmer's ear
    d. *Bacillus cereus* – Pinta disease

29. Pick out the CORRECT pair   [   ]
    a. Pontiac fever – *Clostridium diffcile*
    b. Bejel – *Treponema pallidum*
    c. Gastric adenocarcinoma - *Brucella*
    d. Pyelonephritis – *Moraxella*

30. What is FALSE about *Proteus*   [   ]
    a. Urease positive
    b. Swarming motility
    c. Gram positive rods
    d. Cause urinary tract infections

31. Which of these causes zoonotic infection?   [   ]
    a. *Campylobacter*   b. *Neisseria*   c. *Vibrio*   d. All

32. *Coxiella burnetti is associated with*   [   ]
    a. Q fever
    b. Indicator of milk pasteurization
    c. Granulomatous hepatitis
    d. All

33. Inverted fir tree appearance in gelatin agar butts is shown by   [   ]
    a. *Proteus*
    b. *Bacillus anthracis*
    c. *Corynebacterium diphtheriae*
    d. All

34. Match the columns   [   ]
    A. Drum stick appearance   i. *Corynebacterium diphtheriae*
    B. Kleb-Loeffler's bacillus   ii. *Proteus*
    C. Seminal odour of broth culture   iii. *Vibrio cholerae*
    D. Fish in stream appearance   iv. *Clostridium tetani*

    a. A-iv, B- i, C-ii, D-iii
    b. A-iv, B- ii, C-iii, D-i
    c. A-iii, B- iv, C-ii, D-I
    d. A-ii, B- iv, C-i, D-iii

35. A bluish green water soluble pigment *pyocyanin* is produced by   [   ]
    a. *Proteus mirabilis*
    b. *Pseudomonas aeruginosa*
    c. *Yersinia enterocolitica*
    d. *Bordetella pertusis*

36. Fried egg colonies are characteristic of   [   ]
    a. *Clostridium*   b. *Borrelia*   c. *Mycoplasma*   d. *Nocardia*

37. Weil's disease is associated with   [   ]
    a. *Treponema*   b. *Leptospira*   c. *Mycobacterium*   d. *Listeria*

38. Ogawa, Inaba and Hikojima are strains of   [   ]
    a. *Vibrio cholerae*
    b. *Clostridium tetani*
    c. *Shigella boydii*
    d. *Salmonella typhimurium*

39. Gram negative kidney shaped diplococci   [   ]
    a. *Neisseria meningitides*
    b. *Neisseria gonorrhoeae*
    c. *Sporosarcina spp.*
    d. *Pneumococci*

40. Kauffman-White scheme   [   ]
    a. Phagetyping of *Staphylococcus aureus*
    b. Serotyping of *Salmonella spp.*
    c. Sugar fermentation pattern of Staphylococci
    d. Haemolytic patterns of Streptococci

41. Late lactose fermenter   [   ]
    a. *Shigella flexneri*
    b. *Shigella dysenteriae*
    c. *Shigella sonnei*
    d. *Shigella boydii*

42. V_i Antigens are seen in   [   ]
    a. Shigellae   b. Salmonellae   c. *Neisseria*   d. *Vibrio*

43. What is TRUE about α hemolysis?   [   ]
    a. Partial hemolysis of RBCs
    b. Shown by *Streptococcus pneumoniae*
    c. Shows a greenish zone around the colonies
    d. a, b & c are true.

44. Scientists who work with very dangerous and easily infectious pathogens use   [   ]
    a. Biosafety level 1 labs   b. Biosafety level 2 labs
    c. Biosafety level 3 labs   d. Biosafety level 4 labs

45. The WHO is an international agency based in   [   ]
    a. Switzerland   b. America   c. Africa   d. England

46. Nosocomial diseases   [   ]
    a. Are acquired during medical treatment
    b. Are acquired as result of sexual contact
    c. Are acquired as hypersensitivity reaction to antibiotics
    d. Are acquired as a result of contact with animals.

47. Match the columns   [   ]
    A. Incidence   i. Number of people infected at any one time
    B. Prevelance   ii. Number of cases as a proportion of the population
    C. Morbidity   iii. Number of deaths as a proportion of the population
    D. Mortality   iv. Number of new cases in a period

    a. A-iii, B- ii, C-i, D-iv   b. A-iv, B- i, C-ii, D-iii
    c. A-iv, B- ii, C-iii, D-ii   d. A-i, B- iii, C-iv, D-ii

48. Normal flora of respiratory tract   [   ]
    a. *Enterobacter & Proteus*   b. *Fusobacterium & Moraxella*
    c. Propionibacterium & Acinetobacter   d. *Gardenerella & Ureaplasma*

49. *Klebsiella pneumoniae* is associated with   [   ]
    a. Bovine mastitis and Equine diseases   b. Infant diarrhea
    c. Stitch abscesses & septicemia   d. Laryngitis and carcinoma

50. Match the columns   [   ]
    A. *E.coli*   i. CLED agar
    B. *Vibrio*   ii. BPA medium
    C. *S.aureus*   iii. TCBS agar
    D. *Proteus*   iv. EMB agar

    a. A-i, B- ii, C-iii, D-iv
    b. A-ii, B-iii, C-iv, D-i
    c. A-iv, B- iii, C-ii, D-i
    d. A-iv, B- i, C-iii, D-ii

51. Urethritis is caused by   [   ]
    a. *Neisseria gonorrhoeae*   b. *Mycoplasma genitalium*
    c. *Ureaplasma urealyticum*   d. All

52. Blood agar is NOT used in the cultivation of   [   ]
    a. *Bacillus antracis*   b. *Enterobacter aerogenes*
    c. *Corynebacterium diphtheriae*   d. *Clostridium botulinum*

53. Theyer Martin Agar is employed for selective isolation of [   ]
    a. *Neisseria gonorrhoeae*
    b. *Streptococcus pyogenes*
    c. *Mycobacterium tuberculosis*
    d. All

54. Which of these is NOT cultivable in lab media [   ]
    a. *Corynebacterium diphtheriae*
    b. *Salmonella typhimurium*
    c. *Vibrio parahaemolyticus*
    d. *Treponema pallidum*

55. Mature cyst of *Entamoeba histolytica* [   ]
    a. Contains 8 nuclei
    b. Contains 6 nuclei
    c. Contains 4 nuclei
    d. Contains18 nucleus

56. Which of these is NOT an intestinal amoeba [   ]
    a. *Entamoeba coli*
    b. *Endolimax nana*
    c. *Iodamoeba butschlii*
    d. *Naegleria fowleri*

57. Which of these is a free living amoeba? [   ]
    a. *Naegleria fowleri*
    b. *Acanthamoeba polyphaga*
    c. *Balamuthia mandrillaris*
    d. *All*

58. Pick out the WRONG pair [   ]
    a. Vaginal flagellate-*Trichomonas vaginalis*
    b. Intestinal flagellate-*Giardia lamblia*
    c. Haemoflagellate-*Leishmania donovani*
    d. Neuroflagellate-*Pneumocystis carinii*

59. Which of these trophozoite resembles the *laughing old man?* [   ]
    a. *Giardia lamblia*
    b. *Endolimax nana*
    c. *Trypanosoma bruci*
    d. *Dientamoeba fragilis*

60. Chaga's disease is [   ]
    a. American sleeping sickness
    b. African sleeping sickness
    c. Leishmaniasis
    d. Giardiasis

61. Indian kala azar is caused by [   ]
    a. *Tryopanosoma rhodensiense*
    b. *Leishmania donovani*
    c. *Babesia bovis*
    d. *Toxoplasma gondii*

62. *Glossina species* [   ]
    a. A tsetse fly – Vector for African trypanosomiasis
    b. A sand fly – Vector for Leishmaniasis
    c. A cone nosed bug – Vector for Chaga's disease
    d. A mosquito – Vector for dengue fever

63. *Toxoplasma gondii* infection is transmitted to humans by [   ]
    a. Sexual contact
    b. Intake of contaminated food or water
    c. Mosquito bite
    d. Blood transfusion

64. Malignant Tersian (MT) malaria is caused by [   ]
    a. *Plasmodium vivax*
    b. *Plasmodium malariae*
    c. *Plasmodium falciparum*
    d. *Plasmodium ovale*

65. Parasitic ciliated protozoan [   ]
    a. *Cryptosporidium parvum*
    b. *Balantidium coli*
    c. *Toxoplasma gondii*
    d. *Trypanosoma gambiense*

66. Trematodes are characterized by   [   ]
    a. Elongated, cylindrical, unsegmented body
    b. Tape like, segmented body
    c. Leaf like, undsegmented body
    d. Tape like, unsegmented body

67. What is FALSE about cestodes?   [   ]
    a. Hooks are present
    b. Suckers are absent
    c. Sexes are not separate (monoecious)
    d. Alimentary canal is absent

68. What is TRUE about nematodes?   [   ]
    a. Hooks and suckers are present
    b. Sexes are not separate (monoecious)
    c. Body cavity is known as pseudocoele
    d. Anus is absent

69. Which of these is intestinal nematode?   [   ]
    a. *Necator americanus*
    b. *Loa loa*
    c. *Brugia malayi*
    d. *Trichinella spiralis*

70. Which is NOT an intestinal nematode?   [   ]
    a. *Ancylostoma deodenale*
    b. *Trichuris trichura*
    c. *Enterobius vermicularis*
    d. *Strongyloides stercoralis*

71. Which of these is transmitted through mosquito bite?   [   ]
    a. *Wuchereria bancrofti*
    b. *Brugia malayi*
    c. *Loa loa*
    d. All

72. Pick out the WRONG pair   [   ]
    a. *Ascaris*-round worm
    b. *Enterobius*-Pinworm
    c. *Trichuris*-Ringworm
    d. *Ancyclostoma*-Hookworm

73. Barrel shaped eggs with projecting mucus plug at each pole   [   ]
    a. *Ascaris*   b. *Trichuris*   c. *Ancylostoma*   d. *Enterobius*

74. Intermediate host for *Taenia saginata*   [   ]
    a. Cattle   b. Pig   c. Rat   d. Dog

75. Scolex (head region) of *Taenia solium*   [   ]
    a. Unarmed, four suckers, rostellum absent
    b. Armed, four suckers, rostellum present
    c. Armed, four suckers, rostellum absent
    d. Unarmed, four suckers, rostellum present

76. Match the columns   [   ]
    A. Intestinal fluke
    B. Liver fluke
    C. Lung fluke
    D. Blood fluke

    i. *Fasciola hepatica*
    ii. *Paragonium mexicanus*
    iii. *Fasciolopsis buski*
    iv. *Schistosoma japonicum*

    a. A-iii, B- ii, C-iv, D-I
    b. A-iv, B-i, C-iii, D-ii
    c. A-iii, B- i, C-ii, D-iv
    d. A-i, B- iv, C-iii, D-ii

77. Examination of Peripheral blood smear is useful in the diagnosis of   [   ]
    a. *Enterobius vermicularis*
    b. *Entamoeba histolytica*
    c. *Plasmodium malariae*
    d. All

78. Cellophane tape test is employed in the diagnosis of　　　　　　　　[　]
    a. *Enterobius vermicularis*　　　　　b. *Entamoeba histolytica*
    c. *Plasmodium malariae*　　　　　　d. All

79. Ringworm group of fungi tend to grow in　　　　　　　　　　　　[　]
    a. Epidermis　　　　b. Hair　　　　c. Nail　　　　d. All

80. Which of these is considered subcutaneous mycosis?　　　　　　　[　]
    a. Thrush　　　　b. Maduramycosis　　c. Vaginitis　　d. All

81. *Nocardia braziliensis* is known to cause　　　　　　　　　　　[　]

    a. Folliculitis　　　b. Sporotrichosis　　c. Mycetoma　　d　None of these

82　Severe diabetes may lead to　　　　　　　　　　　　　　　　　[　]
    a. Cutaneous mycoses　　　　　　　b. Subcutaneous mycoses
    c. Systemic mycoses　　　　　　　　d. Superficial mycoses

83. Tinea barbae abd Tinea capitis are caused by　　　　　　　　　　[　]
    a. *Microsporum*　　b. *Trichophyton*　　c. *Sporothrix*　　d. *Epidermophyton*

84. Tinea versicolor and Tinea nigra are　　　　　　　　　　　　　[　]
    a. Cutaneous mycoses　　　　　　　b. Subcutaneous mycoses
    c. Systemic mycoses　　　　　　　　d. Superficial mycoses

85. Cryptococcosis and Histoplasmosis　　　　　　　　　　　　　[　]
    a. Cutaneous mycoses　　　　　　　b. Subcutaneous mycoses
    c. Systemic mycoses　　　　　　　　d. Superficial mycoses

86. Which of these is NOT a dimorphic fungus?　　　　　　　　　　[　]
    a. *Blastomyces dermatitidis*　　　　b. *Histoplasma capsulatum*
    c. *Cryptococcus neoformans*　　　　d. *Coccidioides immitis*

87. Cutaneous fungal pathogens are demonstrated from skin scrapings treated with　[　]
    a. 10% KOH　　　b. 10% HCl　　　c. 10% NaCl　　d. 10% Bile salt

88. *Cryptococcus neoformans* can be isolated from　　　　　　　　[　]
    a. Blood　　　b. Bone marrow　　c. Sputum　　d. All

89. *Malassezia furfur* is the causative agent of　　　　　　　　　[　]

    a. Tinea nigra　　b. Tinea versicolor　c. Pityriasis capitis　d. Tinea capitis

90. Bacterial skin infection　　　　　　　　　　　　　　　　　　[　]
    a. Erythrasma　　　　　　　　　　b. Trichomycosis axillaries
    c. Both a & b　　　　　　　　　　　d. None

91. *Microsporum gypseum* is isolated from　　　　　　　　　　　[　]
    a. Horse skin scrapings　　　　　　b. Soil
    c. Plant roots　　　　　　　　　　　d. Cow dung

92. Parasitic diseases confined to human only　　　　　　　　　　　[　]
    a. Anthraponosis　　　　　　　　　b. Zoonosis
    c. Anthrapozoonosis　　　　　　　　d. None of these

93. Which of these virus is involved in formation of syncytia　　　　[　]
    a. Rabies virus　　　　　　　　　　b. Herpes simplex virus
    c. Polio virus　　　　　　　　　　　d. Pox virus

94. Negri bodies are   [   ]
    a. Intracytoplasmic inclusion bodies of rabies virus
    b. Intranuclear inclusion bodies of rabies virus
    c. Intracytoplasmic inclusion bodies of Herpes simplex virus
    d. Intranuclear inclusion bodies of Herpes simplex virus

95. Polio virus is known for   [   ]
    a. Inhibition of cellular DNA
    b. Disruption of cytoskeleton
    c. Inhibition of cellular protein synthesis
    d. Alteration of cell membrane

96. Flu like symptoms are attributed to   [   ]
    a. T cells
    b. Macrophages
    c. Polymorphonuclear Luecocytes
    d. Interferons

97. Which of these is associated with immunosuppression   [   ]
    a. HIV
    b. Cytomegalovirus
    c. Measles virus
    d. All

98. Human papilloma virus is transmitted through   [   ]
    a. Sexual contact
    b. Oro-fecal route
    c. Maternal-Neonatal mode
    d. Genetic mode

99. Match the columns   [   ]

    A. Respiratory   i. Rabies virus
    B. Feco-oral   ii. Picorna virus
    C. Zoonosis   iii. Rhinovirus
    D. Blood   iv. Hepatitis B virus

    a. A-iv, B- i, C-iii, D-ii
    b. A-iv, B-ii, C-i, D-iii
    c. A-iii, B-ii, C-i, D-iv
    d. A-i, B- iv, C-iii, D-ii

100. Multinucleate giant cells formed by fusion of individual host cells after viral infection are called   [   ]
    a. Schizont
    b. Syncytium
    c. Rosette
    d. Signet

101. Bateriophages produce   [   ]
    a. Plaques
    b. Pocks
    c. Cytopathic effect
    d. None of these

102. Adenovirus associated illness   [   ]
    a. haryngoconjunctival fever
    b. Pneumonia
    c. Acute respiratory disease
    d. All

103. Varicella Zoster virus is   [   ]
    a. Herpes virus
    b. Small pox virus
    c. Measles virus
    d. Cytomegalo virus

104. What is FALSE about coronavirus   [   ]
    a. Enveloped viruses
    b. RNA viruses
    c. Non hemagglutinating
    d. Gastrointestinal pathogens.

Answer Key and Validation

Medical Microbiology

1. a	19. a	37. b	55. c	73. b	91. b
2. c	20. c	38. a	56. d	74. a	92. a
3. d	21. b	39. b	57. d	75 b	93. b
4. b	22. b	40. b	58. d	76. c	94. a
5. a	23. b	41. c	59. a	77. c	95. c
6. c	24. b	42. b	60. a	78. a	96. d
7. b	25. b	43. d	61. b	79. d	97. d
8. a	26. b	44. d	62. a	80. b	98. a
9. d	27. a	45. a	63. b	81. c	99. c
10. a	28. d	46. a	64. c	82. d	100. b
11. a	29. b	47. a	5. b	83. b	101. a
12. c	30. c	48. b	66. c	84. d	102. d
13. a	31. a	49. a	67. b	85. c	103. a
14. d	32. d	50. c	68. c	86. a	104. c
15. d	33. b	51. d	69. a	87. a	
16. c	34. a	52. b	70. d	88. c	
17. a	35. b	53 a	71. d	89. b	
18. a	36. c	54. d	72. c	90. c	

15. *Vibrio parahaemolyticus is primarily a pathogen of the intestinal system*

20. *Cutaneous anthrax is called hide porter's disease and pulmonary anthrax is called wool sorter's disease*

27. *Vibrio are classified under Vibrionaceae on the basis of positive oxidase test and polar flagellae*

28. *Pinta is caused by Treponema carateum. Bacillus cereus causes gastroenteritis*

29. *Treponema pallidum sub spp. Endemicum causes endemic syphilis called Bejel*

30. *Proteus are gram negative rods*

33. *Bacillus anthracis causes gelatin liquefaction in deep gelatin agar tubes and the growth pattern in the tube appears like an inverted fir tree.*

56. *Naegleria fowleri is a free living amoeba*

58. *Pneumocystis carinii is an unclassified unicellular protozoan with a tropism for growth on respiratory surfaces of mammals.*

67. *Suckers are present in cestodes*

70. *Strongyloides stercoralis is a lung parasite*

72. *The common name for Trichuris is whipworm. Ringworm infections are fungal skin diseases.*

11

Pharmaceutical Microbiology

Retrace your subject...
Terminology, Historical account of pharmaceutical Microbiology, spectrum of certain antimicrobial agents, Antibiotic producers and their principal target organisms, Antibiotics: Mechanisms of Action, Drug resistance,

Terminology	
Antibiotic	A substance produced by a microorganism or to a similar substance (produced wholly or partially by chemical synthesis) which, in low concentrations inhibits the growth of other microorganisms.
Broad spectrum antibacterials	Can inhibit a variety of Gram positive and gram negative bacteria
Narrow spectrum antibacterials	Act on only a limited variety of bacteria
Bacteriostatic agent	Inhibits the growth of the test organisms but does not kill them.
Bactericidal agent	Kills the test organisms
Minimum Inhibitory Concentration (MIC)	The lowest concentration of the drug/antimicrobial, that inhibits the growth an organisms
Minimum Bactericidal Concentration (MBC)	The lowest concentration of the drug/antimicrobial, that kills the growth an organisms
Antibiotic synergism	Combination of antibiotics that results in enhanced bactericidal activity when used together as compared with the activity of each antibiotic individually
Antibiotic antagonism	Combination of antibiotics that results in diminished bactericidal activity when used together as compared with the activity of each antibiotic individually; due to interference with activities of each other.
Penicillins	Antibiotics that have a β lactum ring in their chemical structure Natural penicillins, Semisynthetic penicillins and cephalosporins are a part of this group.
Semisynthetic penicillins	
Aminoglycosides	These antibiotics that have amino sugars in glycosidic linkage. The group includes streptomycins, kanamycin, lincomycin, paramomycin, gentamycin, tobramycin and amikacin
Streptomycins	
Macrolides	These antibiotics have a macrocyclic ring to which sugars are linked. The group include serythromycin, oleandomycin and spiromycin.
Tetracyclins	Derivatives of polycyclic naphthacene carboxamide. Antibiotics included are tetracycline, chlortetracycline, oxytetracycline, demeclocycline and minocycline.
Peptide antibiotics	Have peptide linked aminoacids that include D and L forms. Bacitracin, gramicidin and polymyxins are included in this group.
Antifungal antibiotics	The se are of two categories. (a) Polyenes: contain large ring with conjugated double bond system. Nystatin and Amphotericin B fall in this category. (b) Other antifungals : include 5-fluorocytosine, clotrimazole and griseofulvin.
Chloramphenicol	It is a nitrobenzene derivative of dicholoroacetic acid. It is an antibiotic that is distinctly placed in a class of its own.

Table *Contd...*

Drug transport	Drugs must reach at inhibitory levels to the target sites which are usually the cell membranes of cytoplasm of the cells. Most drugs are transported by passive diffusion, while some require the facilitated diffusion mechanism. Formation of cell aggregates, biofilms and drug efflux pupms are some problems encounterd on drug transport.
	In gram negative bacteria, the outer membrane slows down the diffusion of water soluble and lipophilic drugs.
	In gram positive bacteria the strong anionic character of the teichoic acids affect the uptake of ionized molecules.
	Facilitated transport systems enhance transfer of structural analogues
	Multidrug efflux systems reduce intracellular drug concentrations by driving them drug out.

Pharmaceutical Microbiology - Historical	
Early Seventeenth Century	Cinchona bark was used by Indians of Peru for treating malaria. The active principle quinine was isolated in 1820. Quinine still a place in chemotherapy towards treatment of malaria!
1817	Emetine was isolated from root of *Ipecaqcuanha* by Brazilians, who used it in the treatment of amoebiasis
1750	The term *antiseptic* was used by Pringle to describe substances that prevent putrefaction.
Middle age	Mercuric chloride was used by the Arabian physicians to prevent sepsis in open wounds.
1825	Labarraque introduced hypochlorite for the treatment of infected wounds. This was followed by use of tincture iodine in 1839.
1835	Oliver Wendel Holmes reduced the incidence of puerperal fever by washing hands with chloride of lime.
1838	Ignaz Semmelweiss strongly proposed that puerperal fever among mothers and infants was transmitted by physicians and midwives due to unhygienic delivery practices.
1863	Louis Pasteur proposed role of microbes in the process of putrefaction and origin of infections.
1860-70	Joseph Lister introduced use of 2.5% phenol solution for dressing wounds and 5% phenol for washing surgical instruments He also proposed a phenol spray in the environment before surgeries.
1893	Lindgard treated horses suffering from a trypanosome disease *surra*, using arsenious oxide.
1902	The beginning of chemotherapy... Paul Ehrlich worked repeatedly on various arsenicals and finally in 1910, he tasted success with his discovery of *salvarsan* an organoarsenical compound used in the standard treatment of syphilis.
1930s	Ehrlich recognized the problem of resistance of microbes towards chemotherapeutic agents.
1933	Mepacrine was first marketed and gained immense popularity in the Second World War as an antimalarial agent.
1935	Domagk and Trefouel's works introduced the use of sulphonamides against infections.
1928	Alexander Fleming accidentally discovered penicillin, the first antibiotic.
1940-41	Florey and Chain isolated penicillin in its pure and effective form and its great clinical usefulness began.
	Waksman discovered streptomycin another promising antibiotic that gained large reputation.
1960's	Antiprozoal drugs – mefloquine and halofantrine
1970's	Remarkable progresses in understanding the synthesis, structure and functions of various drugs and development of newer techniques in defining the site of action of various drugs
1990's	Serious studies on drug resistance and the underlying mechanisms
Current trends in pharmacy	With the principles of Biochemistry, molecular biology and bioinformatics it is expected that the drug discovery process will be facilitated. Target proteins are sought that bear sequence identity across the major pathogens.

Spectrum of certain Antimicrobial agents

Antimicrobial	Gram Negative Bacteria	Gram positive bacteria	Chlamydias/ Rickettsias	Mycobacteria	Fungi	Protozoa	Helminths	Viruses
Penicillin		*						
Streptomycin	*			*				
Tetracycline	*	*	*					
Isoniazid				*				
Ketoconazole					*			
Mefloquine						*		
Niclosamide							*	
Praziquentel							*	
Acyclovir								*
Sulfonamides	*							
Carbapenems	*	*	*					
Cephalosporins		*						
Rifampicin	*	*	*					*
Nystatin,								*
AmphotericinB								*
Griseofulvin								*
Metronidazole						*		

Antibiotic producers and their principal target organisms		
Antibiotic	**Producer organisms**	**Susceptible organisms**
Penicillin	*Penicilluim chrysogenum, P. notatum*	*Most gram positive bacteria, Nocardia and Actinobacteria*
Streptomycin	*Streptomyces griseus*	*Tubercle bacilli, Gram negative bacteria and limited activity against gram positive cells.*
Bacitracin	*Bacillus licheniformis*	*Several gram positive bacteria*
Griseofulvin	*Penicillium griseofulvin, P.nigricans, P.urticae*	*Pathogenic yeast, plant pathogenic fungi, etc.*
Fumagillin	*Aspergillus fumigatus*	*Bacteriophages and amoebae*
Tetracycline	*Streptomyces aureofaciens, Streptomyces rimosus*	*Broad spectrum*
Chloramphenicol	*Streptomyces venezuelae*	*Broad spectrum, particularly Salmonellae*
Kanamycin	*Streptomyces kanamyceticus*	*Tubercle bacilli, Penicillin resistant staph and some other gram positive bacteria*
Erythromycin	*Streptomyces erythreus*	*Penicillin resistant staph, other gram positive bacteria and limited gram negative cells.*
Nystatin	*Streptomyces noursei*	*Fungal skin infections, moniliasis*
Amphotericin B, Novobiocin, Oleandomycin, Carbomycin, Cycloheximide	*Various species of Streptomyces.*	*Plant parasitic fungi, pathogenic yeast and yeast like fungi*

Antibiotics – Mechanisms of Action

Antibiotic	Target organism	Target area	Mechanism
β-lactams, Glycopeptides, Cycloserine	Antibacterial	cell wall	Inhibit peptidoglycan synthesis
Isoniazid	Antimycobacterial		Inhibits mycolic acid synthesis
Ethambutol	Antimycobacterial		Inhibits arabinogalactan synthesis

Table *Contd...*

Antibiotic	Target organism	Target area	Mechanism
Aminoglycosides	Antibacterial	Ribosome (protein synthesis)	Distort 30S ribosomal subunit
Tetracyclines			Block 30S ribosomal subunit
Chloramphenicol			Inhibits peptidyl transferase
Macrolides & Azalides			Block translocation
Fusidic acid			Inhibits elongation factor
Quinolones	Antibacterial	Chromosome function	Inhibit DNA gyrase
Metronidazole	Antiprotozoal		Breaks DNA strand
Nitrofurantoin	Antibacterial		Breaks DNA strand
Rifampicin	Antibacterial & antimycobacterial		Inhibits RNA polymerase
5-fluorocytosine	Antifungal		Inhibits DNA synthesis
Sulphonamides	Antibacterial, antiprotozoal	Folate metabolism	Inhibit folate synthesis
Trimethoprim	Antibacterial		Inhibits dihydrofolate reductase
Polymyxins	Antibacterial	Cell membrane	Disrupt the bacterial membranes
Polyenes	Antifungal		Disrupt the fungal membranes
Imidazoles, triazoles and Naftidine	Antifungal		Inhibit ergosterol synthesis
Bacitracin	Antibacterial		Inhibits membrane synthesis and movement of precursors

Drug Resistance	
Several mechanisms of drug resistance are shown by microbes. R plasmids in several bacteria impart resistance to certain drugs.	
Chromosomally mediated resistance	Methicillin, quinolones, rifampicin
Plasmid mediated resistance	Aminoglycosides, β lactums, tetracyclins, sulphonamides, trimethoprim, chloramphenicol, erythromycin and fusidic acid
Transposons	Single: ampicillin, choloramphenicol, tetracycline. Multiple: ampicillin+streptomycin+sulphonamide
Mechanisms	Examples
Enzyme inactivation – Hydrolysis of the β lactum ring	Some β lactum antibiotics
Enzymatic trapping – certain penicillin binding proteins	Some β lactum antibiotics
Enzyme modification – alteration of the enzyme by phosphorylation , adenylation or acetylation	Some aminoglycoside antibiotics
Impermeability – Reduction of cell ability to take the drug due to mutational loss of porins	Some β lactum antibiotics, Some aminoglycoside antibiotics
Alteration in binding site – production of altered peptidoglycan precursors, low resistance ribosomes, etc.	Streptomycin, erythromycin, glycopeptides.

Exercises

1. Which of these acts selectively against gram positive bacteria   [   ]

   a. Penicillin   b. Streptomycin   c. Cephalothin   d. Tetracycline

2. Match the Drugs with their target organisms   [   ]

   A. Isoniazid   i. Helminths
   B. Niclosamide   ii. *Plasmodium*
   C. Mefloquine   iii. Fungi
   D. Ketoconazole   iv. Mycobacteria

   a. A – iv, B – i, C – ii, D –iii   b. A – iv, B – ii, C – iii, D –i
   c. A – iii, B – ii, C – iv, D –I   d. A – i, B – ii, C – iii, D –iv

3. Griseofulvin is   [   ]

   a. Antimalarial   b. Antifungal   c. Antiviral   d. Antihelminthic

4. The antibiotic bacitracin is produced by   [   ]
   a. *Streptomyces nodosus*   b. *Cephalosporium*
   c. *Penicillium notatum*   d. *Bacillus subtilis*

5. Broad spectrum antibiotic   [   ]

   a. Tetracycline   b. Streptomycin   c. Isoniazid   d. Praziquantel

6. Gentamycin is obtained from   [   ]
   a. *Micromonospora purpurea*   b. *Streptomyces aureofaciens*
   c. *Bacillus subtilis*   d. *Streptomyces nodosus*

7. Which of these is a product of *Streptomyces* species.   [   ]
   a. Erythromycin   b. Neomycin   c. Amphotericin B   d. All

8. Amoebic infections are tested by   [   ]
   a. Metronidazole   b. Chloroquine   c. Acyclovir   d. Nystatin

9. The drug of choice for *Salmonella* infection   [   ]

   a. Cephalosporin   b. Carbapenems   c. Chloramphenicol   d. Amantadine

10. Erythromycin is the drug of choice for   [   ]

    a. Streptococcal infections   b. Trypanosomiasis

    c. Malaria   d. Diphtheria

11. Which of these is bacteriostatic   [   ]
    a. Sulfonamides   b. Tetracyclines
    c. Chloramphenicol   d. All

12 Tooth discoloration, liver and kidney damage are the toxic effects of   [   ]
    a. Sulfonamides   b. Tetracyclines
    c. Chloramphenicol   d. All

13. Which of these cause inhibition of cell wall synthesis   [   ]
    a. Aminoglycosides   b. Oxacillin
    c. Erythromycin   d. Polymyxin B

14. The mode of action of Chloramphenicol   [   ]
    a. Inhibition of cell wall synthesis
    b. Inhibition of protein synthesis
    c. Inhibition of RNA synthesis
    d. Injury to plasma memebrane

15. Sulfonamides are particularly active against   [   ]
    a. Streptococci   b. *Neisseria meningitis*
    c. *Klebsiella*   d. Chlamydias

16. β lactam ring is seen in   [   ]
    a. Chloramphenicol   b. Gentamicin
    c. Trimethoprim   d. Penicillins

17. Macrocyclic lactone ring is seen in   [   ]

    a. Vancomycin   b. Rifamycin   c. Erythromycin   d. Tetracyclin

18. Polymyxin B is a representative of   [   ]
    a. Macrolides   b. Polypeptide antibacterial antibiotic
    c. Polyene antibiotic   d. Immadazole drugs

19. The most important use of Rifampicin is the treatment of   [   ]

    a. TB & Leprosy   b. Influenza   c. Candidiasis   d. Diptheria

20. Disc diffusion method to study antibacterial sensitivity patterns is called   [   ]
    a. Mueller Hinton test   b. Ehrlich`s test
    c. Kirby-Bauer test   d. Mantoux test

21. The highest dilution of a drug that is capable of inhibiting the growth of the test culture   [   ]
    a. MIC   b. MPN   c. MFT   d. MBC

22. Hereditary drug resistance among bacteria is often carried by   [   ]
    a. D factors   b. M proteins   c. F pili   d. R plasmids

23. Drugs that nullify each others activity when used in combination are   [   ]
    a. antagonistic   b. synergistic   c. mutualistic   d. none of these

24. Antibiotics are   [   ]
    a. microbial products   b. kill or inhibit susceptible microbes
    c. active at low concentrations   d. all

25. Antimetabolite drugs   [   ]
    a. destroy metabolite molecules
    b. destroy key enzymes that synthesize metabolites
    c. competitively inhibit use of metabolites
    d. inhibit absorption of metabolites

26. Pick the WRONG statement   [   ]
    a. Tetracycline disrupts bacterial DNA
    b. Vancomycin inhibits crosslinkage of peptidoglycan layers
    c. Isoniazid inhibits mycolic acid synthesis
    d. Polymyxin inhibits bacterial membranes

27. Match the drugs with their modes of action   [   ]

    | A. Penicillin | i. Antimetabolite |
    | B. Tetracycline | ii. Inhibition of protein synthesis |
    | C. Rifampin | iii. Disruption of cell wall |
    | D. Sulphonamide | iv. Inhibition of nucleic acid synthesis |

    a. A – iv, B – i, C – ii, D –iii   b. A – iv, B – ii, C – iii, D –i
    c. A – iii, B – ii, C – iv, D –i   d. A – i, B – ii, C – iii, D –iv

28. The lowest concentration that kills 99.99% of the population is called   [   ]
    a. Minimum Inhibitory Concentration
    b. Lethal Dose
    c. Minimum Spectral Level
    d. Minimum Bactericidal Concentration

29. Which among these show β lactamase activity   [   ]
    a. Penicillinases   b. Cephalosporinases
    c. Carbapenemases   d. All

30. Narrow spectrum drugs   [   ]
    a. Active only against a limited variety of bacteria
    b. Only inhibit but cannot kill the target organisms
    c. Always require some synthetic modification to be actively antimicrobial
    d. Can easily induce resistance among the target organisms

31. Natural Penicillin   [   ]
    a. Penicillin G   b. Penicillin V   c. Both a & b   d. None

32. Which of these shows enhanced activity against Staphylococci   [   ]
    a. Methicillin   b. Oxacilin   c. Cloxacillin   d. All

33. Drugs that show additional activity against certain organisms other than the usual
    target cells are called   [   ]
    a. Expanded spectrum drugs   b. Broad spectrum drugs
    c. Wide spectrum drugs   d. Extentensive spectrum drugs

34. Streptomycin is an example of   [   ]
    a. β lactum antibiotics   b. Macrolide antibiotics
    c. Aminoglycoside antibiotics   d. Polypeptide antibiotics

35. Bacteria such as *Psuedomonas* are known to gain drug resistance due to   [   ]
    a. development of permeability barriers   b. production of degenerative enzymes
    c. blockage of intracellular target site   d. Production of exogenous inactivation factors

Answer Key and Validation

Pharmaceutical Microbiology

1. a	7. d	13. b	19. a	25. c	31. c
2. a	8. a	14. b	20. c	26. a[*]	32. d
3. b	9. c	15. b	21. a	27. c	33. a
4. d	10. d	16. d	22. d	28. d	34. c
5. a	11. d	17. c	23. a	29. d	35. a
6. a	12. b	18. b	24. d	30. a	

5 *Tetracycline is active against gram positive and gram negative bacteria as well as Chlamydias and Rickettsias. Hence it is a broad spectrum drug.*

26 *Tetracycline prevents polypeptide elongation at 30S ribosome.*

12

Principles of Immunology

Retrace your Subject…
Nonspecific host defenses, Specific host defenses, immune system, Immunity, Antigens, Antibodoes, immunoglobulin Supergene family, Diagnostic Immunology, Hypersensitivity, Vaccines…

Nonspecific host defenses

The general resistance of the body to foreign materials irrespective of the type of antigenic determinant it bears, is called non-specific defense. Eg: Phagocytosis, fever, action of interferons, lysozyme and properdin.

Defense	Nature
First line of defense Skin	Intact skin is a mechanical barrier for microbes to enter, sweat glands ans sebaceous glands posses antimicrobial nature, hair follicles are not easily degradable by microbes, resident skin flora offer competitive inhibition of foreign organisms.
Mucous membranes	This forms the inner lining for entire gastrointestinal, respiratory and genitourinary tracts. Mucous secreted by this membrane traps mobility of the organisms and prevents spread to some extent. Provides a moist surface to prevent drying. Prevents entry of microbes to underlying tissue, but is less effective than skin.

Second line of defense
Phagocytes
  Granulocytes
  Monocytes
	Ingestion of microbial cell or any particulate matter is phagocytosis.
	Neutrophils and eosinophils are phagocytic during initial stages of infection
	Monocytes within tissues are termed macrophages. There are wandering and fixed macrophages. They are phagocytic when the infection advances to tissues. Also phagocytose dead blood cells as infection susides. Also involved in CMI

Inflammation	A defensive response of the body to infections and action of certain physical and chemical agents. It is characterised by redness, pain, heat and swelling. Inflammation destroys the injurious agent, prevents the spread of the agent and repairs the damage caused by the agent.
Fever (pyrexia)	A systemic, generalised defense characterised by body temp above normal, against bacteria, toxins, viruses and fungi that may have established an infection. Interleukin-1 (endogenous pyrogen) causes hypothalamus of brain to release prostaglandins (PGs). The PGs reset the thermostat at a higher temp. Fever destroys pathogens, inactivates toxins, prevents spread of pathogens, activates interferons, decreases available iron and speeds up repair reactions.
Antimicrobial substances Complement system	A group of 20 serum interactive proteins (C_1 to C_9 and others) that follow classical pathway (Ab associated) or alternative pathway (polysaccharide associated). The complement system involves in cytolysis, inflammation and opsonization (complement mediated phagocytosis).

Specific host defenses

Naturally acquired Immunity		Artificially acquired immunity	
Active	**Passive**	**Active**	**Passive**
Antigens enter the body by natural means, body produces Abs and specialised lymphocytes.	Ready Abs are received from mother to fetus, through placenta, or to infant through milk.	Ags are injected through vaccines, body produces specific Abs/ lymphocytes.	Ready Abs are introduced into the body for instant protection.

Immune system

System component	Immunological function
Lymphoid cells T cells	 Express α, β receptors, they are of T_{helper} and $T_{cytotoxic}$ types which are distingushed by expression of CD_4 and CD_8 markers respectively. They carry cytoplasmic *Gall body*. The T cells play a major role in CMI and also render support to humoral immunity.
B cells	The B cells do not have *Gall bodies* but show surface Igs. They act as specific Ag receptors. They mediate HMI by serving as precursors for the Ab producing plasma cells.
NK cells	Natural Killer cells are lymphocytes that contain large number of azurophilic granules. They have a cytotoxic nature. They exhibit *immunological surveilance* by recognizing and killing tumor cells and virus infected cells. They release interferon$_\gamma$ and many lymphokines that regulate hemopoiesis. Some show Ab dependent cellular cytotoxicity.
Mononuclear phagocytes, Antigen presenting cells	These are of two kinds: 1. Phagocytic *macrophages* (tissue monocytes) that remove particulate Ags. 2. Ag presenting cells (*APCs*) take up, process and present the Ag to T cells through MHC.
Polymorphs,	Granulocytes/PMNs consist of, eosinophils and basophils. Neutrophils play a role in *diapedesis* (engulfing foreign particle extravasculary. Eosinophils in atopic reactions and parasitic infestations. Basophils are precursors for mast cells.
Mast cells	Mast cells are tissue eosinophils (mucosalmast cells and connective tissue mast cells). They undergo degranulation on stimulation by an allergen, through of IgE. The degranulation components include: Histamines, serotonins, prolamines, etc.
Platelets	Platelets or *thrombocytes* are derived from *megakaryocytes* that play a key role in the blood clotting cascade. They express class I MHC products for Ag presentation and have receptors for IgG and IgE.

System component	Immunological function
Primary lymphoid organs (Thymus & Bone marrow)	*Thymus* is the site of T cell development. It is a bilobed organ located overlying the heart and major blood vessels in the thoracic cavity. It is made of lobes within which the lymphoid cells (*thymocytes*) are arranged onto outer cortex with immature cells and inner medulla with mature cells.
Secondary lymphoid organs (Spleen & lymph nodes)	*Spleen* is at the upper left quadrant of the abdomen behind the stomach. The splenic tissue is made of the *red pulp* consisting of macrophages, RBCs, platelets, granulocytes, plasma cells and lymphocytes; and *white pulp* consisting of lymphoid tissue with differentiated T abd B cells. *Lymph nodes* occur at brenches of the lymphatic vessels particularly in neck, groin, axillae and abdominal cavity.they are junctions to filter Ags from tissue fluid and lymph during the passage from periphery to thoracic ducts.

Immunity

Tendency of the body to reject, eliminate or counteract against non-self material that enters its tissues. The types of immunity have been enlisted below.

Immunity	Features
Cell Mediated Immunity (CMI)	Specialised cells (Eg. Lymphocytes) are directly involved as *effectors* in imparting the immune reaction. This immunity can be transferred to a non-immune individual through direct transfer of cells.
Humoral (Antibody Mediated) Immunity (AMI)	Immunity derived from body fluid factors such as antibodies and complement. This immunity can be transferred to a non-immune individual through transfer of cell free plasma or serum.
Herd immunity	A form of collective immunity. In a given population when the number of immune individuals is far more than the susceptible ones, the spread of the disease is arrested in the community. Thus the risk to the susceptible individuals is highly reduced. Eg: If in a population, more than 90% members are vaccinated against the disease, the remaining individuals are at very low risk of contracting the disease.
Autoimmunity	An immune response in which the body directs immunological mechanisms against its own components. The resulting disease is autoimmune disease. Autoimmunity is sometimes mediated by Abs (Acquired hemolytic anemias) and sometimes cell mediated (Eg. Multiple sclerosis).

Antigens (Immunogens)

- Any agent (usually proteins or polysaccharides) which can elicit an immune response. Ags are individual macromolecule or homologous/heterologous complex of macromolecules.
- An Ag may be soluble (microbial toxins, extracts, etc.) or particulate (whole cells).
- The antigenicity of a substance depends on its foreignness, chemical nature, route of administration, size, presence of adjuvants, number of available antigenic determinants on the surface, etc.

Antibodies (Immunoglobulins)

- An immunoglobulin molecule which is produced by (plasma cells) the body in response to stimulation by an Ag and which can react specifically, non-covalently and reversibly, with the Ag is called antibody.
- Ab specificity is not absolute. Sometimes an Ab may bind to Ags other than those that caused its production. Such combinations are called *immunological cross reactions.*
- There are five different classes of immunoglobulins. They play a role in humoral immunity and in opsonisation reactions. Artificially raised Abs find use in passive immunity and serological diagnosis.

Antibody class	Features
IgG	Most predominant class of Ig in plasma (75%). It is a bivalent molecule with γ heavy chains. There are 4 subclasses: IgG1, IgG2, IgG3 and IgG4. Plays imp role in transplacental immunity to protect fetus and neonate, as opsonin and antitoxin in extra vascular region and blood.
IgM	First Ig to be formed in humoral response.Contains μ heavy chains. A pentameric molecule, arranged radially with a J chain at the centre. These are complement fixing Abs, cannot cross placenta, and blood vessels. Egs. Wassermann Ab, rheumatoid factor, etc.
IgA	Predominant secretory Ig with α heavy chain, has a protective role in saliva, tears, colostrum and other body fluids. Occur in dimeric forms, linked with a J chain, fix complement via alternate pathway. The subclasses are IgA1 and IgA2.
IgE	A monomeric Ig, found in plasma, contains ε heavy chains. It binds to basophils and mast cells via Fc portion and brings about degranulation in presence of the allergen causing type I hypersensitivity
IgD	Occurs as surface receptor for Ag on the membranes of B cells, contains δ heavy chains. Exact role is nor clearly demonstrated.

Immunoglobulin supergene family

MHC	These are linked genetic loci that dominate the immune response. The MHC genes and their products are designated by prefixes HLA (*Humna Leucocyte Ags)*. The products occur on cell surfaces and serve as markers to differentiate *self* and *non-self* tissues. The genes in the MHC complex occur in three classes I, II and III. Class I and II products recognize target cells for killing and recognition of Ag presenting cells by T cells. Class III genes encode complement components and Factor B.
Interleukins	These are cytokines released by leucocytes and effective on leucocytes.Interleukin (IL) is designated by a number IL1, IL4, IL6, IL18 etc.they have large variety of functions involving: induction of fever, helping killer cells, tumor necrosis factors, stimulate eosinophilia in parasitic infestations, Ab production in B cells, activators of phagocytes. Some ILs (IL10)are anti inflammatory and even inhibit the release of cytokines from leukocytes.
Complement	A group of proteins in plasma, serum and tissues. On activation, they complex with other components to carry out several physiological functions. There are two pathways for complement activation. *Classic pathway* and *Alternative pathway*. Functions of Complement: Opsonisation for phagocytosis, direct killing by cytolysis, processing of immune complexes, activation of leucocytes at sites of inflammation and induction of specific Ab responses.

Diagnostic Immunology

Several Ag-Ab reactions are employed in the field of diagnosis of disorders. These serological tests are advantageous, as they are time saving, specific and safer tests than working with live pathogens. The following tables elucidate typical immunological reactions and their applications.

Test	Features	Applications
Direct Agglutination tests	Reaction of particulate Ags with Abs to form visible aggregates (clump). Direct tests use whole RBCs, bacteria or fungi. Carried out in series of tubes or microtitre plates with serial dilutions of serum.	Widal test, Coomb's test, in diagnosis of brucellosis, in classification of *Salmonella* serotype, , etc.
Indirect (passive) Agglutination tests	Soluble Ags (or Abs) are first adsorbed onto inert particulate surfaces such as bentonite clay, latex spheres or carbon. When these coated particles combine with the corresponding Ab (or Ags), they show clear clumping.	In rapid diagnosis of variety of bacterial and viral diseases, pregnancy tests and several other rapid slide agglutination tests.
Hemagglutination tests	Clumping of RBCs due to any specific Abs against the surface Ags. Some viruses also cause hemagglutination without any Ag-Ab reaction. These are called *viral hemagglutination tests*. Such agglutination is inhibited by use of Abs against the viruses	Blood grouping & Rh typing. In diagnosis of viral infections,by *hemagglutination inhibition tets*
Neutralisation tests	An Ag-Ab reaction in which the harmful effects of a bacterial exotoxin or a virus are blocked by specific Abs..	Cytopathic effects of viruses in cell cultures and embryonated eggs are used in diagnosis of viral diseases. Antitoxins produced in an animal can be injected into humans for therapeutic purposes
ELISA tests (Enzyme Linked Immunosorbent Assay)	A highly sensitive technique to determine the titre of specific Ab/Ag. The Ag-Ab complexes are detected by treating with a *conjugate* (enzyme labelled anti immunoglobulin Ab). The test system is than examined for the bound enzyme by incubating with a substrate, and assaying for the product of the enzyme reaction.	Screening of HIV subjects, ultimate diagnosis of various diseases that may show cross reactions in slide agglutination tests, in determination of Ag/Ab titre to monitor progress of treatment/disease, etc. ELISA Kits, cards and dipstics are designed for clinical labs and bedside tests.
Radioimmunoassay (RIA) tests	A highly sensitive technique to determine the titre of specific Ab/Ag. Using radioactive labelling. In Ab assay, the sample is allowed to react with excess radiolabelled Ag (eg. With ^{125}I). The amount of bound Ag is determined from the level of radioactivity. In Ag assay (*binder-ligand assay*), a competition is created between labelled Ags provided artifically and unlabelled Ags in the sample to bind with a fixed amount of Ab. The conc of Ag is determined from a standard curve	In quantitation of drugs, hormones, tumor markers, IgE and minute quantities of viral Ags, etc.

Test	Features	Applications
Precipitation tests	Involve reactions of soluble Ags with IgG or IgM Abs to form large, aggregates called lattices when the ratio of Ag-Ab is optimum. Precipitation reactions are of various kinds: Ring tests, immunodiffusion tests (Oudin, Outerlony, Oakley Fulthorpe test, SRID,) and immuno-electrophoresis.	To diagnose several diseases, to establish serovars of isolated strains, to establish identity between Ags or Abs. Egs: Ascoli's thermoprecipitin test, C reactive protein test, Kahn test, etc.

Table *Contd...*

Test	Features	Applications
Complement fixation tests	Complement is used up during certain Ag-Ab rections. This is called complement fixation. The test involes 2 steps: complement fixation and indicator (sheep RBCs and their Abs) system. If Ag is present, complement will be used up and hence not available for indicator system and vice versa. Thus hemolysis of the sheep RBC's indicates negative test	Wasserman's test for syphilis was used routinely, but has been replaced by other simpler tests eventually.
Fluorescent Ab (FA) tests (Immuno-fluorescence tests)	These combine fluorescent dyes such as fluorescein isothiocyanate (FITC) with Abs. The labelled Abs fluoresce in UV light. The direct FA tests are performed by using fuorescent labelled Abs against the suspected clinical specimen or Ag. In indirect FA test, the Ag and Ab are allowed to react and the complex is mixed with labelled antihuman immune serum globulin. The test is positive if fluorescence is observed	Can be used to diagnose diseases with increased sensitivity and almost 100% accuracy. Eg; In diagnosis of syphilis and Rabies. Frequent occurrence of nonspecific fluorescence in tissues and other materials is the major problem.

Hypersensitivity

- Hypersensitivity reactions are due to the inappropriate action of immune responses that sometimes cause inflammation and tissue damage.
- Coombs and Gell described four types of hypersensitivity reactions. The first three are Ab mediated and the fourth type is mediated by T cells and macrophages.

Type I	Type II	Type III	Type IV
Immediate hypersensitivity, Caused due to IgE response against pollen, dust, mites, animal dander, certain perfumes etc. IgE binds to mast cells via F_c receptors, and to allergen via F_{ab} regions. Degranulation is induced and meditors that produce allergy are released from the cells.	Complement mediated/cytotoxic hypersensitivity, action against individual's own cells or against foreign cells like transfused RBCs. It leads to cytotoxic action of K cells or complement mediated lysis.	Immune complex hypersensitivity. Some Ag-Ab complexes get deposited within the tissues for longer time than necessary. Complement is activated and phagocytes (PMNs) are attracted to the site and cause local damage and inflammation.	T cell mediated/ Delayed hypersensitivity. The Ag sensitised T cells release lymphokines following secondary contact. Inflammatory reaction is induced and macrophages are also attracted to the site.
Egs: Atopy (asthma, eczema, hay fever, cold allergy), vasodilatation, bronchial smooth muscle contraction, mucosal oedema, eosinophilia, etc	Egs: Hemolytic disease of the Newborn (HDNB), transfusion reactions, hyperacute graft rejection, autoimmune hemolytic anemias, certain drug induced reactions, Pemphigus, Myasthenia gravis, Lambert Eaton syndrome, etc.	Egs: Some patients of malaria, dengue fever, viral hepatitis, staphylococcal endocarditis, RA, SLE and polymistosis. Farmer's lung, pigeon fancier's lung, etc.	Eg: contact with nickel chromate, rubber accelerators, poison ivy, tuberculin type reactions and granulomatous reactions of Ags of *M.tuberculosis and M.leproae*

List of Immunodeficiency Diseases	List of Autoimmune disorders
Humoral immunodeficiencies X-Linked agammaglobulinemia Hypogammaglobulinemia of infancy Selective Ig deficiency diseases	**Hemocytolytic Autoimmune disorders** Autoimmune hemolytic anemia Autoimmune leucopenia Autoimmune thrombocytopenia
Cellular immunodeficiencies (T cell deficiency) DiGeorge syndrome Chronic mucocutaneous candidiasis	**Organ Specific Autoimmune Diseases** Hashimoto's disease (Lymphadenoid Goitre) Addison's disease (adrenal glands)
Combined Immunodeficiencies (B & T cell deficiency) Swis type agammaglobulinemia Ataxia Wiskott-Aldrich syndrome	Autoimmune Orchitis (testes) Myathenia gravis (Myoneural junctions) Autoimmune diseases of the eye Pemphigus vulgaris (skin)
Disorders of Phagocytosis Leucocyte G6PD deficiency Chediak-Higashi Syndrome Job's Syndrome Tuftsin Syndrome Lazy Leucocyte syndrome Hyper IgE syndrome Shwachman's Disease	**Systemic Autoimmune Diseases** Systemic Lupus Erythematosus Rheumatoid Arthritis Polyarteritis nodosa Sjogren's syndrome

Vaccines

Disease	Vaccine employed
Bacterial diseases	
Diphtheria	Purified diphtheria toxoid
Meningococcal meningitis	Purified polysaccharide from *N.meningitidis*
Pertussis (whooping cough)	Killed whole cell/ fragments of *Bordetella pertussis*
Pneumococcal pneumonia	Purified polysaccharide from *Streptococcus pneumoniae*
Tetanus	Purified tetanus toxoid
Diphtheria-pertussis-tetanus	DPT triple vaccine
Viral diseases	
Influenza	Inactivated virus for injection purposes
Measles, Mumps & Rubella	MMR triple vaccine of attenuated viruses
Chicken pox	Attenuated virus
Poliomyelitis	Killed Salk and live Sabin vaccine.
Rabies	Killed virus preparation
Hepatitis B	Antigenic fragments of virus
Hepatitis A	Inctivated virus
Small pox	Live vaccinia virus.

Exercises

1. The ability of a microbe to produce disease   [   ]
   a. Pathogenesis   b. Pathogenicity   c. Communicability   d. Infectivity

2. *Invasiveness* is   [   ]
   a. the ability of a pathogen to survive within the host
   b. the ability of a pathogen to multiply within the host
   c. the ability of a pathogen to spread in the host after establishing the infection
   d. the ability of a pathogen to survive outside the host until it finds a new one

3. Toxoids   [   ]
   a. Nontoxic forms of toxins that retain their ability to induce Ab production
   b. Concentrated forms of toxin that produce toxic effects almost instantly
   c. Toxins released by viruses
   d. Purified toxins

4. What is FALSE about endotoxins   [   ]
   a. They are produced by Gram negative bacteria
   b. They cannot be toxoided
   c. They are heat labile complexes
   d. They are poor Ags

5. Match the columns   [   ]

A. Coagulase	i. Attack PMNs
B. Hyaluronidase	ii. Attack RBCs
C. Hemolysins	iii. Spreading factor
D. Leucocidins	iv. Fibrin barrier

   a. A-iv, B-iii, C-i, D-ii   b. A-iii, B-i, C-ii, D-iv
   c. A-iv, B-iii, C-ii, D-I   d. A-i, B-i, C-iii, D-iv

6. Which of these protect the pathogens from phagocytosis?   [   ]
   a. Capsule of *Klebsiella pnuemoniae*
   b. K Ag of *Escherichia coli*
   c. V_i Ag of *Salmonella typhi*
   d. All of these

7. Which of these help bacteria to withstand the lytic activity of complement?   [   ]
   a. Capsule of *Klebsiella pnuemoniae*   b. K Ag of *Escherichia coli*
   c. V_i Ag of *Salmonella typhi*   d. All of these

8. LD50 refers to   [   ]
   a. The dose of pathogen required to kill 50% of animals tested under standard conditions
   b. The dose of antibiotic required to kill 50% of pathogen population tested under standard conditions
   c. The dose of an antiseptic required to kill 50% of pathogen population tested under standard conditions
   d. The dose of pathogen required to produce infection in 50% of animals tested under standard conditions

9. Match the columns   [   ]
   - A. Endemic
   - B. Sporadic
   - C. Epidemic
   - D. Pandemic

   - i. Affect individuals
   - ii. Spreads rapidly, involves a large population
   - iii. Affects almost the entire globe
   - iv. Present in a confined area

   - a. A-iii, B-i, C-ii, D-iv
   - b. A-ii, B-iii, C-iv D-i
   - c. A-i, B-ii, C-iii, D-iv
   - d. A-iv, B-i, C-ii, D-iii

10. Which of these is a *pandemic* disease?   [   ]
    - a. Cholera
    - b. Plague
    - c. AIDS
    - d. All

11. Which of these is an *epidemic* disease?   [   ]
    - a. Influenza
    - b. Typhoid
    - c. Both
    - d. None

12. Riderpest and Distemper spread   [   ]
    - a. From man to animals only
    - b. From animals to man only
    - c. From man to man only
    - d. From animals to animals only

13. Pick the WRONG statement   [   ]
    - a. Innate immunity is always nonspecific immunity
    - b. Obtaining Abs from mother to foetus is passive natural immunity
    - c. Injecting Abs into an individual is passive artificial immunity
    - d. Vaccination is active artificial immunity

14. A person who once suffers from smallpox never suffers from it again due to   [   ]
    - a. Active natural immunity
    - b. Active artificial immunity
    - c. Passive natural immunity
    - d. Passive artificial immunity

15. Which of these is NOT innate immunity   [   ]
    - a. Spermine in kidneys
    - b. Lactoperxidase in milk
    - c. Autoantibodies against self antigens
    - d. Lactic acid in muscle tissue

16. First line of defense   [   ]
    - a. Skin
    - b. Normal flora
    - c. Epithelial layer
    - d. Acids and bile in the digestive tract

17. Fever   [   ]
    - a. Stimulates production of Abs
    - b. Stimulates production of histamines
    - c. Stimulates production of interferons
    - d. Suppresses action of phagocytosis

18. Many infections result in increase in Acute Phase Proteins (APPs) within the plasma, which   [   ]
    - a. Enhance host resistance
    - b. Promote repair of inflammatory lesions
    - c. Activate alternative pathway of complement
    - d. All

19. In a secondary immune response, when the host is challenged with the same Ag for the second time,   [   ]
    - a. The immune response occurs more quickly than during the first encounter
    - b. The immune response occurs more slowly than during the first encounter
    - c. There is no immune response elicited
    - d. The immune response occurs at the same rate as the first encounter

20. Which is a short lived immunity   [   ]
    - a. Passive artificial immunity
    - b. Active natural immunity
    - c. Active artificial immunity
    - d. All

21. Which of these provides an instant immunity? [   ]
    a. Passive artificial immunity
    b. Passive natural immunity
    c. Both
    d. None

22. Which of these is a killed vaccine preparation? [   ]
    a. BCG vaccine   b. TAB vaccine   c. Sabin vaccine   d. All

23. Substances which are incapable of inducing Ab formation, but can react specifically with Abs [   ]
    a. Epitope   b. Ag   c. Hapten   d. Complement

24. *Horror Autotoxicus* [   ]
    a. A condition in which an individual mounts an immune response against his own normal Ags
    b. A condition in which an individual does not mount an immune response against his own normal Ags
    c. The treatment employed against autoimmune diseases
    d. A vaccine like preparation agains autoimmune diseases

25. Which of these are least antigenic [   ]
    a. Proteins   b. Polysachharides   c. Lipids   d. Glycoproteins

26. Heterophile Antigens [   ]
    a. Same Ags in different biological species
    b. Different Ags amongst same biological species
    c. Different Ags in different biological species
    d. Same Ags amongst same biological species

27. Ab production is by [   ]
    a. B Lymphocytes with the co-operation of T lymphocytes
    b. T Lymphocytes with the co-operation of B lymphocytes
    c. B Lymphocytes
    d. T lymphocytes

28. Ags may be categorized as [   ]
    a. T Independent and T Dependent Ags (TI & TD)
    b. B Independent and B Dependent Ags (BI & BD)
    c. T & B Independent and T & B Dependent Ags (TBI & TBD)
    d. Natural Ags and Artificial Ags

29. In an electrophoretic pattern of human serum, the highest peak is observed for [   ]
    a. α Globulins   b. β Globulins   c. γ Globulins   d. Albumin

30. Abs are [   ]
    a. α Globulins   b. β Globulins   c. γ Globulins   d. Albumin

31. Papain breaks down IgG into [   ]
    a. $2F_c + 1\ F_{ab}$ Portions
    b. $2F_{ab} + 1$ Fc Portions
    c. $2F_{ab}$ Portions
    d. $1F_c + 1\ F_{(ab)}$ Portions

32. In an Ab the *hinge region* is seen on [   ]
    a. The Light chains
    b. The Heavy chains
    c. Both Light and Heavy chains
    d. Variable regions

33. The J chain is seen in [   ]
    a. IgA   b. IgM   c. Both   d. None

34. What is FALSE about IgA [   ]
    a. Secretory Ab
    b. Can fix complement
    c. Second most abundant Ig
    d. Occurs in dimeric form

35. Pick out the WRONG pair   [   ]
    a. IgG – Transplacental Ab
    b. IgM – Earliest Ab
    c. IgE – Defense against helminthic infections
    d. IgD – Reagin Ab

36. Antigenic specificity exclusive to each Ig molecule is called   [   ]
    a. Isotypic specificity   b. Idiotypic specificity
    c. Allotypic specificituy   d. Homotypic specificity

37. Pick out the WRONG statement   [   ]
    a. IgG produces strong Precipitation reaction
    b. IgM produces strong Agglutination reaction
    c. IgA produces strong lytic reaction
    d. IgM produces weak complement fixation reaction

38. The smallest unit of antigenicity   [   ]
    a. Antigenic determinant is called   b. Epitope
    c. Both   d. None

39. Prozone phenomenon occurs due to   [   ]
    a. Ag excess   b. Ab excess
    c. Ag-Ab equivalence   d. Lack of Ab

40. Which of these reactions is easily detected on a slide   [   ]
    a. Precipitation reaction   b. Agglutination reaction
    c. Flocculation reaction   d. All

41. Latex particles may be used to   [   ]
    a. convert precipitation reaction into agglutination reaction
    b. convert agglutination reaction into precipitation reaction
    c. convert precipitation reaction into flocculation reaction
    d. convert agglutination reaction into flocculation reaction

42. VDRL test is   [   ]
    a. a precipitation test   b. an agglutination test
    c. a flocculation test   d. a hemagglutination test

43. Which of these is a quantitative test?   [   ]
    a. Ouchterlony test
    b. Oakley-Fulthorpe test
    c. Single Radial Immunodiffusion test (SRID)
    d. Elek test

44. Oudin test is   [   ]
    a. Single diffusion in one dimension   b. Double diffusion in one dimension
    c. Double diffusion in two dimension   d. Single diffusion in two dimension

45. In the Ouchterlony test, the appearance of a spur indicates   [   ]
    a. Ag or Ab identity   b. Ag or Ab partial relatedness
    c. Ag or Ab unrelatedness   d. Ag relatedness/Ab unrelatedness

46. Rocket Immunoelectrophoresis is used in   [   ]
    a. Identifying Ag
    b. Identifying and quantifying Ag
    c. Differentiating precipitation test from flocculation test
    d. Differentiating precipitation test from agglutination test

47. Coomb's serum contains   [   ]
    a. Incomplete Abs
    b. Anti Immunoglobulin Abs
    c. Ags that cannot elicit the immune response
    d. Erythrocyte Ags

48. Wasserman test is   [   ]
    a. Complement fixation test   b. Slide flocculation test
    c. Immunoelectrophoresis test   d. Slide agglutination test

49. Binder - Ligand Assays are   [   ]
    a. Immunofluorescence test   b. Radioimmunoassay (RIA)
    c. Hemagglutination test   d. Neutralization test

50. The enzyme used in ELISA test is   [   ]
    a. Horse serum peroxidase   b. Serum Oxalacetate transaminase
    c. Staphylococcal coagulase   d. Catalase

51. Thymus is regarded as   [   ]
    a. Birth place of erythrocytes   b. Graveyard of erythrocytes
    c. Primary lymphoid organ   d. Secondary lymphoid organ

52. Bursa of Fabricius in birds is responsible for   [   ]
    a. Secondary immune response   b. Cellular immune response
    c. Humoral immune response   d. Autoimmunity

53. Which of these may be regarded as peripheral lymphoid system   [   ]
    a. Mucosa Associated Lymphoid Tissue (MALT)
    b. Gut Associated Lymphoid Tissue (GALT)
    c. Bronchus Associated Lymphoid Tissue (BALT)
    d. All

54. Lymph nodes   [   ]
    a. Serve as filter for circulating lymph
    b. Phagocytose microbial cells
    c. Help in proliferation and circulation of T cells
    d. All of these

55. The largest lymphoid organ   [   ]
    a. Spleen   b. Thymus   c. Lymph node   d. Bursa

56. Lymphocyte in peripheral blood constitutes   [   ]
    a. Nearly 60% of all blood cells   b. Nearly 40% of all blood cells
    c. Nearly 1% of all blood cells   d. Nearly 90% of all blood cells

57. Majority of activated B cells are transformed into   [   ]
    a. T cells   b. Plasma cells   c. Macrophages   d. Null cells

58. Natural Killer (NK) cells have a role in programmed cell death (apoptosis) in cases of tumours   [   ]
    a. TRUE   b. FALSE

59. Natural Killer cells are   [   ]
    a. Small agranular lymphocytes   b. Large granular lymphocytes
    c. Small granular lymphocytes   d. Large agranular lymphocytes

60. Kupffer cells are macrophages in   [   ]
    a. Lungs   b. Liver   c. Spleen   d. Thymus

61. Macrophages are   [   ]
    a. Circulating B cells
    b. Tissue plasma cells
    c. Tissue monocytes
    d. Blood Monocytes

62. Dendritic cells play a role in   [   ]
    a. Phagocytosis of Ags in blood
    b. Phagocytosis of Ags in tissue
    c. Presentation of Ags to B cells
    d. Presentation of Ags to T cells

63. The HLA (Human Leucocyte Ag) complex is the same as   [   ]
    a. MHC complex
    b. MALT (Mucosa Associalted Lymphoid Tissue)
    c. APC (Ag Processing Cells)
    d. BCDF (B cell Differentiation Factor)

64. The MHC complex was first observed in   [   ]
    a. Human   b. Mice   c. Guinea pigs   d. Chimpanzees

65. Pick out the TRUE statement   [   ]
    a. HLA Class I Ags are involved in graft rejection and cell mediated cytolysis
    b. HLA Class II Ags are found only on cells of the immune system
    c. HLA Ags are 2 chained glycoprotein molecules anchored on the surface membrane of cells
    d. All of these

66. Monoclonal Abs are   [   ]
    a. Abs produced by a single clone of plasma cells
    b. Abs directed against a single antigenic determinant
    c. Abs directed against a single Ag
    d. All

67. Basal culture medium employed in the production of monoclonal Abs   [   ]
    a. HAT medium
    b. HAT deficient medium
    c. HPRT medium
    d. HPRT deficient medium

68. Production of monoclonal Abs is called   [   ]
    a. Monoclonal Technology
    b. Cloning
    c. Hybridoma technology
    d. All

69. Clonal Selection Theory was proposed by   [   ]
    a. Burnet
    b. Ehrlich
    c. Alexander & Mudd
    d. Fenner

70. The theory that shifted immunological specificity to cellular level   [   ]
    a. Direct template theory
    b. Indirect template theory
    c. Side chain theory
    d. None of these

71. Interferons have   [   ]
    a. Antiviral activity only
    b. Antiviral & antitumour activity
    c. Antiviral, antitumour & macrophage activation
    d. Antiviral, antitumour, macrophage activation & Augmentation of Monocyte functions

72. An adjuvant   [   ]
    a. Enhance immunogenicity of an Ag
    b. Confers immunogenicity on non-antigenic substances
    c. Induce or enhance degree of cellular immunity
    d. All of these

73. Sublethal whole body irradiation before Ag stimulus   [   ]
    a. Stimulates Ab response
    b. Suppresses Ab response
    c. Stimulates CMI
    d. Suppresses CMI

74. Which of these does NOT elicit CMI?   [   ]
    a. Trypanosomiasis
    b. Measles
    c. Thyroiditis
    d. Syphilis

75. The most characterisitic immunological functions of CMI   [   ]
    a. Delayed Hypersensitivity
    b. Immune Complex disorder
    c. Stimulation of interferons
    d. Opsonization

76. Which of these is an example of cytokines   [   ]
    a. Interferons   b. Interleukins   c. Lymphotoxins   d. All

77. Interleukins play a role in   [   ]
    a. Differentiation of T & B cells
    b. IgG production
    c. Activation of NK cells
    d. All of these

78. DiGeorge Syndrome   [   ]
    a. An immunodeficiency disorder
    b. A hypersensitivity disorder
    c. An autoimmune disorder
    d. Immune complex disorder

79. Myasthenia gravis is   [   ]
    a. An immunodeficiency disorder
    b. A hypersensitivity disorder
    c. An autoimmune disorder
    d. Metabolic disorder

80. Which of these is NOT an autoimmune disease?   [   ]
    a. Sjogren's syndrome
    b. Guillian Barre syndrome
    c. Job's syndrome
    d. Addison's disease

81. Primary immune deficiency results from   [   ]
    a. Abnormal development of immune mechanisms
    b. First challenge by the immuno suppressive agent
    c. Lack of T cell memory
    d. Underdevelopment of self-Ags.

82. Match the columns   [   ]
    A. Type I Hypersensitivity   i. Immune Complex disorder
    B. Type II Hypersensitivity   ii. Delayed Hypersensitivity 1
    C. Type III Hypersensitivity   iii. Immediate Hypersensitivity
    D. Type IV Hypersensitivity   iv. Cytotoxic/Cytolytic Hypersensitivity

    a. A-iii, B-i, C-ii, D-iv
    b. A-iii, B-iv, C-i D-ii
    c. A-i, B-ii, C-iii, D-iv
    d. A-iv, B-i, C-ii, D-iii

83. Rhesus (Rh) Factor in RBCs was identified by   [   ]
    a. Landsteiner
    b. Landsteiner & Wiener
    c. Peters
    d. Lewis

84. Arthus Phenomenon is   [   ]
    a. Type I Hypersensitivity
    b. Type II Hypersensitivity
    c. Type III Hypersensitivity
    d. Type IV Hypersensitivity

85. Which of these is NOT immediate hypersensitivity   [   ]
    a. Tuberculin reaction
    b. Atopy
    c. Anaphylaxis
    d. All

Answer Key and Validation

Principles of Immunology

1. b	16. a	31. b	46. b	61. c	76. d
2. c	17. c	32. b	47. b	62. d	77. d
3. a	18. d	33. c	48. a	63. a	78. a
4. c	19. a	34. b	49. b	64. b	79. c
5. c	20. a	35. d	50. a	65. d	80. c
6. d	21. c	36. b	51. c	66. d	81. a
7. d	22. b	37. c	52. c	67. a	82. b
8. a	23. c	38. c	53. d	68. c	83. b
9. d	24. b	39.b	54. d	69. a	84. c
10. d	25. c	40. b	55. a	70. d	85. a
11. a	26. a	41. a	56. b	71. d	
12. d	27. a	42. c	57. b	72. d	
13. a	28. a	43. c	58. a	73. b	
14. a	29. d	44. a	59. b	74. d	
15. c	30. c	45. b	60. b	75. a	

4. *They are heat stable polysachharides-protein-lipid complexes*

8. *LD50 is the lethal dose 50. The option d refers to ID50 which is Infecting Dose 50.*

12. *Humans never suffer from rinderpest and distemper*

18. *APPs include C Reactive proteins, alpha-1-acid glycoproteins, serum amyloid P component, mannose binding proteins, etc.*

34. *IgA does not fix complement, but can activate the alternative complement pathway.*

35. *IgE is the ragin Ab. The precise biological role of IgD is not well defined.*

67. *HAT medium contains hypoxanthine, aminopterin and Thymidine which does not permit growth of HPRT enzyme deficient myeloma cells*

70. *The Clonal Selection Theory shifted immunological specificity to cellular level and is widely accepted.*

73. *Irradiation followed by Ag stimulus after 24h suppresses the Ab response while Ag stimulus 2days before irradiation, actually enhances Ab response*

80. *Job's syndrome is an immune deficiency disease characterized by disordered functioning of phagocytosis*

13

Food Microbiology

Principles and Types of Food Spoilage

***Definition*:**   Spoilage is defined as the decay or decomposition of an undesirable nature that affects the physical, chemical or nutritive value of the food adversely. Spoilage may be due to microbes, enzymes in foods, non-enzyme mediated chemical reactions and physical changes during storage by various methods (freezing, pressure holding, drying, burning, etc.

On the basis of spoilage, foods are of three categories:

* *Stable or non-perishable foods:* Sugar, flour, dry beans, pulses…
* *Semi perishable foods:* Potatoes, onions….
* *Perishable foods:* meats, milk, eggs, fruits, vegetables, fish….

***Factors affecting microbial spoilage of foods*:** Extent of contamination, kinds and numbers of flora already present, growth conditions, storage and pretreatments of the food if any.

Biochemistry of food spoilage

Food Component	Spoilage features
Proteins	✓ Proteinases catalyze hydrolysis of proteins to peptides which add bitter taste to foods ✓ Anaerobic protein decomposition results in foul smell, H_2S production, mercaptans, ammonia, amines, indoles and skatols all having obnoxious odor (putrefaction). ✓ *E.coli, Pseudomonas, Bacillus, Clostridium, Desulfotomaculum* are involved in protein degradation.
Carbohydrates	✓ Di- tri- and polysaccharides are broken to simple sugars by microbes first ✓ Carbohydrates are subject to various fates as per the spoilage organism – alcoholic fermentation (yeast), lactic acid fermentation (homofermentative lactic cultures), acid and gas (coliforms), mixed lactate (heterofermentative lactics), propionic acid fermentation, butyric acid fermentation…
Organic acids	✓ Microbes oxidize several organic acids occurring in foods to carbonates, causing the pH to rise to alkaline levels. ✓ Aerobically, the acids may be oxidized completely to CO_2 and water (film yeasts).
Lipids	✓ Microbial lipases hydrolyze fats to glycerol and fatty acids. ✓ Phospholipids and lipoproteins also meet with degradation by microbes to phosphates, glycerol, f.acids, nitrogenous bases, cholesterol esters, etc.
Other compounds	✓ Alcohols are oxidized to organic acids, glycerol is disseminated to products similar to those from glucose, acetaldehyde is converted to acetic acid or ethanol, pectic substances are variously degraded by pectinesterases, polygalacturonases to simple acids and sugars, cyclic compounds in foods are not readily attacked by microorganisms.

Types of food spoilage

Food	Spoilage	Causative agents
Cereals & cereal products.	✓ Do not get easily spoilt if properly stored, due to low water activity ✓ Moisture levels above 12 to 13%, physical damage and temp of storage initiate spoilage by molds. ✓ Bacterial spoilage follows eventually and a result in heavy economic loses.	Fungi: *Aspergillus, Penicillium, Mucor, Rhizopus, Fusarium* Bacteria: *Bacillus cereus, Acetobacter, B.licheniformis, B.subtilis*
Sugar and sugar products	✓ Spoilage is uncommon due to dry and osmotically challenging conditions. ✓ Spoilage may be by osmophilic or xerotolerant microbes.	Fungi: *Saccharomyces rouxii, Candida, Pichia, Hansenula, Rhodotorula* Bacteria: *Leuconostoc,Bacillus, Clostridium, Lactobacillus*
Vegetable & veg. Products	✓ Subject to various spoilages due to damaged peels and improper storage. ✓ Soft rot, sliminess, hard rot, ropiness, discoloration, putrefaction…	Fungi: *Fusarium, Botrytis, Alternaria, Penicillium, Aureobasidium, Sclerotina, Rhizopus* Bacteria: *Pseudomonas, Alkaligenes, Erwinia, Xanthomonas, Bacillus, Micrococcus, Lactic acids bacteria, Coryneforms,…*
Fruits and fruit products	-do-	Fungi: Same as in Veg. + *Phoma, Cladosporium, Trichoderma* Bacteria: *Pseudomonas, Alkaligenes, Erwinia, Xanthomonas,…*
Meat and meat products, Poultry and eggs	✓ Improperly dressed carcasses contaminate the meat from skin, hides and hooves. ✓ High moisture, and ready nutrients lead to rapid growth and spoilage. ✓ Sliminess, brown spots, white spots, ropiness, putrefaction, off odors and off tastes,	Fungi:*Cladosporium, Geotrichum, Penicillium, Mucor, Thamnidium,* Bacteria: *Pseudomonas, Proteus, Clostridium, Alcaligenes, Moraxella, Akinetobacter, Lactobacillus, Leuconostoc, Flavobacterium*
Fish and other seafood	✓ The body slime, scales and fins harbor microorganisms that multiply rapidly causing several kinds of spoilage.	Fungi: *Torula, Rhodotorula,* Bacteria: *Moraxella, Akinetobacter,Flavobacterium, Pseudomonas, Serratia, Sarcina, Clostridium, Escherichia, Proteus*
Milk and milk products	✓ Ideal growth medium for microorganisms. ✓ Spoilages include – Ropiness, rancidity, proteolysis, stormy fermentation, off odor, abnormal colors,…	Bacteria:*Micrococcus, Alkaligenes, Klebsiella, Nitrobacteria, Pseudomonas, Lactobacillus, Streptococcus cremoris, Bacillus, Clostridium,…* Fungi: Alternaria, Penicillium, Cladosporium, Geotrichum, Fusarium, Phoma,..

Typical food spoilages

Spoilage	Causative agent
Vegetables/fruits	
Soft rot	*Erwinia carotovora*
Grey mold rot	*Botrytis cinerea*
Anthracnose	*Colletotrichum lindimuthianum*
Blue mold rot	*Penicillium digitatum*
Downy mildew	*Phytophthora, Bremia*
Watery soft rot	*Sclerotinia sclerotiorum*
Black mold rot	*Aspergillus niger*
Pink mold rot	*Trichothecium roseum*
Green mold rot	*Cladosporium, Trichoderma*
Brown rot	*Sclerotinia*
Meat and Meat products	*Pseudomonas, Akinetobacter, Moraxella, Alcaligenes, Streptococcus, Leuconostoc, Bacillus, Micrococcus.*
Surface slime	*Pseudomonas, Achromobacter,*
Rancidity	*Serratia marsescens*
Red spots	
Milk and Milk products	
Ropy milk	*Alkaligenes viscolactis, Micrococcus freudenreichii, Klebsiella oxytoca…*
Alkaline milk	*Alkaligenes viscolactis, Pseudomonas fluorescens*
Sour milk	*Streptococcus lactis, Lueconostoc, Clostridium…*
Bitter milk	*Proteolytic bacteria, some coliforms, asporogenous yeasts…*
Blue milk	*Pseudomonas syncyanea*
Yellow milk	*Pseudomonas synxantha, Flavobactrium*
Red milk	*Serratia, Brevibacterium erythrogenes*
Brown milk	*Pseudomonas putrefaciens, Ps.fluorescens*

Principles and Methods of Food Preservation

Basis	Approach
Prevention or delay of microbial decomposition	✔ By Keeping out microorganisms ✔ By removal of microorganisms ✔ By hindering growth and activity ✔ By killing microorganisms
Prevention or delay of self decomposition of the food	✔ By destruction of food enzymes ✔ By prevention or delay of certain chemical reactions
Prevention or damage because of macroorganisms: insects, animals…	✔ Out of scope of food microbiologist

Methods of food preservation

Method	Mode of achievement
Asepsis (keeping out microorganisms	Clean and hygienic food collection, transport, handling and processing practices
Removal of microorganisms	Peeling, washing, discarding spoilt or contaminated portions…
Maintenance of anaerobic conditions	Canning in sealed and evacuated containers
Use of high temperature [treatments at, below and above 100^0C]	Boiling, Pasteurization, sterilization, hot sun drying, oven heating
Use of radiations	Radicidation – Low dose of radiation to destroy a particular pathogen selectively Radurization – Low dose of radiation to reduce total counts; serves as pasteurization Radappertization – High dosage of radiation, serves as sterilization process
Inhibition/slowing of growth of microbes	Use of low temperature: Cooling, chilling, freezing Addition of sugar/salt/glycerol Vacuum pack/nitrogen pack to restrict oxygen Acidify/alcoholic fermentation Chemical preservatives: sulphites, nitrites, organic sorbates, benzoates, and parabens…
Preservation by food additives	✓ Benzoic acid, sodium benzoate, methyl and propyl paraben, sorbates, propionates, sulfites Acetates and diacetates, nitrites and nitrates, ethylene oxide and propylene oxide

Commonly used chemical food preservatives

Name of the chemical	Maximum permissible concentration	Characteristically added to Foods
Benzoic acid	0.1%	Carbonated beverages, syrups, fruit drinks, pickles, salads, sauerkraut
Methylparaben	0.1%	
Sodium nitrate	500ppm	All the above items + dried fruits and vegetables
Sodium nirite	200ppm	Sausages
Ethylene oxide	50ppm	Sausages Spices

- ✓ Propionates affect cell membrane permeability although the exact mode of action is not known
- ✓ Exact mechanism is not understood, but the effectiveness of benzoic esters is increased with an increase in the chain length of the ester group.
- ✓ Nitrites decompose to nitric acid, which forms nitrosomyoglobin when it reacts with the heme pigments in meats and thereby forms a stable red color. Also nitrites react with secondary and tertiary amines to form nitrosamines, which are known to be carcinogenic.
- ✓ Nitrates have limited action on limited microbes in foods. Hence they are not regarded as efficient preservatives.
- ✓ SO_2 and sulfites form sulfurous acid in aqueous solutions. Many mechanisms of action have been suggested on microbial cells. Reduction of disulfide linkages, formation of carbonyl compounds, reaction with ketone groups and inhibition of respiratory mechanisms.
- ✓ Ethylene and propylene oxides are known to kill microorganisms. They act as strong alkylating agents attacking labile hydrogens.

Pasteurization

- ✓ The process of heating each and every particle of milk to a predetermined temp for a fixed period of time is generally referred to as pasteurization.

✓ Other liquid commodities such as wine, beer, certain industrial fermentation media and dairy products are also subjected to pasteurization on a routine basis.

✓ **Objectives** – The principal objectives of market milk pasteurization include:

    a. To kill all the milk borne pathogens – the pasteurization temp and holding period are adjusted such that even the most heat resistant pathogen in milk is destroyed completely.

    b. To extend the shelf-life of milk and hence improve the keeping quality

    c. If milk is pasteurized for making butter or cheese, the aim is to destroy microorganisms that would interfere with the activities of the starter culture.

✓ The heat treatment is accomplished without affecting the flavor, appearance, nutritional properties and creaming features of milk.

✓ Heating also destroys lipases that cause deterioration of butter during storage.

✓ **Indicator organism:** The first widely used pasteurization method employed 60^0C for 20min considering that the most heat resistant milk borne pathogen *Mycobacterium*. The time and temp however, were altered after the discovery of more heat resistant milk borne pathogen *Coxiella burnetii*, a rickettsia responsible for Q fever. The new modification was **62.8^0C (145^0F) for 30min**

✓ The initial process was not continuous and was referred to as **vat pasteurization or holding method or LTH – low temp holding process.**

✓ **HTST** (High Temp Short Time) method was introduced to save time and for continuous processing. It employs **71.7^0C (161^0F) for 15seconds.** HTST is the most widely employed method of pasteurization today.

✓ **Ultra High Temperature:** (UHT) Is widely used in Europe and US. It employs **137.8^0C for 2seconds.** The milk is labeled *ultrapasteurized.* the main drawback of UHT is that it causes slight changes in the nutritive and flavor properties of milk. (may give a slight cooked taste and odor). Some suppliers provide milk treated at **148.90C (300^0F) for 1sec** and label the product as *commercially sterile.* UHT for butter employs 87.7 to 93.3^0C for 2 to 3 seconds.

✓ The common methods of heating include:

    a. Steam injected into milk process – **Steam injection technique**

    b. Milk injected into steam – **stem infusion technique**

✓ In both the cases, the excess water/ steam is removed in a sterile vacuum chamber.

✓ Conventional pasteurization kills all yeast ant mold and several vegetative bacterial cells. Those that survive are termed thermodurics. The various pasteurization surviving bacteria include:

    a. High temp lactics: *Streptococcus thermophilus, Lactobacillus bulgaricus, Microbacterium*

    b. Aerobic spore formers: *Bacillus cereus, B. licheniformis B. subtilis, B. coagulans, B. polymyxa*

    c. Anaerobic spore formers: *Clostridium butyricum, C. sporogenes.*

✓ Other miscellaneous bacteria also survive in milk but do not grow well in milk.

Fermented Foods

Desirable activities of microorganisms in certain foods or raw materials result in fermented foods.

Fermented foods have multiple applications and merits over the unfermented ones.

Importance

✓ The non edible raw material is converted to **edible food**

✓ The palatability and **acceptability** of the product is improved several folds after fermentation by way of texture, structure and color after the fermentation.

✓ The **nutritive value** of the food is enhanced by way of simplified, easily digestible components and addition of growth factors such as vitamins in the finished product

✓ The organoleptic value of the food increases as several **flavoring compounds and aroma producing substances** are released in the final product.

✓ Fermented foods have a longer shelf-life over the raw material, and hence are a means of **natural preservation** in most cases.

Fermented food	Key organism(s)	Features
Bread	Baker's yeast. Strains of high CO_2 producing capacity. Generally *Saccharomyces spp*	✓ Soft fermented flour product subjected to baking after fermentation. ✓ Yeast contribute the flavor and leavened texture ✓ Major steps include – mixing the dough, leavening by bread yeast, baking and setting, drying slicing and packing. ✓ Modified products include: bread from homofermentative lactic cultures, leavening by chemicals, rye bread, *Torulopsis* bread, flavored breads with milk, spices or fruits.
Sauerkraut	**Lactic acid bacteria:** *Leuconostoc mesenteroides, Lactobacillus plantarum, L.brevis.*	✓ A clean, sound product of characteristic krauty flavor obtained by lactic fermentation of shredded quality cabbage in the presence of two to three percent salt. ✓ Anaerobic conditions develop rapidly in salted, shredded cabbage due to plant cell and bacterial cell respiration that encourage growth of lactic acid bacteria. ✓ Final product has about 0.7 to 1.0 percent lactic acid. ✓ Flavor of good sauerkraut is due to formation of lactic acid, acetate, ethanol, methanol, mannitol, dextran, esters and CO_2 all released in right proportions.
Fermented dairy products [cheese, yogurt, kefir, cumiss]	*Lactobacillus delbrueckii, Streptococcus thermophilus, Leuconostoc cremoris, Propinibacterium freudenreichii, etc.*	✓ Lactic acid fermentation is involved in almost all of them ✓ The acidity is sufficient to prevent spoilage by proteolytic and other undesirable bacteria. ✓ Most of them involve curdling the milk by lactic fermentation followed by flavoring/ripening/de-watering and such other procedures that lead to a variety of finished products. ✓ Some kinds of cheese are subjected to mold treatment as part of ripening
Beer	Brewer's yeast. Strains of bottom type yeast. Generally *Saccharomyces carlsbergens*	✓ A fermented malt beverage of low alcohol content ✓ Yeast contributes to flavor, color and aroma in conjunction with the contributions of several other ingredients such as malt and hops. ✓ Major steps include: Malting, mashing, and boiling of wort with hops, fermentation of wort, aging, maturing and finishing. ✓ Barley grains are soaked, sprouted, mashed and filtered. The filtered wort is boiled with hops and fermented with yeast. The green beer is stored and 'lagered' for weeks to months and mildly carbonated if desired. ✓ Types of beer: Pilsner, bock, Weiss, Porter, Stout, ales, and lager beer. Related beverages include Sake, Pulque and Ginger beer.
Wine	Natural inoculum (yeast present on grapes) or special wine yeast *Saccharomyces cerevisiae* var *ellipsoideous.*	✓ Alcoholic fermented product from grape juice by yeast followed by aging process. ✓ Wines made from black grapes are called red wines and those from green grapes, or peeled black grapes are white wines. ✓ Major steps include: Preparation of must, fermentation of must, racking, storing & aging. ✓ Grapes at a right stage of ripening are subjected to natural fermentation the fermented product is separated from the culture and clear wine is stored and aged. ✓ Types: Still wines, Sparkling wines (Champagne), dry wines, Fortified wines, Table wines and dessert wines. ✓ Distilled liquors: Rum, Whiskey and Brandy
Vinegar	Dual fermentation: **a. Alcoholic fermentation** by yeast. *Saccharomyces cerevisiae* var *ellipsoideous*	✓ A condiment made from sugary or starchy material by an alcoholic fermentation followed by acetous fermentation. Typical vinegar should contain at least 4g of acetic acid per 100ml or (40 grain) ✓ Vinegars are made from fruit juices/malted cereals/molasses/honey/spirits or alcohols.

	b. Acetic acid fermentation by acetic acid bacteria *Acetobacter* and *Glucanobacter spp*	✓ Fermentation begins with alcoholic reaction by yeast. The alcohol is then subjected to acetate fermentation by slow or quick methods. ✓ Slow methods are let-alone fermentations such as the Orleans's surface fermentation ✓ Quick methods make use of filter beds such as Fring's generator or submerged aerated fermentors. ✓ Finished vinegar is aged to improve body and taste and pasteurized in bulk. Strength of vinegar is expressed as grains, which is 10 times the no. of Gms of acetic acid per 100ml of vinegar.

Oriental Fermented Foods

Oriental fermented food	Key organism(s)	Features
Soy sauce	*Aspergillus soyae/oryzae, lactic acid bacteria, Hansenula, Saccharomyces rouxii, Bacillus subtilis..*	✓ Koji or chou (starter culture) is first prepared by growing the fungus on autoclaved soybeans ✓ Koji is inoculated into bulk autoclaved soybeans and incubated for three days. ✓ Brined for 21/2 months to 1 year in 24% NaCl. ✓ Fermentation is lactic and alcoholic.
Miso	*Aspergillus oryzae, Saccharomyces rouxii, Zygosaccharomyces spp, Lactic acid bactera.*	✓ Koji is grown on steamed polished rice and mixed with crushed soybeans ✓ Salt is added and fermented for a week to 2 months ✓ The final product is ground to a paste to be used in combination with other foods.
Tempeh (Indonesian food)	*Rhizopus stolonifer, R. oryzae, R.oligosporus or R. arrhizus.*	✓ Split soybeans are boiled, dried and inoculated with starter culture, packed in banana leaves and incubated for 1 day. ✓ The mash is sliced, dipped in salt and fried in vegetable fat (ghee).
Ang-Khak (Chinese red rice)	*Monascus purpureus*	✓ Culture is grown on autoclaved polished rice. ✓ Used in coloring and flavoring fish and other products.
Natto	*Bacillus natto*	✓ Boiled soybeans are wrapped in rice straw and incubated for 1 or 2 days until the package becomes slimy from outside.
Idli (Indian fermented food)	*(Lactic acid bacteria) Leuconostoc mesenteroides, Streptococcus faecalis, pediococcus cerevisiae*	✓ Rice and black gram are soaked separately, ground, mixed and allowed to ferment overnight until the batter has risen enough. ✓ Batter is cooked by steaming and served hot with chutney and sambar.
Tofu or to-fu-ju (Chinese soy bean cheese)	*Mucor*	✓ Soaked soybeans are ground and filtered. ✓ The filtrate is curdled by calcium salt. ✓ The curd is pressed into blocks which are left for fermentation by *Mucor*. ✓ Final ripening is in brine or wine.

Exercises

1. The correct way to represent water activity of a food is __________   [   ]
   a. A_W b. W_a c. w_a d. a_w

2. *Hansenula, Kloeckera and Brettanomyces* are __________   [   ]
   a. Molds significant in food microbiology
   b. Yeast significant in food microbiology
   c. Starter cultures for dairy product
   d. Spoilage causing bacteria in foods

3. Match the following   [   ]
   A. *Alcaligenes*   i.   Saccharolytic
   B. *Clostridium*   ii. Proteolytic
   C. *Bacillus*   iii. Pectinclytic
   D. *Erwinia*   iv. Lipolytic

   a. A- iii, B-ii, C-i, D-iv   b. A-iv, B-iii, C-ii, D-i
   c. A-iv, B-i, C-ii, D-iii   d. A-ii, B-iii, C-iv, D-i

4. Spices are best preserved by use of   [   ]
   a. Ethylene oxide   b. Nitrites   c. Sulfites   d. Propionates

5. Wood smoke   [   ]
   a. Is more effective on vegetative cells than against spore bearers
   b. Is more effective against spores that against vegetative cells
   c. Is equally effective against spores and vegetative cells
   d. Is only for flavoring and not useful for preservation purposes

6. Principle of microwave oven   [   ]
   a. Microwaves pass through the food and evolve heat as they leave the food, thus heating from center to the surface of the food.
   b. Heat is produced when microwaves pass through the food, due to rapid oscillation of food molecules to align themselves with the electromagnetic field
   c. A combination of a & b
   d. None of the these

7. Which of these delays microbial decomposition   [   ]
   a. Refrigeration   b. Chilling   c. Cold storage   d. All

8. Which of these causes *removal* of microbial contamination   [   ]
   a. Washing   b. Canning   c. Freezing   d. Adding benzoate

9. The temp/time employed in UHT processing of milk is   [   ]
   a. 137.8^0C / 2 Seconds   b. 148.9^0C / 3 Seconds
   c. 161^0 F / 5 Seconds   d. 145^0F / 3 Seconds

10. Rancidity is related to   [   ]
    a. Lipid spoilage   b. Saccharine spoilage
    c. Protein spoilage   d. all

11. Bitter taste to milk is due to   [   ]
    a. Proteolytic bacteria   b. Coliforms
    c. Asporogenous yeast   d. all

12. Lactic acid fermentation is observed in   [   ]
    a. Sauerkraut   b. Pickling   c. a & b   d. None

13. Sauerkraut is   [   ]
    a. fermented cabbage in water   b. fermented cabbage in brine
    c. fermented cabbage in sugar syrup   d. fermented cabbage in alcohol

14. Which is the correct sequence for cheese manufacture   [   ]
    a. salting – curdling – draining – ripening
    b. curdling – draining – salting – ripening
    c. curdling – salting – draining – ripening
    d. curdling – draining – ripening – salting

15. Cheese with characteristic 'eyes'   [   ]
    a. Cheddar cheese   b. Swiss cheese
    c. Roquefort cheese   d. Limburger cheese

16. Which of these exhibits dual fermentation   [   ]
    a. Wine   b. Bread   c. Vinegar   d. Idli

17. Fermented cabbage   [   ]
    a. Sausage   b. Sauerkraut   c. Ang- khak   d. b and c

18. What is FALSE in bread making   [   ]
    a. There is very little growth of yeast during the first 2h in the dough.
    b. Fermentation continues until the temp of oven inactivates the yeast enzyme.
    c. Dough conditioning results mainly from action of saccharolytic enzymes on gluten.
    d. Optimum temp for fermentation is 23 to 24^0C

19. *Mucor* is involved in the making of   [   ]
    a. Tofu   b. Beer   c. Natto   d. Miso

20. In beer making, mashing causes   [   ]
    a. hydrolysis of malt starch   b. hydrolysis of malt proteins
    c. hydrolysis of malt adjuncts   d. None of these

21. Hops in beer contribute to   [   ]

    a. antiseptic property   b. Flavour

    c. final colour   d. all

22. Match the pairs correctly   [   ]
    A. Whiskey   i) Distilled grape wine
    B. Brandy   ii) Distilled fermented sugar cane juice
    C. Champagne   iii) Distilled fermented grain mash
    D. Rum   iv) Carbonated beer

    a. A-i, B-ii, C-iii, D-iv   b. A-iii, B-i, C-iv, D-ii
    c. A-iii, B-ii, C-iv, D-i   d. A-iv, B-i, C-iv, D-iii

23. Dill pickles are   [   ]
    a. Pickles flavoured with Dill herb
    b. Pure cucuber pickles
    c. Mango (Avakaya) pickle flavoured with jaggery
    d. Mixed vegetable pickle

24. Found in vegetable pickles   [   ]
    a. *Leuconostoc mesenteroides*   b. *Pediococcus cerevisiae*
    c. *Lactobacillus plantarum*   d. All

25. Kumiss, kefir and emmantal   [   ]
    a. All are fermented dairy products   b. All are non-fermented foods
    c. All are oriental fermented foods   c. all are Soy fermented foods

26. Choose the WRONG pair   [   ]
    a. Tea - fermentation of leaves by self-enzymes
    b. Coffee - Lactic fermentation of beans
    c. Cocoa - *Candida krusei* fermentation of beans
    d. Tofu - hard cheese.

27. *Enology* is   [   ]
    a. Study of tea and coffee production
    b. Study of wine production
    c. Study of cheese production
    d. Study of oriental fermented foods

28. Izushi and katsuabushi are   [   ]
    a. Fish pickles from Japan   b. Chicken pickles from china
    c. Fish sauces from Japan   c. Rice pickles from china

29. Salami and Bologna are _______   [   ]
    a. Fermented meat products   b. Fermented crab foods
    c. Fermented Prawn pickles   d. Fermented soft drinks.

30. The strength of vinegar is expressed in _______   [   ]
    a. grams   b. ppm   c. grains   d. ppb

31. Key organism(s) in vinegar production   [   ]
    a. *Acetobacter acetigenum*   b. *Glucanobacter oxydans*
    c. both   d. None

32. Most oriental fermented foods involve fermentation by   [   ]
    a. Yeast   b. Molds   c. Lactics   d. Algae

33. *Koji* means   [   ]
    a. Starter for most oriented fermented foods
    b. A Plant leaf in which the grain mash is wrapped for fermentation
    c. NaCl brine used for in the making of oriental fermented foods
    d. A kind of oriental fermented food

34. What is FALSE about Idli?   [   ]
    a. It employs rice and blank gram
    b. The ingredients are soaked separately, ground, mixed together and left for fermentation
    c. Overnight fermentation causes rising of the batter without change in pH
    d. It is cooked only by steaming.

35. Pick out the WRONG statement (if any) about Mysore Bajji   [   ]
    a. It is a combination of fermentation + caustic soda action
    b. The quality & rate of fermentation is improved by adding curd to the batter
    c. The batter is subjected to deep frying
    d. a, b, c statements are Correct

36. Which of these requires fermentation of batter   [   ]
    a. Dosa   b. Utappam   c. Upma   d. a and b

37. The purpose of food preservation   [   ]
    a. To improve the palatability of the food
    b. To extend the shelf-life of the food
    c. To encourage favourable fermentations in food
    d. All

38. An example of removal of microbes from food   [   ]
    a. Removing the upper peels of cabbage   b. Heating milk at 70^0C for 5seconds
    c. Adding buttermilk to pasteurized milk   d. Adding turmeric to rice

39. Which of these is a natural method of preservation   [   ]
    a. Filtering water through micron filter
    b. Adding sodium benzoate to fruit juices
    c. Adding a spoon of curd to coconut chutney
    d. Adding red chilly powder to mango pickle.

40. Which of these is a WRONG method for preservation   [   ]
    a. Smoking of meat   b. Washing of eggs in potable water
    c. Overlaying pickle with edible oil   d. Addition of tamarind juice to rice

41. Blanching   [   ]
    a. Destroys spoilage causing organisms in foods
    b. Adds antimicrobial oils to foods
    c. Destroys pathogenic bacteria in foods
    d. Delays self-decomposition of foods.

42. Match the columns   [   ]

    | A. Microbicidal | i) Washing knives & tables |
    | B. Microbistatic | ii) Refrigeration |
    | C. Removal of microbes | iii) Pasteurization |
    | D. Restriction | iv) Filtration |

    a. A-iii, B-ii, C-iv, D-I   b. A-i, B-ii, C-iii, D-iv
    c. A-i, B-ii, C-iv, D-iii   d. A-iv, B-iii, C-ii, D-I

43. Addition of glycerol in foods   [   ]
    a. Reduces available O_2 for microbes   b. Reduces water activity of the food
    c. Kills vegetative cells in the food   d. Creates harmful pH for microbes.

44. HACCP is   [   ]
    a. Health Analysis Council For Corporate Products
    b. Heat And Chill For Clarity And Preservation
    c. Hazard Analysis Critical Control Point
    d. Howard And Crowford Chill Proof.

45. TDT (Thermal Death Time)   [   ]
    a. Time required to kill the given pure culture held at a fixed temp.
    b. Time required to kill 75% of the given pure culture at a fixed temp.
    c. Time required to kill the given pure culture when held at 75^0C.
    d. None of these

46. TDP (Thermal Death Point) [   ]
    a. Temp required to kill a given pure culture when held for 15min.
    b. Temp required to kill the given pure culture when held for a fixed time
    c. Temp required to kill 50% of the given pure culture when held for a fixed time
    d. None of these.

47. When treating food with high temp, which of these improves the process of killing microbes. [   ]
    a. Lowering the pH of the food
    b. Increasing the pH of the food
    c. Decreasing the surface area of the food
    d. Adding chemical preservation to the food.

48. Retort chambers are employed in [   ]
    a. Radiation preservation   b. Pasteurization
    c. High steam pressure sterilization   d. Appertization

49. Hot water spray on vegetables followed by drying is [   ]
    a. Scalding   b. Flash spraying   c. Spray drying   d. Pinning.

50. Which wavelength of UV light is preferred for food preservation processes [   ]
    a. 260nm   b. 200nm   c. 300nm   d. 210nm

51. Smoking renders [   ]
    a. Preservation activity   b. Flavoring activity
    c. Drying activity   d. All

52. Wood that is NOT recommended for smoking meat [   ]
    a. Walnut   b. Neem   c. Oak   d. Apple

53. Smoking temp for meat ranges from [   ]
    a. 43 to 71 °C   b. 75 to 90 °C   c. 60 to 90 °C   d. 50 to 70 °C

54. Wood smoke contains [   ]
    a. Formaldehyde   b. Phenols   c. Cresols   d. All

55. Which of these spices is most effective against yeast. [   ]
    a. Cinnamon   b. Clove   c. Black pepper   d. Mustard

56. Which of these is most antibacterial [   ]
    a. Black pepper   b. Clove   c. Bay leaf   e. Curry leaf

57. D and Z values are related to [   ]
    a. High temp treatment of culture   b. Treatment of cultures with organic acids
    c. Treatment of cultures with alkali   d. None

58. Which of these has highest a_w [   ]
    a. Cashew nut   b. Strawberry   c. Fresh meat   d. Orange

59. Sun drying is employed for [   ]
    a. raisin   b. salted fish   c. figs   d. all

60. *Sweating* is [   ]
    a. Equalization of moisture   b. Reduction of moisture
    c. Both   d. None

61. Pick the WRONG pair [   ]
    a. Film / aerosol drum drying– milk   b. Freeze-drying – coffee
    c. Rotary drying – tea   d. Conveyer band drying – grains

62. Decimal reduction time   [   ]
    a. Time of heating at a temp to cause 90% reduction in viable counts
    b. Time of heating at temp to cause 10 % reduction in viable counts
    c. Temp of heating at a fixed time to reduce fixed no of viable cells
    e. None

63. Butter is manufactured by   [   ]
    a. churning pasteurized milk   b. churning pasteurized curd
    c. churning pasteurized cream   d. churning pastuerised butter milk

64. The key compound that imparts flavour and aroma to butter is   [   ]
    a. Acetoin   b. Diacetyl   c. Lactic acid   d. Indole

65. Starter cultures for butter making   [   ]
    a. *Streptococcus cremoris + Leuconostoc dextranicum*
    b. *Pediococcus lactis + Brevibacterium linen*
    c. *Oidium lactis + Lactobacillus helveticus*
    d. *Streptococcus thermophilus + Lactobacillus bulgaricus*

66. Bulgarian milk is   [   ]
    a. Milk with High SNF (Solids Not Fat) content
    b. Yogurt
    c. Sour cream
    d. A mixture of cow and buffalo milk

67. Pickling of cucumbers in salt stock includes   [   ]
    a. Lactics + *Aerobacter* + Yeast   b. Lactics + yeast
    c. *Aerobacter* + Yeast   d. Lactics + *Aerobacter*

68. Dill pickles   [   ]
    a. Pickles made by fermenting Dill herb
    b. Pickles that are flavored by Dill herb
    c. Salt stock cucumbers + Dill + Indian spices
    d. Pickel of Dill herb + garlic + Indian spices

69. Fermentation of tea leaves is by   [   ]
    a. *Aspergillus*   b. *Penicillium*
    c. *Rhizopus*   d. Enzymes of the leaves

70. Fermentation of coffee beans is by   [   ]
    a. Lactics
    b. Pectinolytic coliforms
    c. Pectinolytic coliforms followed by lactics
    d. Yeast + *Pseudomonas*

71. Pick out the WRONG pair   [   ]
    a. Soy sauce   - *Aspergillus oryzae*
    b. Tamari sauce   - *aspergillus tamari*
    c. Tempeh   - *Mucor*
    d. Idli   - *Leuconostoc mesenteroides*

72. Pick out the FALSE statement   [   ]
    a. Tamari sauce is soy sauce to which rice is added
    b. Tofu is soybean cheese prepared by curdling soy paste filtrate
    c. Ang-Khak is a soybean paste steamed and fermented by *Apergillus spp.*
    d. Pidan are preserved duck eggs fermented with colifoms and *Bacillus* species.

73. Slimy sauerkraut is due to   [   ]
    a. excessive growth of *Leuconostoc mesenteroides*
    b. *Torulopsis* contamination
    c. *Bacillus polymyxa* contamination
    d. very high concentration of salts

74. What is/are the demerits with use of algae as SCP   [   ]
    a. They require warm temp & sunlight
    b. Their cell wall is indigestible by humans
    c. They require CO_2 for rapid mass production
    d. All

75. Match the following with respect to SCP.   [   ]

    A. *Candida utilis*   i) Poor public acceptance
    B. *Pseudomonas*   ii) Indigestible cell wall
    C. *Scenedesmus*   iii) Utilize pentose & hexose sugars
    D. *Saccharomyces*   iv) Utilize only hexose sugars

    a. A-iv, B-i, C-ii, D-iii   b. A-iii, B-ii, C-i, D-iv
    c. A-i, B-ii, C-iii, D-iv   d. A-iii, B-i, C-ii, D-iv

76. Which of these cannot be subjected to smoking for preservation   [   ]
    a. Poultry meat   b. mutton & beef   c. cheese   d. eggs

77. Optimum temperature for yogurt fermentation   [   ]
    a. 42 to 44^0C   b. 50^0C   c. 37^0C   d. 25^0C

78. Lactic acid bacteria are   [   ]
    a. Gram positive only
    b. Gram negative only
    c. Some are gram positive & some are gram negative
    d. Acid fast bacteria

79. Clostridial stormy fermentation is seen during spoilage of   [   ]
    a. Dill pickles   b. Idli batter   c. Milk   d. Soya sauce

80. Spices can be best preserved by using   [   ]
    a. Benzoic acid   b. Methylparaben   c. Sodium nitrite   d. Ethylene oxide

Answer Key and Validation

Food Microbiology

1. d	14. b	27. b	40. b	53. a	68. b
2. b	15. b	28. a	41. d	54. d	69. d
3. c	16. c	29. a	42. a	55. d	70. c
4. a	17. b	30. c	43. a	56. b	71. c
5. a	18. c	31. b	44. c	57. a	72. c
6. b	19. a	32. c	45. a	58. d	73. a
7. d	20. a	33. a	46. b	61. c	74. d
8. a	21. d	34. c	47. a	62. a	75. a
9. a	22. b	35. d	48. d	63. c	76. c
10. a	23. b	36. d	49. a	64. b	77. a
11. a	24. d	37. b	50. a	65. a	78. a
12. c	25. a	38. a	51. d	66. b	79. c
13. b	26. d	39. c	52. b	67. a	80. d

14

Fermentation Biotechnology

Retrace your subject...

Terminology, Types of Fermentors, Major gropus of Fermentation products and their starter cultures (Antibiotics, Organic acids, enzymes beverages, glycerol, acetone butanol. Amino acids, Vitamins, Growth factors, Single cell proteins, Microbial insecticides), Microorganisms used in bioremediation...

Terminology	
Fermentation	A process in which an energy rich substrate is oxidized without an exogenous electron acceptor. Usually organic molecules are oxidized to form other organic molecules. Fermentative reactions may be in the presence of oxygen (aerobic) or in the absence of oxygen (anaerobic). Fermentations are brought about by microorganisms or their enzymes at the level of industries.
Fermentor vessel	The container used in the industries to carry out fermentative production of commercial goods is called a fermentor vessel. The term has been modernized as *Bioreactor*. Fermentor vessels may be of various sizes ranging from as small as 20 litre capacity for pilot studies on the lab to as large as 5lac litre capacity.
Impellers	Impellers are agitating device used for stirring the contents of a fermentor. They bring about even distribution of nutrients and oxygen within the vessel. Impellers are fitted iside the fermentor vessel from the centre usually as a shaft with vertical pain blades. The shaft is rotated by means of a motor.
Sparger	A device fitted in the fermentor vessel, to allow aeration. Sparger comprises of a perforated metal tube through which sterile air is allowed to pass into the vessel.
Baffles	Short vertical metal plates on the inner surface of the bioreactor. Baffles are meant to create turbulence and prevent formation of vortex in the centre of the fermentor.
Starter Culture	The pure culture of industrially potent microorganisms that is used in the large scale cultivation and production of a particular commodity is called as the starter culture for that fermentation.
Primary screening	The process of preliminary isolation and identification of a potentially useful microorganism in the environment is called primary screening.
Secondary Screening	The qualitative and quantitative tests carried out to select the best starter culture from among those identified by primary screening is called secondary screening. Such tests as optimal growth conditions, optimal production conditions, rate and yield of production, genetic stability of the culture being used, etc comprise of secondary screening.
Strain improvement	The methods employed to genetically modify the starter cultures inorder to improve the fermentation conditions by way of increasing yield, simplifying growth requirements of the organism, etc. Some of the strain improvement techniques include: inducing mutation and selection of mutants, conjugation between valuable strains, protoplast fusion, etc.
Down stream processing	The steps involves in recovery, harvest and purification of the product from the fermentation broth is called downstream processing.
Immobilization techniques	The starter microbial cultures or their enzymes can be trapped (immobilized) in a suitable manner so that they can be used for fermentations with ease. Immobilized starters can be separated from the finished fermented broth very conveniently and can also be reused in the next batch. Immobilization may be brought about by trapping the cells in polymer matrices such as in alginate beads or cross linking to reagents or by directly binding to water insoluble carriers such as ion exchange resins.

Types of Fermentors

Anaerobic fermentor	A closed vessel that allows anaerobic fermentative reaction. Eg. Huge tanks employed in the production of biogas. These vessels do not have sophisticated accessories. They bear an inlet, outlet and pressure control or gas escape valves if required.
Aerobic fermentor	A typical aerobic bioreactor is a deep vessel fitted with impellers, sparger, pH & antifoam controls, etc. inlets and outlets are as usual, but several valves are fitted to enable variety of operations even as the fernetation is in progress.
Bubble column fermentor	Energy saving fermentor modification comprising of inner special compartment for aeration. Solids can be left suspended in these fermentors effectively. Liquid media circulation is exclusively driven by the gas.
Fluidized bed fermentor	A fermentor in which the culture exists as biofilm on solid rings of plastic or other inert material. The solid articles are allowed to move freely in the fermentor by agitation. The organisms on the surfaces of the solid material use the substrate in the liquid broth and release the product.
Packed bed fermentor	A kind of trickling filter bed in which the fermentor vessel is packed with material such as stones, corn cobs, wood chips etc. the substrate solution is allowed to trickle down the material slowly. During this process, the film of starter culture developed on the packing material acts on the substrate to form the product. The product formed is released from the outlet at the bottom of the fermentor. Eg: Oxidation of sewage, vinegar manufacture employ packed bed fermentors.
Airlift fermentor	A fermentor consisting of two compartments, one is sparged with air (riser part) and the other is kept off from the air (downcomer part). Media ingredients are mixed and risen in the riser column while they settle downwards in the other column. Air lift fermenter is an energy saving fermentor vessel and has several other advantages such as mass heat transfer capacity. Eg: Production of biopharmaceutical proteins employs this fermentor.

Antibiotic	Starter culture
Penicillin	*Penicillium chrysogenum*
Streptomycin	*Streptomyces griseus*
Tetracyclin	*S. aureofaciens, S. rimosus*
Griseofulvin	*Pencillium platulum, P.greseofulvin, P.nigricans P.utricase.*
Polymixin –B	*Bacillus polymixa*
Bacitracin	*B.lichiniformis*
Fumigellin (Cephalosporin C)	*Aspergillus fumigatus*
Oxytetracyclin	*Streptomyces aureofaciens, S.rimosus.*
Chloramphenicol	*S.venuzuelae*
Kanamysin	*S.kanamycetius*
Erythromycin	*S.erythreus*
Carbomycin	*S.halstedii*
Oleandomycin	*S.antibioticus*
Nystatin	*S.noursei*
Amphotericin-B	*S.nodosus*
Cycloheximide	*S.griseus*
Novobiocin	*S.nineus, S.spheroides*
Neomycin-B	*S.fradiae*

Organic Acids	
Citric acid	*Aspergillus niger, Aspergillus clavatius* *Pencillium luteum, Pencillium citrinum* *Paecilomyces divaricatum, Mucor piriformis* *Ustilina vulgaris. Candides lipolytica*

Lactic acid	*Lactobacillus delbruckii, L.bulgaricus* *L.leichmannii, L.casei, L.pentosas, Streptococcus lactis, Rhizopus oryzae*
Acetic acid	*Aspergillus terreus, Acetobacter aceti*
Gluconic acid	*Aspergillus niger*
Kojic acid	*A.oryzae*
2-keto gluconic acid	*Pseudomonas fluorescens*
5-keto gluconic acid	*Gluconobacter suboxydans*
D-Araboasecorbic acid	*Penicillum notatum*
Propionic acid	*Propioni bacterium shermanii*
Pyruvic acid	*Pseudomonas auruginosa*
Succinic acid	*Bacterium sucinicum*
Tartaric acid	*Gluconobacter suboxydans*
Fumaric acid	*Rhizopus delemar*
Malic acid	*Lactobacillus brevis*
Alpha ketogluratic acid	*Candida hydrocarbofumarica*
L-isocitric acid	*Candida brumptii*
L-allocitric acid	*Pencillium purpurogenum*

Enzymes	
Amylase	*Aspergillus oryzae, A.niger* *Bacillus subtitis, B.amyloliquefaciens*
Protease	*Aspergillus oryzae, Aspergillus niger* *Aspergillus saitoi, Mucor pusillus* *Bacillus subtilis, Streptomyces grisecs* *Aspergillus phoenias*
Lipase	*Aspergillus niger, Rhizopus spp* *Candida cylindracea*
Cellulase	*Aspergillus niger, Trichoderma viridi* *Trichoderma koningi*
Pectinase	*Aspergillus niger, Aspergills oryzae* *Aspergillus flavis*
Invertase	*Sacharomuces cervisiae* *Aureobasidium pullulans*
Aspartase	*E.Coli*
Catalase	*Aspergillus niger*
Lactase	*Aspergillus oryzae*
Rennet	*Mucor pusillus*
Takadiastase	*Aspergillus oryzae*

Bewerages	
Ethyl alchol	*Saccharomyces spp, Torulopsis spp* *Kloekera spp, Candida spp*

Glycerol	Certain yeast spp
Acetone and butanol	*Clostridium acetobutylicum S25* *C.acetobutylicum B14, Butyl culture* *C.aurianticum, C.liquefaciens*

Amino acids	
L-glutamic acid	*Corynebacterium glutamicum* *Brevibacterium falvim* *Brevibacterium thigentalis* *Bacillus megaterium* *Brevibacterium pentosminacidicum*
L-Lysine	*Brevibacterium flavum* *Corynebacterium glutamicum* *Brevibacterium lactofermentam* *Nocardia spp*
L-threonine	*E.Coli*
Leucine	*Corynebacterium glutamicum*
Isoleucine	*Serratia marcescens*
Arginine	*Bacillus subtilis*
Histidine	*Corynebacterium glutamicum*
Tyrosine	*Corynebacterium glutamicum*
Phenylalanine	*Corynebacterium glutamicum*
Tryptophan	*Corynebacterium glutamicum*

Vitamins	
Vitamin B 12	*Streptomyces griseus* *S.olivaceous* *Bacillus megaterium* *Bacillus coagulans* *Pseudomonas denitrificans* *Propionibacterium freudenreichii* *P.shermanii* *Proteus spp* *Pseudomonas spp*
Riboflavin B2	*Eremothecium ashbyii* *Ashbya gossypii*
L-ascorbic acid/Vitamin C	*Acetobacter suboxydans*
Beta carotene/Vitamin A	*Phycomyces blakesluanus* *Chomephora cucurbitarum* *Blakeslea trispora*

Growth Factors	
Gibberellins	*Gibberella fujikuroi*

Single Cell Proteins	
Bacterial SCP Hydrogen utilising organisms	*Pseudomonas facilis* *P.sacccharophila* *P.falva* *P.pelleronii* *Alkaligenes autrophes* *Alkaligenes paradoxus*
Methane or methanol utilising bacteria	*Methanomonas methanica* *M.methanooxidans* *Methylococcus capsulatus* *Pseudomonas methanica* *Methylomonas methylovora* *M.clara* *M.methanolica* *Methylophilus*
n-paraffins utilising bacteria	*Brevibacterium insectiphilum* *Corynebacterium spp* *Corynebacterium paurimetabolum* *Pseudomonas ligustri* *Pseudomonas pseudomalli* *Pseudomonas orvilla* *Alcaligenes spp* *Cellulomonas galba* *Micrococcus aurificians*
Actinomycetous proteins	*Nocardia paraffinica, Mycobacterium phlei,* *Thermomonospor fusca, Streptomyces spp*
Yeast proteins	*Saccharomyces cereviciae* *Saccharomyces carlsbergensis*
Fungal proteins	*Trichoderma viridae, Aspergillus oryzae* *Rhizopus arrhizes*
Algal proteins	*Scenedesmus, Chlamydomonas* *Spirulina, Anabaena, Chlorella* *Mosocales chysophyta*

Microbial Insecticides

Bacterial insecticides	*Bacillus thuringensis* *B.popilliae* *B.leutimorbus* *B.morital*
Viral insecticides	*Tobacco budworm* *Heliothis* *Biotrol VH2*
Fungi insecticides	*Beauveria bassiana* *Meturrhizum anisopliae*
Nematodes insecticides	*Acrostalagmus aphidum*
Protozan insecticides	*Neoaplectania carpolapsae*
Rickettsiae insecticides	*Biotrol NCS*

Microorganisms used in bioremediation

Hydrocarbon pollution	*Pseudomonas*
Bioleaching of metal	*Thiobacillus ferrooxidans*
Arsenic and antimony compounds	*Alcaligens utrophus, Corynebacterium haccumfacieus*
Cupper compounds	*E.Coli*
Chromium compounds	*Pseudomonas aeroginosa*
Silver compounds	*Pseudomonas stutzerikk*
Camphor and toluene compounds	*Pseudomonas putida*
Alkyl benzene sulfonate	*Pseudomonas testosterone*
Biphenyl compounds	*Beijerinckia*

Exercises

1. The concept of fermentation was scientifically simplified by   [   ]
   a. Louis Pastuer   b. Robert Koch   c. D.A. Micklos   d. G.A. Freyer

2. Growth of culture without presence of added free water is followed in a   [   ]
   a. Fixed bed reactor   b. Lift tube fermentor
   c. Solid state fermentation   d. Fluidized bed reactor

3. Molasses serve as   [   ]
   a. Antifoam agent   b. Carbon and energy source
   c. Nitrogen source   d. Source of growth factor

4. Corn steep liquor source   [   ]
   a. Antifoam agent   b. Carbon and energy source
   c. Nitrogen source   d. Source of growth factor

5. Whey is a fermentation medium that comes as a waste product from   [   ]
   a. Diary industry   b. Paper industry   c. Sugar industry   d. Oil industry

6. Which of these cultures was significantly improved for better yields in the early days of fermentation technology   [   ]
   a. *Aspergillus niger*   b. *Streptococcus thermoacidophilus*
   c. *Penicillium notatum*   d. *Saccharomyces cervisiae*

7. Which of these is employed for agitation in a bioreactor   [   ]
   a. Impeller   b. Sparger
   c. Dissolved oxygen probe   d. Biosensor unit

8. In a fluidized bed reactor   [   ]
   a. the microbes are immobilized in alginate beads
   b. the microbial cells are freely suspended in the broth medium
   c. the artificial addition of inoculum is avoided
   d. the microbes are cultured as biofilms on surfaced of suspended solid particles

9. Which of these occurs in a dialysis culture unit   [   ]
   a. Waste products diffuse out of the culture medium
   b. substrates diffuse across the membrane into the culture medium
   c. both a & b
   d. neither a nor b

10. Which of these organic acids can be produced at the level of a fermentation industry   [   ]
    a. gluconic acid   b. fumaric acid   c. kojic acid   d. all

11. The growth and physiology of yeast differs depending on whether it is grown under aerobic or anaerobic conditions.   [   ]
    a. Kluyer hypothesis   b. Hesse's observation
    c. Koch's postulates   d. Pastuer effect

12. The head space is   [   ]
    a. The total internal volume of the bioreactor
    b. The broth holding capacity of a bioreactor
    c. The depth of the fermentor
    d. The space left between the surface of the fermentation medium and the lid of the bioreactor

13. Which of these serve as antifoam agents   [ ]
    a. Silicones
    b. Fertilizer grade phosphate
    c. Crude carbonates
    d. Nitrates

14. Cultivation of microbial cells using gasous substrates can be best brought about by   [ ]
    a. Trickling filter bed
    b. Dialysis culture unit
    c. Bubble cap fermentor
    d. Solid state fermentor

15. Identification & isolation of potential cultures from natural samples such as air, water and soil is called   [ ]
    a. Strain improvement
    b. Primary screening
    c. Secondary screening
    d. Crowd plating

16. Growth along a central streak to test inhibition of test organisms is called   [ ]
    a. Crowded plate technique
    b. Replica plating
    c. Kirby-Baeur technique
    d. Giant colony technique

17. Turbulence is due to presence of   [ ]
    a. Impeller
    b. Sparger
    c. Baffles
    d. Head space

18. A quick and widely used method to determine cell content in a bioreactor   [ ]
    a. Packed cell volume (PCV)
    b. Haemocytometry
    c. Direct microscopic count
    d. Petroff-Hausser chamber count

19. Toxic effect of heavy metals on microbial cells is referred to as   [ ]
    a. Oligodynamic action
    b. Mac Farland reaction
    c. MIC
    d. End-point assay

20. The main advantage of invert molasses over black strap molasses is that it is   [ ]
    a. relatively cheaper
    b. higher in sugar content
    c. lower levels of toxic subtstances
    d. higher content of N source

21. Molasses resulting from the manufacture of crystallillne dextrose from corn starch is called   [ ]
    a. Hi-test molasses
    b. Refinery black strap molasses
    c. Hydrol
    d. Starch molasses

22. Sulfite waste liquor must be subjected to stream stripping before use as fermentation media inorder to   [ ]
    a. remove free sulfur dioxide & sulfurous acid
    b. concentrate the sugars
    c. remove toxic salts
    d. digest cellulose to simpler sugars

23. Lard oil in fermentation technology finds use as   [ ]
    a. Antifoam agent
    b. Buffer
    c. PH regulator
    d. agent of anaerobiosis

24. Photosynthetic microorganisms are useful in the production of   [ ]
    a. vitamins
    b. biofuel
    c. somatostatin
    d. alkaloids

25. *Rhizopus* and *Arthrobacter* find use in   [ ]
    a. production of gibberllins
    b. organic acid production
    c. steroid biotransformations
    d. all

26. Which of these is employed in the industrial production of vitamins   [ ]
    a. *Ashbya*
    b. *Exmothecium*
    c. *Blakeslea*
    d. all

27. Pick out the CORRECT pair   [   ]
    a. *Claviceps-perpurea* – Production of alkaloids
    b. *Zymomonas* – Production of insulin
    c. *Xanthomonas* – nucleotide biosynthesis
    d. *Corynebacterium glutamicum* – Interferon production

28. Which of these is involved in the production of polysaccharides as food additives   [   ]
    a. *Kluyveromyces fragilis*   b. Clostridium acetobutyricum
    c. Xanthomonas   d. Giberlla fujikuroi

29. Which of these productions is an outcome of microbial r-DNA technology   [   ]
    a. Ethanol   b. Methane   c. Citric acid   d. Human growth hormone

30. Accumulation of secondary metabolites occurs during   [   ]
    a. Trophophase   b. Idiophase   c. Secondary phase   d. Diauxic phase

31. The antibiotic Bacitracin is produced from   [   ]
    a. *Bacillus licheniformis*   b. *Aspergillus fumigatus*
    c. *Streptomyces griseus*   d. *Streptomyces aureofaciens*

32. *Streptomyces venezuelae* is used in the production of the antibiotic   [   ]
    a. Oxytetracyclilne   b. Chlortetracycline
    c. Cycloheximide   d. Chloramphenicol

33. The antifungal antibiotic griseofulvin is produced by   [   ]
    a. *Penicillium* spp.   b. *Streptomyces* spp.
    c. *Bacillus* spp.   d. *Cephalosporium*

34. Which of these organic acids is NOT synthesized by *Aspergillus* spp.   [   ]
    a. citric acid   b. fumaric acid   c. gluconic acid   d. kojic acid

35. *Curvularia lunata* finds use in the production of   [   ]
    a. Steriod prednisolone   b. Amino acid L-lysine
    c. Antibiotic cephalosporin   d. Coenzyme A

36. Which of these spp. is employed in environmental management   [   ]
    a. *Saccharomyces*   b. *Streptomyces*
    c. *Leuconostoc*   d. *Pseudomonas*

37. In the fermentative production of ethanol from molasses, the molasses medium is
    acidified with   [   ]
    a. H_2SO_4 b. HCl   c. Citric acid   d. Acetic acid

38. Which of these is employed in producing artificial honey?   [   ]
    a. Lipase   b. Pectinase   c. Amylase   d. Invertase

39. Pick the FALSE statement   [   ]
    a. Rennin is used in cheese production
    b. Amylase is used in clarifying fruit juices
    c. Lipase is used in coffee bean fermentation
    d. Taka diastase is used in starch dextrinization

40. Mash tub is required in the manufacture of   [   ]
    a. Butter   b. Wine   c. Beer   d. Cheese

41. Sake is   [   ]
    a. Japanese alcoholic beverage   b. Chinese sauce
    c. British dairy fermented product   d. Russian pickle

42. Kuass, Chicha & Jaette are   [ ]
    a. Undistilled alcoholic beverages
    b. Fermented dairy products
    c. Oriental fermented foods
    d. Flavoured vinegar preparations

43. The aminoacid valine is fermentatively produced from   [ ]
    a. *Escherichia* mutants
    b. *Brevibacterium* spp.
    c. *Claviceps* spp,
    d. *Corynebacterium* spp.

44. The fermentor vessel is packed with wood chips, corn cobs and other material in the production of   [ ]
    a. Vinegar
    b. Ethanol
    c. Lysine
    d. Cheese

45. Which of these is produced in Russia by yeasty fermentation of potatoes   [ ]
    a. Gin
    b. Brandy
    c. Rum
    d. Vodka

46. A bioreactor is   [ ]
    a. the agent used to bring about biological reactions
    b. the vessel in which microorganisms grow
    c. the vessel in which a fermented product is collected
    d. the vessel in which microbes or their metabolites are used to produce industrially viable products

47. *Arthrobacter* simplex finds application in   [ ]
    a. rennet production
    b. lipase production
    c. steroid biotransformation
    d. dextran production

48. *Streptomyces* is employed in the production of   [ ]
    a. Neomycin
    b. Amphotericin B
    c. Streptomycin
    d. All

49. Majority of hormones are produced industrially from   [ ]
    a. *E.coli* (via rDNA technology)
    b. *Xanthomonas*
    c. *Bacillus subtilis*
    d. *Mycobacterium*

50. Cell free fermentation was started by   [ ]
    a. Buchner
    b. Pasteur
    c. Koch
    d. Liebig

51. Screening is the process of   [ ]
    a. preserving industrial culture
    b. selecting industrial cultures
    c. separating culture from product
    d. All

52. Downstream processing is   [ ]
    a. harvest and recovery of industrial product
    b. draining the fermentation broth from below the fermentor
    c. mixing the lower contents of a fermentor
    d. removal of contaminants in a fermentor

53. Match the media with the product   [ ]
    A. Molasses   i. Streptomycin
    B. Whey   ii. Penicillin
    C. Soy meal   iii. Alcohol
    D. Corn steep liquor   iv. Lactic acid

    a. A-iii, B-iv, C-ii, D-i
    b. A-iii, B-iv, C-I, D-ii
    c. A-ii, B-iii, C-iv, D-i
    d. A-iv, B-ii, C-I, D-iii

54. Distillers soluble serves as   [ ]
    a. Starchy medium
    b. Saccharine medium
    c. Hydrocarbon medium
    d. Nitrogenous medium

55. Which of these is not a saccharine source medium for industrial fermentation   [   ]
    a. Fruit juices   b. Molasses   c. Whey   d. Sulfite Waste Liquor

56. Pharma media is raw material for fermentation media, obtained from   [   ]
    a. embryo of cotton seed   b. maize pulp
    c. paper industry   d. hydrocarbon extractions

57. Vegetable oils are used   [   ]
    a. in production of lipases   b. in production of antibiotics
    c. as antifoam agents   d. in production of $VitB_{12}$

58. The yeast used as animal feed   [   ]
    a. *Saccharomyces*   b. *Torula utilis*
    c. *Candida albicans*   d. *Kloekera*

59. Which of these is an aerating device   [   ]
    a. sparger   b. impeller   c. baffles   d. all

60. Autotitrations are fitted to a fermentor to automatically control   [   ]
    a. temp. changes   b. PH changes   c. foam production   d. All

61. The top yeast used in ethanol production   [   ]
    a. *Saccharomyces cervisiae*   b. *S. carlbergensis*
    c. *Kloekera utilis*   d. *Torulopsis*

62. Pitching is   [   ]
    a. Distillation of ethanol from fermented broth
    b. Removal of inoculum from fermented broth
    c. Addition of inoculum to fermentation medium
    d. Ageing of ethanol after purification

63. What is 'TRUE' about ethanol fermentation   [   ]
    a. Medium used is cane molasses   b. Optimum pH is 4.8 to 5 & opt. temp is 28 to $30^{\circ}C$
    c. Duration of fermentation is 30 to 72h   d. a, b, & c are true

64. Which of these is used for agitation   [   ]
    a. Sparger   b. Impeller   c. Butterfly valve   d. Baffle

65. The fermentation medium for ethanol production must contain sugar conc. up to   [   ]
    a. 5%   b. 10 to 18%   c. 35%   d. 50%

66. The fermentation medium for ethanol production must be   [   ]
    a. filtered before yeast is added   b. autoclaved before yeast is added
    c. steam sterilized before yeast is added   d. pasteurized before yeast is added

67. The byproduct (distillery effluent) of ethanol fermentation, fusel oil contains   [   ]
    a. acetates   b. aldehydes   c. fats   d. higher alcohols

68. Fusel oil is used in the manufacture of   [   ]
    a. acetates   b. perfumes   c. liquors   d. all

69. Pick out the WRONG pair of solvent – culture   [   ]
    a. Glycerol   -   *Corynebacterium glutamicum*
    b. Acetone butanol   -   *Clostridium acetobutylicum*
    c. 2,3 – Butanedial   -   *Bacillus polymyxa*
    d. Ethanol   -   *Saccharomyces*

70. Which of these serves as immuno suppressor   [   ]
    a. Bestatin   b. Piericidin   c. Cyclosporin   d. Avermectin

Answer Key and Validation

Fermentation Biotechnology

1. a	13. a	25. c	37. a	49. a	61. a
2. c	14. c	26. d	38. d	50. a	62. c
3. b	15. b	27. a	39. c	51. b	63. d
4. c	16. d	28. c	40. c	52. a	64. b
5. a	17. c	29. d	41. a	53. b	65. b
6. c	18. a	30. b	42. a	54. d	66. d
7. a	19. a	31. a	43. b	55. d	67. d
8. d	20. b	32. d	44. a	56. a	68. d
9. c	21. c	33. a	45. d	57. c	69. a
10. d	22. a	34. b	46. d	58. b	70. c
11. d	23. a	35. a	47. c	59. a	
12. d	24. b	36. d.	48. d	60. b	

20. *Invert molasses contains approx. 70-75% sugar. Whole cane juice is partially inverted to prevent crystallization. Hence it has lower levels of non-fermentable sugars. Invert molasses is also called Hi-test molasses*

24. *Hydrogen is commercially produced as a biofuel using photosynthetic microorganisms*

39. *Pectinases are employed in coffee bean fermentation*

49. *Genetically engineered strains of E.coli are used for production of insulin, human growth hormone, somatostatin & interferon*

68. *Fusel oil is the by product of ethanol fermentation. It contains isoamyl alcohol & butyl alcohol which are variously useful.*

69. *Glycerol is fermentatively obtained from yeast*

15

Air, Water and Soil Microbiology

<table>
<tr><td colspan="2">Retrace your subject…</td></tr>
<tr><td colspan="2">Air: Atmospheric layers, Air microflora, Air sampling devices, Air spores and Air borne diseases.
Water: Aquatic environments, Water related terminology, Treatment process for drinking water, Treatment process for waste water, Common water borne diseases.
Soil: Soil microflora, Soil related terminology, Methods for detecting soil microflora, Biological Nitrogen fixation, Biofertilizers.</td></tr>
</table>

Atmospheric Layers

Troposphere	The first layer closest to the earth is the troposphere. This is where all plants and animals live and breathe. This is where *weather* takes place.
Stratosphere	The next layer is the stratosphere. Ozone in this layer stops many of the sun's harmful rays from reaching the earth. People can not breathe in this layer.
Mesosphere	The third layer is the mesosphere. It is very cold. This is where we see "falling stars" (rocks and other materials burning as they rush through the atmosphere from space).
Exosphere	The exosphere is the highest layer of the atmosphere. The air is very thin here.
Thermosphere	The thermosphere is directly above the mesosphere and below the exosphere. Within this layer, ultraviolet radiation causes ionization. It is the fourth atmospheric layer from earth.

Air Microflora

Bacteria in Air	*Bacillus, Micrococcus, Mycobacterium , Corynebacterium,*
Fungi in Air	*Alternaria, Cercospora, Helminthosporium, Puccinia, Aspergillus, Cladosporium, Penicillium, Epicoccum, Tricothecium, Mucor, Rhizopus, etc.*
Algae & BGA in Air	*Chlorella, Clorococcum, Chlamydomonas, Aulosira, Nostoc, Phormidium, Protococcus, Spirogyra, Oscillatoria, etc.*
Lichens in Air	*Cladonia, Heteroderma, Parmelia, Usnea, etc..*
Flora common in Hospitals/ Pharmaceuticals	*Staphylococcus aureus, Cladosporium cladosporioides, Alternaria, Penicillium, Epicoccum, Aspergillus flavus, A. fumigatus, Candida albicans, Mycobacterium, etc.*
Plant pathogens in air	*Alternaria brassicae, A. solani, Cercosporaarachidicola, Colletotrichum gleosporioides, Helminthosporium oryzae, Puccinia graminis, etc.*

Air sampling devices

Traditional air sampling	Holding greasy kites or greasy gas balloons up in the air and collecting the washing from their surfaces, ir the traditional method of sampling air flora.
Settling plate technique	A simple method to stydy air flora. Agar Petri dishes are exposed to the air for 15min followed by incubation. Colonies of different flora are obtained that can be characterized and identified.
Low volume air samplers	These are devices that collect small volumes of air in polythene bags or metal canisters with the help of motor pumps. The air thus collected is released into sterile saline and the organisms are isolated.

Table Contd…

High volume air samplers	These are sophisticated units the collect air from large areas such as the entire hospital or a large field. The air is blown directly into sterile saline. The saline is then centrifuged and processed for the flora in the labs.
Personal air samplers	A small kit that can be attached to the individual's shirt. It collects air spora from all the places the individual moves about on a daily basis. These samplers help in identifying allergens for individual personnel.
Biological air samplers	Several sampling devices cause impingement of the air spora directly onto the surfaces of agar media. Eg. Slit to agar air sampler, membrane filter air sampler, portable air blower.
Anderson air sampler	A sampling device consisting of 8 serially arranged agar petridishes, each covered by a porous grid with decreasing pore sizes from top to bottom. Air particles settle on the plates as per their size and can be easily studied.
Burkhard air sampler	A special spore trap air sampler device.
Rotorod air samplers	Consists of a rotating pair of rods that cause impingement of air spora on the rods. It is useful as an open field sampling device.

Air borne diseases

Bacterial	Diphtheria, psittacosis, whooping cough, tuberculosis, scarlet fever, Q fever, pneumococcal pneumonia, pneumonic plague, pulmonary anthrax.
Fungal	Aspergillosis, Gilchrist's disease (caused by *Blastomyces spp)*, Cryptococcosis, Histoplasmosis, Candidiasis, Blastomycosis.
Viral	Influenza, Measles, Mumps, Chicken pox, Small pox, polio, common cold, Swine flu.

Aquatic Environments

Water bodies	There are three kinds of water bodies. Fresh water, Marine water and Estuarine water bodies.
Fresh water bodies	Lotic Habitats: Comprise of running waters. Eg. Spring, falls, running streams, etc. Lentic habitats: Standing water bodies. Eg. Pond, lake, standing river, etc.
Estuaries	The sea river union. Show zones of high salt water, zone of high fresh water and the middle zone of dispersion.
Ground water	Water in the water table underground, formed as a result of surface water percolation over the years. Eg. Well water, bore water.
Zones of fresh water bodies	Upper limnetic zone, middle profundal zone, lower benthic zone. The upper limnetic zone is called littoral zone along the shore and the pelagic zone away from the sore. The profundal zone is divided to photic and aphotic zones one below the other depending on the availability of light in that zone.
Zones of marine water bodies	Upper pelagic, middle Photic and Lower Aphotic zone. The water immediately along the shore is the littoral zone. Further to the land is the intertidal zone that is underwater only during high tides.

Water related terminology

Planktons	Microorganisms that are found in the surface layers of the sea and other aquatic environment are collectively known as planktons. The algal mass and other photosynthetic flora floating on the water surfaces are called phytoplankons while Zooplanktons comprise of the floating protozoa and minute animal members.
Benthos	Microorganisms present in the bottom regions of aquatic bodies are designated as benthos. The benthos comprise of anaerobes, extreme heterotrophs, halophile (salt loving) and barophile (pressure loving) flora.
Coliform bacteria	Coliforms are gram negative, aerobic to facultatively anaerobic, coccobacillary to short rod shaped bacteria that can ferment lactose producing acid and gas.
Biological Oxygen Demand (BOD)	The amount of oxygen required by microorganisms to oxidize organic matter in the water sample when held at 20^0C for 5 days. BOD is expressed in parts per million (mg/Lt).

Treatment process for drinking water

Pumping	Raw water is pumped from a river/lake by means of a cleanly maintained pumping station
Flocculation	The collected water is mixed with alum, colloidal silicate or bentonite in a mixing tank. This facilitates removal of many coarse and fine impurities.
Sedimentation	The supernatant water is transferred to a settling basin for further clarification
Filtration	The water is then passed through tall columnar sand filters to remove ultra fine particulate matter. Many microbes that remain adhered to fine particles are filtered by this process.
Disinfection	The filtered water is then chlorinated to enable killing of pathogens. The extent of chlorination and the net residual chlorine depends on individual standards maintained by the treatment plant.
Storage	The treated drinking water is transported to large storage reservoir tanks from where the water is supplied to the consumers.

Treatment process for wastewater

Sewage	The liquid waste emanating from domestic establishments is called sewage. Raw untreated sewage consists of more than 99% water and the rest is suspended solid matter. Fecal bacteria, coliphage viruses, protozoa and parasites constitute the microflora of sewage. Sewage is treated aerobically in three different stages. Primary, Secondary and Tertitiary treatment processes.
Primary treatment	The sewage is allowed to pass through a series of mesh grids of decreasing pore sizes. This removes coarse solids such as glass pieces, plastic bags, stones, dust, shells etc.
Secondary treatment	This is the actual biological treatment process. It may be accomplished by use of Lagoons or trickling filter bed or activated sludge process. These processes oxidize the organic matter in the sewage as fully as possible so that the biological oxygen demand of the sewage is reduced drastically. The lesser the BOD value, the safer it is to dispose the sewage into natural water bodies. Among the various methods, the activated sludge process is considered fastest and best method of sewage treatment.
Tertiary treatment	This is an optional step for sewage treatment. After secondary treatment the sewage may be safely disposed into natural water bodies. However, if the water is to be re-used, the sewage must be subjected to tertiary treatment. In this the sewage is subjected to coarse sand filtration followed by disinfection with chlorine. The treated sewage can be employed as swimming pool water, animal washing, industrial coolers, etc. it is even possible to recycle sewage water into pure drinking water, but the process is highly expensive and not practically feasible.
Sludge digestion	The solid slurry that remains after sewage treatment is sludge. Sludge is digested aerobically by composting in open air. The finished compost is used as a fertilizer. Sludge may be digested anaerobically for production of biogas (methane).
Anaerobic treatment of sewage	Small amounts of sewage from individual dwelling units can be treated underground, by construction of septic tanks. Sewage water undergoes partial oxidation by action of methanogens and thermophiles. The process is convenient for small establishments, but the treated sewage is still high in BOD value and may not be always safe for disposal. Imhoff tanks are similar to septic tanks. However, they are much large and suited for anaerobic treatment of large volumes of sewage emanating from large establishments such as complexes, buildings, malls, etc.

Common water borne diseases

Bacterial	Cholera, Typhoid, paratyphoid, Shigellosis.
Protozoa	Amoebic dysentery, *Giardia*sis.
Viral	Poliomyelitis, Hepatitis

Soil Microflora

Soil Bacteria	*Achromobacter, Agrobacterium, Arthrobacter,, Azotobacter Clostridium, Desulfovibrio Gallionella, Nitrobacter, Nitrosomonas, Methanobacillus, Pseudomonas, Rhizobium, Thiobacillus, Ferrobacillus, , Hydrogenomonas, Carboxydomonas, Xanthomonas, etc.*
Soil Actinomycetes	*Actinoplanes Actinomyces, Micromonospora,Nocardia, Streptomyces, Streptosporangium Thermoactinomyces, etc.*
Soil Fungi	*Alternaria, Aspergillus, Botrytis, Cephalosporium, Cercospopra, Cunninghamella, Fusarium, Monilia, Mucor, Penicillium, Phoma, Rhizoctonia, Rhizopus, Saprolegnium,, Sclerotium, Trichoderma , Pesta, Pullularia, Helminthosporium, Pyricularia, Zygorrhynchus, etc.*
Soil Protozoa	*Balantiophorus, Bodo, Biomyxa, Euglypha, Tetramitus, Cercobodo, Naegleria, Uroleptus, Colpoda, Hartmannella, Trienma, Euglyphia, Colpidium, Halteria, etc.*
Soil Algae	*Chladophora, Chlamydomonas, Chlorella, Clorococcum, Cymbella, Hormidium, Navicula, Pinnularia, Scenedesmus, Stichococcus,, Surirella, Synedra,, Ulothrix, Vaucheria etc.*
Soil Cyanobacteria	*Chroococcus, Cylindrospermum, lyngbya, Nodularia, Anabena, Nostoc, Oscillatoria, Scytonema, Tolypothrix, Phormidim, etc.*

Soil related terminology

Soil horizons	The vertical layers of soil in a given area, from uppermost surface to the lower ones are the soil horizons. Usually the horizons are named A, B, C, D from surface to subsurface gradually. Each horizon differs by physical characteristics of the soil. i.e. from fine soil to coarse, gravel, stones and rock area. The geological, biological and chemical reactions occurring in the soil determine the depth of each horizon. The layer of organic matter above the soil surface is sometimes referred to as O layer. Followed by A horizon of surface soil, B (sub surface zone) and C (parent rock zone). The number of soil microorganisms usually decrease from top to bottom.
Mineralization	The process of release of simple organic compounds during decomposition of dead animal or plant matter in soil. Mineralization contributes significantly in improving soil fertility.
Humus	The dark brown degraded organic matter in soil is called humus. Humus is a stable material and not easily degradable further. It is very rich in easily absorbable nutrients for plants and serves as a natural fertilizer.
Rhizosphere region	The immediate area of soil that surrounds the roots of the plant is called rhizosphere. Rhizosphere region extends upto the area where the plant root exudates can reach. Soil in this region is nutritious for microbial growth as many micronutrients are released by the plant root. The sloughed plant root matter also serves as food for bacteria. Hence the average count of the microflora in the rhizospehre region is far higher than that of the non rhizosphere regions. Rhizosphere regions encourage growth of nitrogen fixing bacteria and stimulate germination of spores of many mycorrhiza.
Rhizoplane region	The surface of the plant root that is in direct contact with the soil is called rhizoplane. Many microorganisms remain adhered to the rhizoplane. Study of rhizoplane flora is important in the process of understanding nodule formation, symbiotic associations and also plant root infection processes.
Biogeochemical cycles	The cyclic pathway that shows the movement of a particular chemical element through the biotic and abiotic componets in nature is a biogeochemical cycle of that element. Eg. Nitrogen cycle, Carbon cycle, etc. careful studies of these cycles enable to understand how living organisms assimilate naturally available elements into their body and how these elements reach back to the envt after the death of the organisms.
Biological Nitrogen fixation	Microbial reduction of atmospheric nitrogen to ammonia by the enzyme nitrogenase. BNF is either symbiotic (*Rhizobium spp*) or non-symbiotic (*Klebsiella spp*).

Methods for Detection of Soil microflora

Method	Particulars
Light Microscopy	Direct examination of smears prepared from soil suspension.
Rossy Colodny Buried Slide Technique.	Slides are buried in the soil close to the roots for a week and one side is stained.
Direct observation of root surfaces	Small roots are observed directly.
Impression slide technique	Roots are pressed against nitrocellulose coated slides and pulled away. The impression is stained.
Direct Culture method	Fine soil is directly sprinkled on selective agar plates for bacterial, actinomycetes and fungal growth
Immunofluorescence methods	Specific fluorescent tagged Abs are added to minute amount of soil sample. After washing, fluorescence will be seen only if the particular organism is present in the sample.
ELISA assay technique	The technique has been applied mainly to *Rhizobia* in soil and in roots of legumes The major difficulty is removal of the microbial cells from the substrates.
Gene probe and nucleic acid hybridization	Detecting specific sequences of nucleic acids in the organisms under study. This technique can find specific organisms in soils and other environmental samples. The gene probe is a short segment of nucleotides that binds specifically with the homologous sequence in the target microorganism. If the segment is labelled with radioactive ^{32}P, any binding to the target nucleotides can be detected by the presence of the radioactivity after reaction.
Polymerase Chain Reaction (PCR)	PCR has recently been applied to soil microbial ecology. In this technique, extracted DNA is melted to form single strands, annealed with primers, and the DNA is extended from the primers by nucleotide addition using DNA polymerase enzyme. The primers are chosen to link to regions of DNA of interest.
Respiration measurements	Soil is incubated at the required temperature, moisture content, in the main flask and air can be added through the CO_2 absorber in the tube on top of the main flask. During incubation the CO_2 produced by respiration is absorbed into the alkali in the side tube. This is then titrated to discover the amount of alkali neutralized by the evolved CO_2. To obtain reproducible results from assays of respiration rate the soils are usually sieved to remove plant materials, insects, etc.
Other Methods	Measurement of rate of cell division, Mycelial extension, Enzyme activity, Autoradiography, Determination of ATP content of soil and Dilution plate technique.

Biological nitrogen fixation

BNF is the process of conversion of atmospheric nitrogen to ammonia by nitrogen fixing microorganisms. It is an extremely important and most studied biological phenomenon first discovered by Martinus Beijerinck. Plants cannot utilize the amply available molecular nitrogen directly from the atmosphere. Nor can they convert it to absorbable forms. Hence they are entirely dependent on BNF microflora for nitrogen requirement. These organisms are called *diazotrophs* and can fix nitrogen with or without symbiotic association with plant root. Formula for BNF: $N_2 + 6\,H^+ + 6\,e^- \rightarrow 2\,NH_3$

Nitrogenase Enzyme: This is an enzyme complex that catalyses the BNF reaction. It requires energy in the form of ATP. The enzyme complex is composed of heterotetrameric protein (MoFe protein) and a homodimeric protein (Fe Protein). During ATP hydrolysis, the nitrogenase enzyme uses the transferred electrons to break the bonds between N_2 molecule. Three such cycles are required to form ammonia as the N_2 molecule has three bonds between the nitrogen atoms. The ammonia so formed is bonded to glutamate to from glutamine. The enzyme complex is irreversible damages by oxygen which oxidises the Fe-S co-factors of the enzyme.

Major groups of Nitrogen fixers	Examples of Bacterial genera
Proteobacteria	*Rhizobium, Bradyrhizobium, Azospirillum, Beijerinckia, Azotobacter*
Actinobacteria	*Frankia*
Cyanobacteria	*Anabaena, Nostoc, Trichodesmium, Calothrix, Phormidium, Scytonema, Oscillatoria*
Endospora	*Some species of Clostridium*
Important species of *Rhizobium*	*R. leguminosarum, R melilotii, R trifoli, R. phaseoli, R. lupinii, R, japonicum*
Genera of Non-symbiotic nitrogen fixers	*Rhodobacter, Rhodospirillum, Azotobacter, Azotomonas, Derxia, Methylomonas, Clostridium, Bacillus, Desulfotomaculum, Beijerinckia, Alkaligens, Arthrobacter, Azospirillum, Flavobacterium, Pseudomonas, Enterobacter, Klebsiella.* (note that only some species in each case are nitrogen fixers).

Biofertilizers

Microorganisms that serve as agents to improve soil fertility are called biofertilizers. They are produced in large scale at the industry level and applied to the soil by various means as dust sprays, with irrigation water, direct seed inoculation, mixing with carriers such as soil, peat, lignite or compost, etc.

Eg: *Rhizobium, Azotobacter, Anabena, Nostoc, etc* are used as biofertilizers for bringing about biological nitrogen fixation. *Mycorrhiza and bacteria such as Thiobacillus* are used for phosphate solubilization in the soil.

Biofertilizers are relatively less expensive than chemical ferlitlizers. They are also ecofriendly and show long term sustainability in the fields. However, the results of the yields using biofertilizers are not rapid and extraordinary. This has limited the use of biofertilizers in a large way.

Exercises

1. Tilak air sampler consists of   [   ]
   a. a slowly rotating drum with cello tape
   b. a series of agar plates arranged in a row
   c. a metal grid placed on the agar plate
   d. a sticky microscope slide

2. Air borne microorganisms can cause   [   ]
   a. immediate hypersensitivity
   b. delayed allergic reactions
   c. both a& b
   d. immune complex hypersensitivity

3. Which of these are seen in air   [   ]
   a. Microalgae   b. Microfungi   c. Protozoa   d. all

4. Which of these is air borne disease   [   ]
   a. Measles
   b. Q fever
   c. Gilchrist's disease
   d. All

5. Which of these is caused by air borne fungi   [   ]
   a. Cryptococcosis   b. Influenza   c. Psittacosis   d. Scarlet fever

6. HEPA filters are fitted in   [   ]
   a. Laminar air flow units
   b. Andersons air sampler
   c. Kulkarni's air sampler
   d. Slit sampler devices

7. Simplest method to study air flora   [   ]
   a. Tilak air sampling technique
   b. Slit sampling technique
   c. Settling plate technique
   d. Air sieve technique

8. Which of these is effective in chemical air sanitation   [   ]
   a. Propylene glycol spray
   b. DDT spray
   c. Ethanol spray
   d. Bleaching powder sprinkle

9. Typical air flora   [   ]
   a. *Nostoc*   b. *Micrococcus*   c. *Halobacter*   d. HIV

10. The atmospheric layer closest to the earth's surface is   [   ]
    a. Troposphere   b. Stratosphere   c. Exosphere   d. Thermosphere

11. Slit to agar air sampler is   [   ]
    a. Personal air sampling device
    b. Biological air sampler
    c. high volume air sampler
    d. Rotorod air sampler

12. Swine flu virus that caused the 2009 pandemic   [   ]
    a. H1N1   b. H2N2   c. H1N2   d. H3N3

13. Coliforms are   [   ]
    a. indicators of fecal pollution
    b. indicators of industrial pollution
    c. indicators of air pollution
    d. all

14. What is FALSE about coliforms   [   ]
    a. They are pathogenic
    b. All coliforms are bacteria
    c. They are found in the colon
    d. They are found in sewage waters

15. Coliforms   [   ]
    a. ferment lactose
    b. end products of sugar fermentation are acid and gas
    c. are aerobes to facultative anaerobes
    d. all of the above

16. Which of these is NOT a coliform   [   ]
    a. *Enterobacter*   b. *E.coli*   c. *Klebsiella*   d. *Pseudomonas*

17. Ideally, water in springs & deepwells is   [   ]
    a. Sterile   b. Unsafe for consumption
    c. Hard water   d. heavily contaminated with bacteria

18. Which of these is NOT included in routine examination of water   [   ]
    a. Test for anaerobic sulphite reducers   b. Test for coliforms
    c. Test for *Pseudomonas aeruginosa*   d. Test for *Giardia*

19. Metallic sheen is produced on EMB agar by   [   ]
    a. *E.coli*   b. Coliforms   c. *Proteus*   d. None

20. Presumptive coliform test requires   [   ]
    a. Lactose fermentation broth   b. EMB agar
    c. Salt mannitol agar   d. Peptone water

21. Dark centered faint pink colonies on EMB agar   [   ]
    a. *E.coli*   b. Coliforms   c. *Enterobacter*   d. *Klebsiella*

22. Brilliant green lactose bile (BGLB) broth is used in   [   ]
    a. Presumptive coliform test   b. Confirmed coliform test
    c. Completed coliform test   d. MPN test

23. Which of these is a quantitative coliform test   [   ]
    a. Completed coliform test   b. MPN test
    c. Eijckman's test   d. All of these

24. Pick out the WRONG pair   [   ]
    a. *S. aureus* - Salt mannitol agar   b. *Proteus*   - Cetrimide agar
    c. *Salmonella* - Bismuth sulfite agar   d. *Vibrio*   - TCBS agar

25. Aerobic plate count in water are most accurately determined by   [   ]
    a. Serial dilution technique   b. Membrane Filteration Technique (MFT)
    c. Spread plate technique   d. Pour plate technique

26. The ONPG & MUG coliform test is dependent on   [   ]
    a. lactose fermentation   b. production of β galactosidase
    c. production of β galacturonidase   d. both b & c

27. Killing effect of chlorine in water on   [   ]
    a. *E.coli* > Viruses > *Giardia*   b. Viruses > *Giardia* > *E.coli*
    c. *E.coli* > *Giardia* > Viruses   d. *Giardia* > Viruses > *E.coli*

28. Which of these is considered as nutrient cycle   [ ]
    a. Hydrological cycle   b. Gaseous cycle
    c. Sedimentary cycle   d. All

29. Pick out the FALSE statement   [   ]
    a. Solar energy evaporates water from hydrosphere to atmosphere
    b. Evaporation from oceans leads to formation of maritime air mass
    c Evaporation snow, plant transpiration and animal respiration, do not contribute to the formation of clouds.
    d. The amount of water vapor in the atmosphere is greatest near the equator

30. Hydrological cycle is dependent on   [   ]
    a. Sun   b. Wind   c. Ocean current   d. All

31. Denitrification   [   ]
    a. Nitrites are reduced to molecular nitrogen
    b. Nitrates are reduced to molecular nitrogen
    c. Nitrites or nitrates are reduced to molecular nitrogen
    d. Nitrites or nitrates are reduced to molecular nitrogen or nitrous oxide

32. Ammonification   [   ]
    a. Organic nitrogenous compounds are converted to ammonium ions
    b. Atmospheric nitrogenous compounds are converted to ammonium ions.
    c. a and/or b
    d. None of these

33. Non-symbiotic N_2 fixers   [   ]
    a. *Rhodospirillum*   b. *Azotobacter*   c. *Nitrosomonas*   d. All

34. Nitrogenase enzyme is inhibited by accumulation of   [   ]
    a. Nitrates in the soil   b. Ammonia in the soil
    c. a and b   d. None of these

35. In the carbon cycle, maximum amount of CO_2 is returned to the atmosphere by   [   ]
    a. Respiration (exhalation)
    b. Combustion of fossil fuel (aeroplanes)
    c. Volcanic eruptions
    d. Anaerobic bacteria, in the form of methane

36. At 12noon the CO_2 concentration around tree tops   [   ]
    a. Reaches the maximu   b. Reaches the minimum
    c. Reaches zero   d. Remains unaltered

37. Which of these algal members are rarely found in soil except in highly saline soils?   [   ]
    a. Bacillariophyceae   b. Cholorophyceae
    c. Chrysophyceae   d. Rhodophyceae

38. Which of these bacteria are dominant in cellulose rich soils?   [   ]
    a. *Cytophaga & Sporocytophaga*   b. *Arthrobacter & Achromobacter*
    c. *Micrococcus & Flavobacterium*   d. *Myxococcus and Chondrococcus*

39. Which of these are responsible for soil texture   [   ]
    a. *Tolypothrix*   b. *Actinomyces*   c. *Saccharomyces*   d. *Polyangium*

40. The main contribution of fungi in soil is   [   ]
    a. Mineralization   b. Parasitizing plant roots
    c. Phosphate solubilization   d. Production of antibiotics

41. Aggregation of soil particles is aided by [   ]
    a. Protozoa   b. Algae   c. Fungi   d. Viruses

42. Which of these flora is reduced when soil remains shaded under huge trees or
    heavy surface litter [   ]
    a. Algae   b. Fungi   c. Protozoa   d. Viruses

43. *Bodo, Heteromita and Tetramitus* in soil are commonly occurring [   ]
    a. Algae   b. Fungi   c. Protozoa   d. Viruses

44. Water permeability of soil is increased best by [   ]
    a. Humus   b. Mushrooms   c. Wood shavings   d. Saw dust

45. Ammonium salts are converted to nitrites in soil by [   ]
    a. *Nitrobacter*
    c. *Nitrosomonas*
    b. *Nostoc*
    d. *Bacillus denitrificans*

46. Which of these processes is significant in the carbon cycle [   ]
    a. Ammonification
    b. Anaerobic decay with production of H_2S
    c. Breakdown of cellobiose to glucose
    d. Combustion of fossil fuel

47. Which of these releases H_2SO_4 [   ]
    a. *Desulfotomaculum*
    c. *Rhodopseudomonas*
    b. *Desulfovibrio*
    d. *Thiobacillus thiooxidans*

48. Phosphate solubiliser [   ]
    a. *Aspergillus*   b. *Penicillium*   c. *Fusarium*   d. All

49. Which of these is a soil remediation technique [   ]
    a. Land farming   b. Terrace farming   c. Composting   d. Grafting

50. Hemicellulose is known to be decomposed by [   ]
    a. *Chaetomium*   b. *Polyangium*   c. *Cellfalcicula*   d. *Micromonospor*

51. Heterocyst in cyanobacteria is site for [   ]
    a. photosynthesis and energy production
    b. nitrogen fixation and sexual reproduction
    c. nitrogen fixation and photosynthesis
    d. nitrogen fixation

52. Some non-heterocystous cyanobacteria can fix nitrogen in vegetative cells
    when grown anaerobically and vigorous bubbling of nitrogen [   ]
    a. TRUE   b. FALSE

53. In increased amount of ammonia in the environment, the rate of nitrogen fixation
    by cyanobacteria [   ]
    a. Remains unaffected
    b. Increases
    c. Decreases
    d. Increases only when supplemented with nitrogen gas

54. Highest level of nitrogen fixation occurs in root nodule when the color of the nodule is [   ]
    a. Pink   b. Deep brown   c. Green   d. Pale yellow

55. The *dinitrogenase-dinitrogenase reductase* complex of nitrogenase enzyme is made of [   ]
    a. MoFe-Fe   b. Fe-MoFe   c. MgFe-Fe   d. Fe-MgFe

56. Which of these is a non-symbiotic nitrogen fixer [   ]
    a. *Azotobacter*   b. *Azotococcus*   c. *Azomonas*   d. All

Answer Key and Validation

Air, Water and Soil Microbiology

1. a	11. b	21. c	31. d	41. c	51. d
2. c	12. a	22. b	32. a	42. a	52. a
3. d	13. a	23. b	33. d	43. c	53. c
4. d	14. a	24. b	34. c	44. a	54. a
5. a	15. d	25. b	35. a	45. c	55. a
6. a	16. d	26. d	36. b	46. d	56. d
7. c	17. a	27. a	37. d	47. d	
8. a	18. d	28. d	38. a	48. d	
9. b	19. a	29. c	39. b	49. a	
10. a	20. a	30. d	40. a	50. a	

14. *Coliforms are indicators of fecal pollution, but not necessarily pathogenic. Only some strains or species may be pathogenic.*

38. *Cytophaga & Sporocytophaga are cellulose degrading bacteria.*

39. *Actinomyces are filamentous bacteria that contribute to the texture of the soil by holding together the ultra fine particles of soil.*

42. *Counts of Algae are very low in shaded soils due to decreased availability of direct sunlight required for their photosynthesis*

50. *b, c and d are cellulose degraders*

16

Ecology

Prologue	
Ecology	Ernst Haeckel a German Biologist coined the term ecology in 1866. Haeckel defined ecology as *the comprehensive science of the relationship of the organism to the environment*
Eugenius Warming	A Danish botanist presented the first significant textbook on Ecology. Hence Warming is regarded as the founder of Ecology.
Autecology	Study of individual organism or species in relation to environment.
Synecology	The study of interactions of groups of organisms among themselves and with their environment.
Realm of Ecology	Ecology is a multi-disciplinary branch of Biology that involves principles of Geology and Geography, Meteorology, Pedology, Botany, Zoology, Microbiology, Genetics, Chemistry, and Physics.
Scope of Ecology	Ecology undertgakes the challenging task of identification and solutions to problems of environmental pollution, overpopulation, human survival, pest control and conservation of natural resources and events of ecological calamities. Ecology is often studied at the level of populations, communities, and ecosystems.
Ecosystem	A defined locality within which, the biological interactions are studied. The organisms are interrelated in an ecosystem through *Food webs* and *Food chains*. An Ecosystem is the functional and structural unit of ecology.

Types of ecology	
Habitat Ecology	Includes fresh water ecology, desert ecology, forest ecology, grassland ecology, cropland ecology and marine ecology etc.
Population Ecology	Study of interrelationships of different groups of organisms. Also the study of number of organisms and what determines the number.
Ecosystem Ecology	Analysis of structural and functional ecosystems including the interrelationship of biotic and abiotic components.
Conservation Ecology	Deals with proper management of natural resources like land, water, forest, and sea etc.
Production Ecology	Studies the gross and net production of different ecosystems so that proper management is possible to get maximum yield.
Radiation Ecology	Studies radioactive substances and the environment.
Ecological energetics	Energy conservation and flow of energy in an organism of an ecosystem.
Palaecology	Traces gaps in the evolutionary lines and deals with the organism of past ecological environment.
Gene Ecology	Genetic make-up of species or populations in relation to environment.

Table *Contd…*

Systems Ecology	It is the research of community dynamics at a mathematical level, where complex formulae and computer programmes summarize and model a particular dynamic system being examined.
Behavioral Ecology	The behavioral relationship between individuals of a species — Example, the study of the queen bee, and her interactions with the worker bees and the drones
Community Ecology	The organized activity of a species in a community. Example, the activity of bees assures the pollination of flowering plants. Bee hives produce additional honey that is consumed by other species, such as bears.
Environmental ecology	The relationship between the environment and a species — The environment affects the activity of a species and is thus linked with the survival of the species.

Ecosystem	Features
1) Arid and semi-arid zone ecosystems	
Grassland ecosystems	• Dominated by grasses, grass-like plants and woody plants and are further characterized by periodic drought, fire and large herbivore grazing. • Soils are usually marked by low fertility. • The earth's land surface is composed of 20 percent grasslands with temperate grasslands representing one forth of the area and savannas the remainder.
Savannahs	• Characterized by a dominant grassy ground layer which can be present in treeless plains, open woodlands, and closed-canopy woodlands.
Warm desert, semi-desert and Hyper-arid desert ecosystems	• Have irregular rainfall with periodic droughts lasting several years. The Sahara desert represents nearly 70 per cent of the global hyper-arid area. Semi-desert areas are dominated by succulents and other xerophytic plants and are often characterized by sparse or absent tree cover. Rainfall in arid and semi-arid areas varies from less than 200 to a maximum of 800mm annually.
Tundra communities and cold desert biomes	• Non-humid areas are composed of cold regions and dry lands and represent 61 per cent of the Earth's total land surface. Cold regions include polar and tundra areas as well as certain high mountains and plateaus and compose 14 per cent of the non-humid area. Temperatures remain below freezing for a long period which restricts plant growth. Liquid water is usually unavailable for a significant part of the year.

Ecosystem	Features
2) Coastal, marine and freshwater ecosystems	
Mangroves (A hybrid terrestrial ecosystem)	• Mangroves are composed of shrubs, trees, and ferns living in or adjacent to the inter-tidal zone. Mangroves are predominantly tropical with the most diverse systems located in Southeast Asia.
Coral reefs	• Corals are most abundant in shallow marine environments usually characterized by warm low-nutrient waters. Coral reefs are among the most diverse ecosystems in the world with over 450 different coral species identified in Southeast Asia alone. They are composed of calcium carbonate.
Seagrasses	• Found in shallow coastal areas between the Arctic and Antarctic, seagrasses are flowering plants that live submerged in seawater. Entire ecosystems are founded on their high productivity as they provide an important source of food for many organisms.
Pelagic or open-ocean communities	• Based on plankton, which supports a large number of free-swimming organisms, pelagic communities occupy a greater area than any other major community on earth. Characterized by vertical gradients that fluctuate seasonally or as often as daily, the community responds to the physical and chemical variations of seawater.
Deep-sea communities	• Exist in permanent darkness at depths greater than 3000 meters. Marked by low biomass.

Table *Contd...*

Rivers	• Freshwater bodies draining from elevated land toward sea level. Rivers show considerable variation in water flow, underlying geology and altitude. There can be a wide range of habitats within an individual river. Some rivers are marked by continual change while others have been in continuous existence for millions of years.
Lakes	• Lakes are mainly glacial in origin. As a rule, they are geologically short-lived, typically changing or disappearing within 100,000 years. As a result, lake ecosystems are often less complex and less evolved than others.
Wetlands	• Permanent or seasonal shallow waters and large aquatic plants and are usually categorized into bogs, fens, marshes, and swamps. Bogs are peat-producing wetlands in most climates in which organic matter has accumulated over long periods. Fens, also peat-producing wetlands, are usually supplied by mineral-rich groundwater. Marshes and swamps are inundated areas with herbaceous emergent vegetation or forested wetlands where little or no peat production occurs.

Ecosystem	**Features**
3) Forest ecosystems	
Boreal and temperate needle leaf forests	• Covers a larger area than any other. Dominated by a small number of tree species. Lichens and Mosses are usually relatively high. Serving as major reservoirs of organic carbon, these forests are significant in the carbon cycle both above and below ground.
Temperate broadleaf and mixed forest	• Structurally more complex than needle leaf forests with considerably higher species diversity. The temperate forests are generally representative of warmer latitudes. Biomass is lower than that of temperate needle leaf.
Tropical moist forests	• These are the most diverse ecosystems on earth. Contain an estimated 60 per cent of all species. Diversity is extremely high with as many as 300 tree species per hectare in areas such as the western Amazon. Such forests account for nearly one third of the annual global terrestrial net primary production.
Tropical dry forests	• Characteristic of tropical areas affected by seasonal drought, Composed of a deciduous forests. Sparse trees and parkland occupy areas in transition from forested to non-forested landscapes and most tropical dry forests have lower species diversity than tropical moist forests.
4) Mountain ecosystems	• Highlands represent one quarter of the earth's land surface. They are susceptible to changes from landslides, avalanches, lava flows and torrents and are slow to recover. Mountain ecosystems are often highly biodiverse, with large numbers of endemic species.
5) Agricultural ecosystems	• Agricultural land represents 38 per cent of earth's land area and is characterized by permanent pastures. Agricultural ecosystems contribute to soil structure protection and air and water quality.

Food Chains & Food Webs	
Food Chains	• The picture of the unidirectional flow of energy in an ecosystem, consisting of the series of members eating and being eaten. • *Grazing Food Chain:* Flows across Producers, Primary consumers, Secondary consumers and Tertiary consumers. • *Detritus Food Chain:* Begins with decomposers that feed on dead organic matter and flows ahead through their consumers. • Food chains are usually short (not more than six levels). The longer the chain, the less stable it is. • The main demerit of food chains is thet they do not consider the interrelations of the organisms with those from a different food chain. Hence the food chains cannot give a true picture of the flow of energy in the ecosystem.

Table *Contd...*

Food Webs	<ul><li>A picture of all the food chains in the ecosystem and their interrelations is referred to as a Food Web.</li><li>Food webs give a better understanding of the ecosystem. However, it is a complicated figure and can be designed only when all the groups of organisms in the ecosystem are known.</li><li>Keen observation, analysis of fecal pellets and use of radiolabelled isotopes in the nutrients are some of the methods employed to trace the food chain/web.</li><li>Food webs are more realistic than food chains. However, they are unable to focus on the quantitative significance of each group of organism. Hence, ecological pyramids are a better choice to understand the state of affairs in an ecosystem.</li></ul>
Trophic levels	The organisms on an ecosystem are categorized into different trophic levels. Each level indicates a common mode of nutrition. Example: Producers (Green plants), Primary consumers (Herbivores), Secondary consumers (Lower Carnivores), Tertiary consumers (Higher carnivores) are all examples of trophic levels. Identification of trophic levels aids in easy construction of ecological pyramids and in the calculation of energy transfers.
Abiotic Factors	*Environmental Factors*: Temp, Light, Hydrostatic Pressure and osmotic pressure. *Chemical Factors*: pH, Oxygen and Carbon dioxide.

Ecological Pyramids	
Pyramids of Numbers	The diagrammatic representation of the numbers of organisms at the different trophic levels in a community. The term *pyramid of number* was coined by Elton (1927). This may quite **often be inverted.** Eg. A single tree supports many smaller organisms or insects and produces an inverted pyramid. The unit of pyramids of number is *individuals/Area.*
Pyramids of Biomass	These are calculated by determining for a given unit area, the biomass of a producer level, biomass of herbivores, biomass of primary carnivores, etc. very few pyramids of biomass are developed so far as they are very difficult to determine. These are **most often upright** unlike pyramids of numbers which are often inverted (However, in a pond the pyramid of biomass is inverted in shape). The unit of pyramids of Biomass is *Mass/ Area.*
Pyramids of Energy	This is a diagrammatic representation of the flow of energy from a trophic level of a community to the ones above. The unit of pyramids of energy are *Energy/area/time.* Eg. KJ/ha/yr, i.e. pyramids of energy show the rate at which energy flows up the food web and they are measured over a stated period of time. [Note: The pyramids of biomass cannot be converted to Pyramids of energy by using any conversion factor. Pyramids of energy **are always upright.** They can never be inverted.]
Major Ecological Calamities	Global Warming, Ozone layer hole issue, Increase in desertification, disappearance of many species, 1986 Nuclear meltdown at Chernobyl and Exxon Valdez oil spill off Alaskan coast in 1989

Exercises

1. The term ecology was first used by [   ]
   a. Reiter in 1885 followed by Ernst Haeckel in 1886.
   b. Ernst Haeckel in 1885 followed by Reiter in 1886
   c. Warming in 1885 followed by Reiter in 1886
   d. Reiter in 1885 followed by Warming in 1886.

2. Ecology is the study of [   ]
   a. Interactions of living being among themselves.
   b. Interactions of living beings with their physical environment
   c. Both a & b
   d. None

3. The population genetically adapted a particular environment is called. [   ]
   a. ecotype   b. ecocline   c. guild   d. ecophene

4. An Ecocad is [   ]
   a. Change in organisms due to inbreeding.
   b. A population adapted to its environment
   c. A population having the same genetic makeup but different morphological characteristics due its particular environment
   d. A population, unable to adapt itself to its environment.

5. A Community is [   ]
   a. Assemblage of populations   b. Disintegration of populations
   c. Division of populations   d. None of these

6. Study of individual species or population in relation to its environment is called. [   ]
   a. Autecology   b. Species ecology
   c. Autecology *or* Species ecology d. Synecology

7. Growth, natality, mortality, competition etc are dealt within [   ]
   a. Community ecology   b. Population ecology
   c. Ecosystem ecology   d. genecology.

8. Study of different communities living in a particular climate or area [   ]
   a. Synecology   b. Ecosystem ecology
   c. Biome ecology   d. Cytoecology.

9. Ecological energetics [   ]
   a. Conservation of energy   b. Net productivity of ecosystems
   c. Flow of energy in an ecosystem   d. All

10. Interpreting ecologocial concepts and processes in terms of mathematical formulae and models. [   ]
    a. Ecological energetics   b. Algebraic ecology
    c. System ecology   d. Calculus ecology

11. Pollution ecology, forest ecology, agronomy, conservation ecology are all part of [   ]
    a. Zoogeographical ecology   b. Applied ecology
    c. Palaeocology   d. Phytogeographic ecology.

12. The portion of the habitat occupied by a species based on the tolerance volume is its [   ]
    a. Ecological niche   b. Ecological equivalent
    c. Character displacement   d. Edaphic factor

13. Two birds living in the same nest, one feeding on insect and the other on seeds are said to differ by their   [   ]
    a. Fundamental niche
    b. Habitat niche
    c. Trophic niche
    d. Ecosystems

14. Which of these is a microecosystem   [   ]
    a. Rhizosphere region
    b. oral cavity
    c. a test tube of contaminated peptone water.
    d. All

15. All the ecosystems on earth comprise of   [   ]
    a. Holosphere
    b. Geosphere
    c. Biosphere
    d. Ecosphere.

16. The first paper published on ecology, written by Hippocrates was called.   [   ]
    a. *On air, water and places*
    b. *We and Our Environment*
    c. *Ecology*
    d. *Population Regulation*

17. The concepts of *food chains* and *population regulation* in modern ecology was first put forth by   [   ]
    a. Aristotle
    b. Theophrastus,
    c. Anton van Leeuwenhoek
    d. Darwin

18. Early Indian ecologists   [   ]
    a. Misra, Tiwari, Puri, Mal
    b. Sharma, Sarma, Mukherjee,
    c. Shah, Patel, Johnson, Kumbham
    d. Kumaran, Maheswari, Tilak, Khan.

19. Soil formation, nutrient cycles, energy flow and productivity are studied under   [   ]
    a. Gynecology
    b. Physiological ecology
    c. Ecosystem ecology
    d. Community ecology.

20. Interpretation of animal behavior under natural conditions is   [   ]
    a. Sociology
    b. Ethnology
    c. Systems ecology
    d. evolutionary ecology

21. Pedology is the branch of ecology that deals with study of   [   ]
    a. Soils
    b. humans
    c. ecological energetics
    d. Paleoecology

22. Demecology is another word for   [   ]
    a. Paleoecology
    b. Space ecology
    c. Gene ecology
    d. Population ecology

23. The most dominant ecological factor affecting each and every micro and macro ecosystem is   [   ]
    a. Light
    b. Water
    c. Temperature
    d. Soil

24. Select the Correct sequence of Water Cycle   [   ]
    a. Water bodies → rain → evaporation → water bodies.
    b. Evaporation → rain → Condensation→ water bodies.
    c. Rain → Water bodies → evaporation → Condensation → Rain.
    d. Condensation → evaporation → water bodies →Rain

25. An association in which one organism is harmed and the other is neither benefited nor harmed is   [   ]
    a. Mutalism.
    b. Commensalism
    c. Amensalism
    d. Mullerian mimicry

26. Interaction which is beneficial to both partners, but is not obligatory [   ]
    a. Bayesian mimicry
    b. Amensalism
    c. Protocooperation
    d. Neutralism.

27. In *Competition* [   ]
    a. Each organism favours the other
    b. Each organism inhibits the other
    c. Neither organism affects the other
    d. One organism exploits the other

28. Ant-aphid relation is [   ]
    a. Commensalism.
    b. Mutuatism
    c. Amensalism
    d. Parasitism

29. Production of antibiotics is an example of [   ]
    a. Amensalism
    b. Commensalism
    c. Parasitism
    d. Predation

30. Which of these is NOT predation [   ]
    a. Plants feeding on insects
    b. Cats feeing on rats
    c. Leeches feeding on blood
    d. Wolves feeding on sheep

31. Ecosystem is the basic functional unit of organisms and their
    environment – This definition was proposed by [   ]
    a. E. P. Odum
    b. A. G. Tansely
    c. S. Mathavan
    d. Misra

32. The term *Ecosystem* was proposed by [   ]
    a. Varnadsky (1944)
    b. A G Tansley (1935)
    c. Koestler (1969)
    d. Friederichs (1930)

33. Which of these is an ecosystem [   ]
    a. A laboratory aquarium
    b. *E. coli* growing in peptone water in a tube
    c. Courtyard garden
    d. All

34. Pick up the FALSE statement [   ]
    a. An ecosystem is a unit that has achieved the stage of a *stable system*
    b. All ecosystems are connected with other ecosystem around them
    c. When the ecosystem is getting completely destroyed, the stage is called ecological climax
    d. An ecosystem is the level of biological and physical and organization in a particular unit

35. The major energy circuit in an ecosystem [   ]
    a. The grazing circuit
    b. The detritus circuit
    c. The carnivore circuit
    d. The photosynthetic circuit

36. In a stable ecosystem, the exchange of materials follows a [   ]
    a. Circular path
    b. A linear forward path
    c. A linear backward path
    d. A vertical upward path

37. Any factor in an ecosystem that tends to determine growth is called [   ]
    a. Abiotic factor
    b. Growth factor
    c. Limiting factor
    d. Stability factor

38. In an ecosystem, the stratum in which intense heterotrophic metabolism
    occurs is called [   ]
    a. Green belt
    b. Brown belt
    c. Black belt
    d. White belt

39. Phytoplanktons occupy the position of [   ]
    a. Producers
    b. Primary consumers
    c. Secondary consumers
    d. Decomposers

40. The transfer of energy from one trophic level to the next, in a food
    chain is 100% efficient   [   ]
    a. TRUE   b. FALSE
    c. True only in certain ecosystems   d. True only in forest ecosystems

41. The fourth trophic level in a food chain is occupied by   [   ]
    a. Detrivores   b. Carnivores that consume herbivores
    c. Carnivores that consume carnivores   d. Photosynthetic biotic factors

42. A diagram of all trophic relations among themselves is   [   ]
    a. A food web   b. A Food chain   c. A pyramid   d. A trophic galaxy

43. The loss of energy between trophic levels is in the form of   [   ]
    a. Heat   b. Light   c. Moisture   d. All

44. Vultures play the part of   [   ]
    a. Omnivores   b. Scavengers   c. Producers   d. None of these

45. The disadvantage of food chain   [   ]
    a. Flow of energy is not depicted
    b. Overall picture of interrelations among biotic factors is not clear
    c. Shows only unidirectional picture
    d. Cannot be unrstood easily

46. The disadvantage of a food web   [   ]
    a. Does not show individual food chains
    b. Does not give significance to the relative number of population depicted
    c. Does not highlight the role of detrivores
    d. Does not differentiate trophic levels

47. Ecological Pyramid is best explained as   [   ]
    a. A diagram of directional relations   b. A diagram of quanlitative relations
    c. A diagram of energy relations   d. A diagram of quantitative rerlations

48. The Pyramid of biomass is inverted   [   ]
    a. Always
    b. Only if turnover rate of products is more rapid than that of consumers
    c. Only if turnover rate of producers is much less than that of consumers
    d. Never

49. Which of these Pyramids is always erect   [   ]
    a. Pyramid of biomass   b. Pyramid of Energy
    c. Pyramid of Numbers   d. None of these

50. The shallow water areas where sunlight penetrates is   [   ]
    a. Littoral zone   b. Limnetic zone
    c. Profundal zone   d. Lotic zone

51. Pick out the FALSE statement in water body   [   ]
    a. The open water zone, full of microbial population is the limnetic zone
    b. The zone where light does not penetrate and has no producers is profundal zone
    c. Dystrophic lakes are very poor in nutrition but rich in acids due to which plants
       and animals are inhabited
    d. Stable water bodies are called lotic water

52. The lower most zone in oceans   [   ]
    a. Neritic zone   b. Intertidal zone   c. Abyssal zone   d. Benthic zone

53. Pick out the WRONG pair   [   ]
    a. Zooplankton – BGA, *Chlorella*   b. Nektons – Fishes
    c. Benthos – Crabs and Snails   d. Decomposers – Saprophytic bacteria and fungi

54. What is common in Kangaroos, Hawk and owl   [   ]
    a. All are forest animals   b. All are desert animals
    c. All are secondary consumers   d. All are never seen in food chain

55. Which of these is considered as nutrient cycle?   [   ]
    a. Hydrological cycle   b. Gaseous cycle
    c. Sedimentary cycle   d. All

56. Pick out the FALSE statement   [   ]
    a. Solar energy evaporates water from hydrosphere to Atmosphere
    b. Evaporation from oceans lead to formation of maritime air mass
    c. Evaporation, snow, plant transpiration and animal respiration do not
       contribute to the formation of clouds
    d. The amount of water vapor in the atmosphere is greatest near the equator

57. Hydrological cycle is dependent on   [   ]
    a. Sun   b. Wind   c. Ocean currents   d. All

58. Denitrification   [   ]
    a. Nitrites are reduced to molecular nitrogen
    b. Nitrates are reduced to molecular nitrogen
    c. Nitrites or Nitrates are reduced to molecular nitrogen
    d. Nitrites are reduced to molecular nitrogen or nitrous oxide

59. Ammonification   [   ]
    a. Organic nitrogenous compounds are converted to ammonium ions
    b. Atmospheric nitrogen compounds are converted to ammonium ions
    c. a & b
    d. None of these

60. Non symbiotic nitrogen fixers   [   ]
    a. *Rhodospirillum*   b. *Azotobacter*   c. *Nitrosomonas*   d. All

61. Nitrogenase energy is inhibited by accumulation of   [   ]
    a. Nitrates in the soil   b. Ammonia in the soil
    c. a & b   d. None of these

62. In the Carbon cycle, maximum amount of CO_2 is returned to
    the atmosphere by   [   ]
    a. Respiration (Exhalation)   b. Combustion of fossil fuels (aeroplanes)
    c. Volcanic eruptions   d. Anaerobic bacteria in the form of methane

63. At 12.00noon, the CO_2 conc. Around tree tops   [   ]
    a. Reaches the maximum   b. Reaches the minimum
    c. Reaches zero   d. Remains unaltered.

Answer Key and Validation

Ecology

1. a	12. a	23. c	34. c	45. a	56. c
2. c	13. c	24. c	35. b	46. b	57. d
3. a	14. d	25. c	36. a	47. d	58. d
4. c	15. c	26. c	37. c	48. b	59. a
5. a	16. a	27. b	38. b	49. b	60. d
6. b	17. c	28. b	39. a	50. a	61. c
7. b	18. a	29. a	40. b	51. d	62. a
8. c	19. c	30. c	41. c	52. d	63. b
9. d	20. b	31. a	42. a	53. a	
10. c	21. a	32. b	43. a	54. b	
11. b	22. d	33. d	44. b	55. d	

17

Botany

Retrace your subject...

Development of Botany, Categories under Kingdom Plantae, Plant tissues, Plant growth substances, Bentham and Hooker's system of classification of seed plants, Plant ecology, Plant pathology...

Botany is the study of eukaryotic, photosynthetic non-motile living organisms called **plants**. A vast and interesting subject as it is, Botany is one of the elementary branch of Biology. The botanist covers the entire plant Science under external and internal Morphology, Taxonomy, Metabolism, Physiology, Reproduction, Nutrient Transport, Coordination and control, Evolution and Ecology of Plants.

Much of the Microbiologist's knowledge originates from the Principles of Botany particularly in the study of Algae and Fungi.

Although comprehensive know-how of plant science may not be mandatory, it **is nevertheless essential to have a clear understanding of the basics of Botany before attempting the competitive exam of Microbiology.**

This Chapter covers Exercises that range from basic plant morphology to histology, taxonomy, transport, coordination and control in plants. Some of the primary facts are presented in the following tables. The reader is further directed to read a standard book of Botany for advanced acquaintance of the subject.

Historical

Scientist	Contribution(s)
Socrates, Plato, Aristotle	Responsible for development of Botany as a distict branch of Biology
Theophrastus	Father of Botany. Authored the Book *Historia Plantarum*.
Brunfels, L'Obel, Fuchs	Herbalists. Described medicinal plants in the herbals.
Robert Hooke	Observed Cells. Authored *Micrographia.*
Camerarius	First explained sexual reproduction in plants
Carolus Linnaeus	Development of Taxonomy
Stephan Hales	Water is transported implants through xylem
Joseph Priestly	Plants absorb bad air and release good air
Gregor Mendel	Father of Genetics. Proposed Law of inheritance and carried out hybridization experiments on garden pea.
Amici	Discovered pollen tube
Hooker	Father of Phytogeography – Distribution of plants in different regions of earth
Darwin	Proposed theory of Natural Selection
Hatch & Slack	Described C_4 pathway – dark reactions of photosynthesis in certain plants
Calvin, Benson and Basham	Described C_3 pathway– dark reactions of photosynthesis.
Kurasawa	Discovered gibberellins

Branches (*Exclusively related to Plant science are included here*)

Branch	Study of..	Branch	Study of..
Phycology Mycology	Algae Fungi	Agriculture	Methods of cultivation, harvest, storage and protection of crops
Agronomy	Field crops	Horticulture	Improvement of ornamental plants
Palaeobotany	Fossil plant	Olericulture	Vegetable yielding plants
Pharmacognosy	Medicinal plants	Floriculture	Cultivation of ornamental flower yielding plants
Dendrology	Trees	Arboriculture	Cultivation of trees
Plant pathology	Plant diseases	Dendrochonology	Estimating age of plants on the basis if the annular rings
Pomology	Fruit yielding plants	Bryology	Bryophyte – Plant Amphibians
Agrostology	Grasses	Pteridology	Pteridophytes – Vascular Cryptogams or Snakes of the plant Kingdom
Anthology	Flowers	Ethnobotany	Relation between tribals and plants
Limnology	Fresh water ecosystem		

Kingdom Plantae

Phylum			
Bryophyta	**Filicinophyta**	**Coniferophyta**	**Angiospermophyta**
Liverworts & Mosses	*Ferns*	*Conifers*	*Flowering Plants*
Dominant Gametophyte	Dominant Sporophyte	Dominant Sporophyte	Dominant Sporophyte
Defined vascular tissue absent	Defined vascular tissue present	Defined vascular tissue present	Defined vascular tissue present
Only one type of spore produced	Only one type of spore produced	Two types of spores produced	Two types of spores produced
No seeds	No seeds	Seeds produced	Seeds produced
Gametophyte is free living	Gametophyte is free living	Gametophyte is protected within the sporophyte	Gametophyte is protected within the sporophyte
No fruits	No fruits	No fruits	Fruits are produced
Motile male gamete	Motile male gamete	Non-motile male gamete	Non-motile male gamete
Class Hepaticae: Liverworts *Eg. Pellia, Marchantia, Lophocolea* Class Musci: Mosses *Eg. Funaria, Mnium, Sphagnum*	Eg. *Dryopteris filix* (male feren), *Pteridium* (Bracken)	Eg. *Pinus sylvestris*	Class: Dicotyledonae: Pea, Rose, Buttercup, Dandelion. Class: Monocotyledonae: Grassess, Iris, lilies.
Live in damp shady places. No true roots, stems and leaves. Rhizoids serve as roots. Sporophyte is attached to dominant gametophyte. Spores are within a capsule at the tip of a slender stalk that arises from the gametophyte.	Gametophyte is a simple prothallus. Xylem and phloem are present in sporophyte. True roots, stem and leaves are present in sporophyte. Spores produced in sporangia are called *Sori*	Produce cones on which sporangia, spores and seeds develop. No ovary, no fruits.	Produce flowers in which sporangia, spores and seeds develop. Ovary is present which develops to fruit after fertilization.

Plant tissues

One type of cells			More than one type of cells	
Parenchyma	**Collenchyma**	**Sclerenchyma**	**Xylem**	**Phloem**
Packing tissue	Support tissue	Support tissue	Translocation of water and minerals	Translocation of organic solutes
Living tissue	Living tissue	Dead tissue	Living tissue with mixture of living and dead cells	Living tissue with mixture of living and dead cells
Cell walls are made up of cellulose, pectins and hemicelluloses	Cell walls are made up of cellulose, pectins and hemicelluloses	Cell walls are mainly made up of Lignin. Cellulose, pectins and hemicelluloses	Cell walls are mainly made up of Lignin. Cellulose, pectins and hemicelluloses	Cell walls are made up of cellulose, pectins and hemicelluloses
Spherical to elongated cells	Elongated, polygonal cells with tapering ends	Elongated, polygonal cells with tapering, interlocking ends	Elongated and tubular cells	Elongated, tubular/narrow cells
Present in cortex, pith, packing tissue in xylem & phloem	Present in outer regions of cortex: mid rib of leaves and angles of stems	Present in pericyle of stems, outer regions of cortex.	Constitute the vascular system	Constitute the vascular system
--	--	Occur as fibres and Sclereids	Contains fibres, parenchyma, tracheids and vessels	Contains fibres, sclereids, Sieve tubes and Companion cells

Plant Growth Substances

Growth Substance	Characteristics
Auxins	**Promote:** Cell enlargement in stems, Promotes callus, Promotes fruit growth, apical dominance. **Inhibits:** Root growth at very high conc, Delays leaf senescence, inhibits abscission.
Gibberellins	**Promote:** Cell enlargement in stems in presence of auxins, Promotes leaf growth, Breaks bud dormancy and seed dormancy. **Inhibits:** Root growth at very high conc, Delays leaf senescence in few species.
Cytokinins	**Promote:** Cell division in apical meristem and cambium, Promotes leaf growth, Promotes fruit growth, promotes lateral bud (antagonistic to auxins), Promotes stomatal opening. **Inhibits:** inhibits primary root growth, Delays leaf senescence.
Abscisic acid	**Promote:** Seed dormancy, Abscission, Closing of stomata during wilting, Promotes bud dormancy in some spp. **Inhibits:** Root growth, inhibits stem growth during physiological stress
Ethene	**Promote:** Flowering in pineapple, Promotes fruit ripening. **Inhibits:** Root growth, inhibits stem growth during physiological stress

Bentham and Hooker's system of classification of Phaenerogams

<table>
<tr><th colspan="6">Phaenerogams (seed Plants)</th></tr>
<tr><td colspan="2" align="center">Dicotyledonae</td><td rowspan="3" colspan="2" align="center">Gymnospermae</td><td colspan="2">Monocotyledonae</td></tr>
<tr><td colspan="2" align="center">Sub-class</td><td colspan="2">Series: Coronariae</td></tr>
<tr><td align="center">Polypetalae</td><td align="center">Gamopetalae</td><td rowspan="1">Monochlam ydae</td><td colspan="2">Family: Liliaceae</td></tr>
<tr><td>Series   Thalamiflorae:
Family   Malvaceae</td><td>Series   Inferae
Family Asteraceae</td><td></td><td></td><td></td><td></td></tr>
<tr><td>Series   Disciflorae</td><td>Series   Heteromerae</td><td></td><td></td><td></td><td></td></tr>
<tr><td>Series   Calyciflorae
Family   Fabaceae</td><td>Series   Bicarpellatae
Family Solananceae</td><td></td><td></td><td></td><td></td></tr>
</table>

Plant Pathology

Plant disease	Causative agent
Potato Blight Disease	*Phytophthora infestans* (fungus)
Neck blast or black neck of rice	*Pyricularia oryzae* (fungus)
Red rot of Sugarcane	*Colletotrichum falcatum* (fungus)
Grain smut of Sorghum	*Sphacelotheca sorghi* (fungus)
Citrus canker	*Xanthomonas axonopodis* (bacterium) *[Formerly – Xanthomonas camperstris]*
Rust	*Puccinia graminis (fungus)*
Tikka disease of groundnuts	*Circospora arachidicola*
Smut of maize	*Ustilago maydis (fungus)*

Exercises

1. Liverworts belong to Phylum   [   ]
   a. Bryophyta   b. Filicinophyta   c. Coniferophyta   d. Angiospermophyta

2. Which of these produce seeds   [   ]
   a. Angiosperm   b. Conifers   c. Both a & b   d. Ferns

3. Match the following   [   ]
   A. Liverwort   i. *Pteridium*
   B. Mossess   ii *Ranunculus*
   C. Fern   iii. *Pellia*
   D. Angiosperms   iv. *Funaria*

   a. A-iii, B-iv, C-i, D-ii   b. A-ii, B-i, C-iii, D-iv
   c. A-ii, B-i, C-iv, D-iii   d. A-i, B-iv, C-ii, D-iii

4. What is FALSE about Bryophytes?   [   ]
   a. No xylem or phloem is present
   b. Sporophyte is the dominant generation over gametophyte
   c. Body is a thallus
   d. No true roots, leaves or stem present

5. Conifers are different from angiosperms because   [   ]
   a. In conifers seeds are not enclosed in the ovary
   b. In conifers there is no ovary
   c. In conifers there is no fruit
   d. All

6. In monocotyledonous plants   [   ]
   a. Leaves are thin, long and show parallel venation
   b. Flowers are often insect pollinated
   c. Roots show secondary growth
   d. All

7. Vascular cambium   [   ]
   a. A layer of cells found between xylem and phloem
   b. Disperse cells found within xylem and phloem
   c. A type of cell system in xylem tissue
   d. A type of cell system in phloem tissue

8. The driving force in xylem is   [   ]
   a. Transpiration   b. Osmosis   c. Active transport   d. Root pressure

9. Plants lose water through   [   ]
   a. Stomata   b. Stomata and Lenticels
   c. Stomata, Lenticels and Cuticle   d. Stomata and Cuticle

10. Water potential and Solute potential are representaed as   [   ]
    a. δ & δ_s b. ψ & ψ_s c. λ & λ_s d. ξ & ξ_s

11. Apoplast pathway, Symplast pathway and Vascular pathway   [   ]
    a. Explain movement of water from root to stem
    b. Explain movement of water in the leaf
    c. Explain movement of water in the stem
    d. Explain movement of water in the root

12. Cabbage (*Brassica*) is placed under dicots   [   ]
    a. TRUE   b. FALSE

13. Which of these tissues contain only one type of cells   [   ]
    a. Parenchyma   b. Collenchyma   c. Sclerenchyma   d. All

14. Photosynthetic tissue   [   ]
    a. Collenchyma   b. Pericyle and Epidermis
    c. Mesophyll   d. Sclerenchyma

15. Xylem is made up of   [   ]
    a. Tracheids, Vessels, Fibres and Parenchyma
    b. Sieve tubes, Companion cells, Fibres and Sclereids
    c. Tracheids, Sieve tubes, Sclereids and Fibres
    d. Parenchyma, Companion cells, Tracheids and Sclereids

16. Phloem is made up of   [   ]
    a. Tracheids, Vessels, Fibres and Parenchyma
    b. Sieve tubes, Companion cells, Fibres and Sclereids
    c. Tracheids, Sieve tubes, Sclereids and Fibres
    d. Parenchyma, Companion cells, Tracheids and Sclereids

17. Which of these stores starch in a dicotyledenous leaf?   [   ]
    a. Lower epidermis   b. Palisade mesophyll
    c. Spongy mesophyll   d. Vascular tissue

18. The DNA is circular in   [   ]
    a. Chromosomes   b. Mitochondria   c. a & b   d. Nucleus

19. A chlorophyll is made up of   [   ]
    a. Flat hydrophilic head with Mg & long hydrophobic tail.
    b. Flat hydrophilic head with Mn & long hydrophobic tail.
    c. Flat hydrophobic head with Mg & long hydrophilic tail.
    d. Flat hydrophobic head with Mn & long hydrophilic tail.

20. The C_4 pathway was first observe in   [   ]
    a. Maize   b. Millet   c. Sugarcane   d. Sorghum

21. The C_4 pathway is also called   [   ]
    a. GP pathway   b. Kranz pathway
    c. Hatch-Slack pathway   d. Malate Shunt

22. C_4 are more efficient in taking up CO_2 from the atmosphere than C_3 plants   [   ]
    a. TRUE   b. FALSE

23. What is FALSE about C_4 Plants?   [   ]
    a. their leaves contain two types of cloroplasts
    b. The first product of photosynthesis is oxalacetate
    c. CO_2 fixation occurs in bundle sheath cells in place of mesophyll cells
    d. More efficient in photosynthesis than C_3 plants, but use more energy.

24. Sucrose in plants is transported through   [   ]
    a. Cambium   b. Xylem   c. Phloem   d. All

25. In plants, the waste gases are eliminated through   [   ]
    a. Stomata   b. Lenticels   c. Epidermis   d. All

26. When water enters inside the plant cell… [   ]
    a. The pressure potential inside the cell decreases
    b. The pressure potential inside the cell increases
    c. The pressure potential inside the cell is unaffected
    d. The pressure potential instantly becomes negative

27. Water potential is given by [   ]
    a. (Solute potential) + (Pressure potential)   b. (Solute potential) - (Pressure potential)
    c. (Pressure potential) - (Solute potential)   d. ½ (Pressure potential)

28. A potometer is used [   ]
    a. To measure water potential within a cell
    b. To measure pressure potential within a cell
    c. To measure rate of water uptake by a cut shoot or young seedling
    d. To measure solute potential within a cell.

29. What is FALSE about transpiration? [   ]
    a. Low humidity outside the leaf favours transpiration
    b. Rate of transpiration increases with total leaf surface area
    c. Rate of transpiration increases with the number of stomata per unit area
    d. Rate of transpiration increases with the number of stomata per unit area

30. Guttation is [   ]
    a. Loss of water as drops of liquid from the plant surface
    b. Loss of water as vapour from the plant surface
    c. Absorption of water as liquid drops through the plant surface
    d. Absorbtion of water as vapour through the plant surface

31. Movement of plant parts in response to touch is called [   ]
    a. Geotropism   b. Thigmotropism   c. Haptotropism   d. b or c

32. Movement against water current or air current [   ]
    a. Rheotaxis   b. Aerotaxis   c. Magnetotaxis   d. Geotaxis

33. Which of these is positively phototactic? [   ]
    a. Cockroach   b. Earthworm   c. *Euglena*   d. All

34. Auxins do not have any effect on [   ]
    a. Stem growth   b. Root growth   c. Fruit growth   d. Stomatal mechanism

35. Which of these is inhibitory to stem growth? [   ]
    a. Cytokinins   b. Gibberellins   c. Abscissic acid   d. Ethene

36. Which of these promotes fruit ripening [   ]
    a. Cytokinins   b. Ethene   c. Gibberellins   d. Abscissic acid

37. Gibberellins [   ]
    a. Enhance action of auxins in apical dominance
    b. Encourage bud dormancy
    c. Encourage seed dormancy
    d. Promote root initiation

38. Stomatal opening is promoted by [   ]
    a. Auxins   b. Cytokinins   c. Gibberellins   d. Ethene

39. Stomatal closing is promoted by [   ]
    a. Auxins   b. Cytokinins   c. Gibberellins   d. Abcsissic acid

40. One of the earliest books on Botany, *Historia Plantarum* was written by [   ]
    a. Plato   b. Aristotle   c. Theophrastus   d. Fuchs

41. Who first observed dead cells and published the book *Micrographia*?　　　[　]
    a. Zacarius, Janssen
    b. Robert Hooke
    c. Marcello Malpighi
    d. L'Obel

42. Strassbeger is known for　　　[　]
    a. Discovery of nucleic acids
    b. Descroption of cell division in plants
    c. Proposal of theory of Natural Selection
    d. Proposal of Laws of Inheritance

43. Mutation in plants were discovered and explained by　　　[　]
    a. Gregor John Mendel
    b. MacLeod & MacCarty
    c. Jacob & Monod
    d. Hugo de Vries

44. Hargovind Khorana　　　[　]
    a. Synthesized artificial genes
    b. Crystallised enzyme urese
    c. Described photoperidism
    d. Described light reactions of photosynthesis

45. Study of structure of pollen grains and spores　　　[　]
    a. Embyology
    b. Polynology
    c. Palaeobotany
    d. Pteridology

46. Match the following　　　[　]
    A. Study of flowers　　　i. Agrostology
    B. Study of grasses　　　ii Pomology
    C. Study of fruit yielding plants　　　iii. Anthology
    D. Study of vegetable yielding plants　　　iv. Olericulture

    a. A-iii, B-i, C-ii, D-iv
    b. A-ii, B-i, C-iii, D-iv
    c. A-iv, B-ii, C-i, D-iii
    d. A-i, B-iv, C-iii, D-ii

47. Ethnobotany is　　　[　]
    a. Study of life on other planets and outer space
    b. Study of environmental pollution
    c. Study of relation between tribals and plants
    d. Study of plants that are of religious concern

48. Estimating the age of trees by counting the number of annual rings is called　　　[　]
    a. Dendrochronology
    b. Phytometry
    c. Phytophysiology
    d. Arborimetry

49. Indian Agricultural Research Institute is situated at　　　[　]
    a. New Delhi
    b. Mumbai
    c. Lucknow
    d. Chennai

50. Father of Botany　　　[　]
    a. Theophrastus
    b. Amici
    c. Hooker
    d. Linnaeus

51. What is the correct sequence of units in classifying plants?　　　[　]
    a. Kingdom, Family, Class, Order, Series, Genus & Species
    b. Kingdom, Order, Family, Class, , Series, Genus & Species
    c. Kingdom, Class, Series, Order, Family, Genus & Species
    d. Family, Kingdom, , Class, Series, Order Genus & Species

52. All these families are of dicotyledonous plants except　　　[　]
    a. Liliaceae
    b. Solanaceae
    c. Malvaceae
    d. Fabaceae

53. Match the following　　　[　]
    A. Malvaceae　　　i. Pea Family
    B. Fabaceae　　　ii. Datura Family
    C. Asteraceae　　　iii. Sunflower Family
    D. Solanaceae　　　iv. Hibiscus Family

    a. A-iii, B-i, C-ii, D-iv
    b A-iv, B-i, C-iii, D-ii
    c. A-iv, B-ii, C-i, D-iii
    d. A-i, B-iv, C-iii, D-ii

54. Cryptogams are   [   ]
    a. Flowerless and Seedless spore plants
    b. Flowering plants without ovary and fruit
    c. Flowering plants with ovary but no fruit
    d. Flowering plants with ovary and fruit

55. Phaenerogams are classified into   [   ]
    a. Thallophytes & Bryophytes
    b. Thallophytes, Bryophytes and Pteridophytes
    c. Thallophytes & Pteridophytes
    d. Angiosperms & Gymnosperms

56. Cycas belongs to   [   ]
    a. Bryophyta   b. Pteridophyta   c. Gymnospermae   d. Angiospermae

57. The *Great Bengal Famine* of India (1942) was due to the plant disease   [   ]
    a. Brown leaf spot of rice   b. Tikka disease of ground nut
    c. Blast disease of paddy   d. Grain smut of sorghum

58. The *Great Irish Famine* (Ireland 1845) was due to   [   ]
    a. Red rot of sugarcane   b. Grain smut of sorghum
    c. Potato blight disease   d. Brown leaf spot of rice

59. Pick out the WRONG pair   [   ]
    a. Potato blight disease - *Phytophthora infestans*
    b. Paddy blast – *Pyricularia oryzae*
    c. Red rot of sugarcane – *Sphacelotheca sorghi*
    d. Citrus canker – *Xanthomonas campestris*

60. Which of these is the CORRECT pair?   [   ]
    a. *Agaricus bisporus* – White button mushroom
    b. *Volveriella volvaceae* – Paddy straw mushroom
    c. *Pluerotus sajor caju* – Oyster mushroom
    d. All three pairs are correctly matched

61. In mushroom production spawn is   [   ]
    a. The seed material used for inoculation
    b. The bed of raw material prepared for mushroom cultivation
    c. Each crop of mushroom buttons
    d. The material used to cover the mycelial network so as to encourage fruiting

62. The Cohesion Tension Theory that explains Ascent of Sap, was proposed by   [   ]
    a. F W Went   b. Sumner   c. Dizon   d. M S Swaminathan

63. Sheperd's crook   [   ]
    a. A typical fungal disease of pulses
    b. Typical curling of legume root hair when infected with *Rhizobium*
    c. The arrangement of vascular bundles in Xerophytes
    d. None of these

***

Answer Key and Validation

Botany

1. a	12. a	23. c	34. d	45. b	56. c
2. c	13. d	24. c	35. c	46. a	57. a
3. a	14. c	5. d	36. b	47 c	58. c
4. b	15. a	26. b	37. a	48. a	59. c
5. d	16. b	27. a	38. b	49. a	60. d
6. a	17. c	28. c	39. d	50. a	61. a
7. a	18. c	29. b	40. c	51. c	62. c
8. a	19. c	30. a	41. b	52. a	63. b
9. c	20. c	31. d	42. b	53. b	
10. b	21. c	32. a	43. d	54. a	
11. b	22. a	33. c	44. a	55. d	

59. *Red rot of sugarcane is caused by Colletotrichum falcatum. Sphacelotheca sorghi causes grain smut of sorghum*

18
Zoology

Retrace your subject...
Principles and outlines of Classification of Kingdom Animalia, Histology, Animal Diseases...

Zoology (Gr. *zoion* = animals, *logos* = study) deals with the study of structure, organization and function of animal organism. It is a vast subject with several specializations such as endocrinology, zoogeography, parasitology, helminthology, ichthyology, ornithology, to name few. Knowledge of Zoology is a must for all biologists because animals pose as models to understand fundamental principles of morphology, physiology, psychology and even sociology. Moreover several animals are subjects of experimentation in the labs.

On the practical front, Zoology is necessary for preliminary studies of anatomy, dissection skills and techniques of histology. Day-to-day application of animal science includes in fishery, poultry, piggery, aquaculture, veterinary, medicine, wild life conservation, dietetics, dentistry and medicine. The Micobiologist's interest in Zoology lies in study of zoonoses (diseases transmitted to man through animals) veterinary microbiology, raising antisera, preparation of vaccines, use of transgenic animals in molecular pharming and parasitology.

The present Chapter focuses on outlines of classification of animal kingdom, principles of animal histology, reproduction, respiration and animal diseases. Not many of the exercises may appear directly in competitive exams of Microbiology, but these are useful to the student in acquiring reasonable authority on Life Sciences.

Major Branches Of Zoology	
Endocrinology	Study of endocrine systems and function of hormones in animals
Embryology	Study of development of organism from the stage of fertilization to birth.
Paleoecology	Study of life as it existed in the past
Zoogeography	Study of distribution of animals on the surface of earth
Entomology	Study of insects
Malacology	Study of mollusks
Ichthyology	Study of fishes
Ornithology	Study of birds
Animal husbandry	Cultivation of domesticated animals (cattle)
Piggery	Cultivation of pigs
Poultry	Cultivation of fowl
Aquaculture	Cultivation of fish and other aquatic animals
Apiculture	Cultivation of honeybees
Sericulture	Cultivation of silkworms
Helminthology	Study of parasitic worms (helminthes)
Parasitology	Study of parasites in animals

<table>
<tr><td colspan="6" align="center">Kingdom Animalia</td></tr>
<tr><td rowspan="2" align="center">Subkingdom Parazoa</td><td colspan="5" align="center">Subkingdom: Eumetazoa</td></tr>
<tr><td rowspan="2">Grade: Radiata (or) Diploblastica
Phylum: Cnidaria</td><td colspan="4" align="center">Grade: Bilateria (or) Triploblastica</td></tr>
<tr><td rowspan="3">Phylum:
Porifera</td><td colspan="3" align="center">Div: Protostomia</td><td>Div: Deuterostomia
Phyla:
Echinodermata
Chordata</td></tr>
<tr><td></td><td>Subdiv: Acoelomata</td><td>Subdiv: Pseudocoelomata</td><td>Subdiv: Schizocoelomata</td><td></td></tr>
<tr><td></td><td>Phylum:
Platyhelminthes</td><td>Phylum:
Nematoda</td><td>Phyla:
Annelida
Arthropoda
Mollusca</td><td></td></tr>
</table>

Examples of Some typical animals with common names

Phylum	Zoological name	Common name
Porifera (Sponges)	*Hyalonema*	Glass rope sponge
	Euplectella	Venus flower basket
	Chalina	Dead man's fingers
	Euspongia	Bath sponge
Cnidaria	*Physalia*	Portuguese man of war
	Adamsia	Sea anemone
	Gorgonia	Sea fan
	Pennatula	Sea pen
Platyhelminthes	*Fasciola hepatica*	Liver fluke
	Schistosoma haematobium	Blood fluke
	Taenia solium	Pork tapeworm
	Taenia saginata	Beef tapeworm
	Echinococcus granulosum	Dog tapeworm
Nematoda	*Ascaris*	Common round worm
	Wuchereria bankroftii	Filarial worm
	Ancylostoma duodenale	Hookworm
	Enterobius vermicularis	Pinworm
Annelida	*Neries*	Sand worm
	Aphrodite	Sea mouse
	Hirudinaria	Leech
Arthropoda	*Limulus*	King crab/Hoerseshoe crab
	Heterometru	Scorpion
	Aralea	Spider
	Sarcoptes	Mite

Table Contd...

	Periplaneta	Cockroach
Mollusca	*Pila*	Snail
	Aplysia	Sea-Hare
	Unio	Fresh water Mussel
	Pinctada	Pearl oyster
	Sepia	Cuttle fish
	Loligo	Squid
Echinodermata	*Salmacis*	Sea Urchin
	Echinodiscus	Sand dollar
	Clypeastor	Sea biscuit (Cake urchins)
	Holothuria	Water polyp

Classes Of Phylum Chordata

Chondricthyes	**Osteithyes**	**Amphibia**	**Reptilia**	**Aves**	**Mammalia**
Cartilaginous fish	Bony fish	Amphibians	Reptiles	Birds	Mammals
Sharks, Skates, rays..	Herring	*Rana* (frog) *Bufo* (toad)	*Natrix* (grass snake), *Crocodylus* (crocodile)	*Columba* (Pigeon) *Aquila* (eagle)	*Homo* (Human). *Canis* (Dog)

Animal Tissues

1. Epithelial tissue		2. Connective tissue
Simple epithelium (One cell thick layer)	Compound epithelium (More than one cell thick layer)	Loose connective tissue – Areolar tissue
		Fibrous connective tissue White Fibrous, Yellow elastic
		Adipose tissue – Fat cells dominate
- Squamous - Cuboidal - Columnar - Ciliated - Pseudostratified	- Stratified - Transitional	Dentine (ivory) – Similar to bone, but harder
		Skeletal Cartilage – Hyaline, Yellow elastic, White fibrous Bone - Spongy, Memnrane, Compact
		Blood cell making tissue – RBCs and WBCs

Animal Tissues

3. Muscle tissue	4. Nervous tissue
Striated - Skeletal Smooth - Unstriated Cardiac - Heart	Neuron – Nerve cells
	Nerves – Bundles of nerve fibres in a connective tissue sheath called the **epineurium.** Inward extensions of einerium is perineurium and easch fibre is surrounded by connective tissue called **endoneurium**

Major Animal Diseases	
Bird diseases	Fowlpox, Plague, Avian malaria, Newcastle disease, Avian sarcoma, Avian tuberculosis
Cattle diseases	Bovine viral diarrhea, Bovine haemorrhagic syndrome, Bovine mastitis, Bovine tuberculosis, Bovine brucellosis, Bovine ephemeral fever, Bovine trypanosomiasis, Lumpy skin disease
Sheep Diseases	Blue tongue, Scrapie, Sway back, Pneumonia,
Swine disease	Swine pneumonia, Oedema disease, African swine fever, Classical swine fever
Horse disease	African horse sickness, Equine infectious anaemia, Glanders
Rodent diseases	Murine acquired immunodeficiency syndrome, Monkeypox, Ectromelia,
Dog diseases	Distemper, Canine hepatitis, Canine dysplasia,
Cat disease	Feline acquired immunodeficiency syndrome, Feline panleucopenia, Feline leukemia
Animal Parasitic diseases	Helminthiasis, Protozoan infection
Fish disease	Viral hemorrhagic septicemia, Bacterial body ulcers, Skin and gill flukes, Parasitic white spots, Bacterial gill diseases, Argulkus (louse) infestation

Exercises

1. Which phylum contains the largest number of species of animals? [   ]
   a. Arthropoda   b. Porifera   c. Cnidaria   d. Platyhelminthes

2. The extensive space that separates the animal gut from the body wall is [   ]
   a. Peritoneum   b. Periplasmic space
   c. Coelom   d. Buccal cavity

3. Which of these is NOT found inside the nucleus of a typical animal cell [   ]
   a. Heterochromatin   b. Euchromatin
   c. Centriole   d. Nucleolusy

4. Secretary vesicles arise from [   ]
   a. Endoplasmic reticulum   b. Golgi apparatus
   c. Microvilli   d. Nuclear envelope

5. If cell components are subjected to fractional centrifugation, which is the correct order of organelles settled from bottom to top of the tube? [   ]
   a. Nuclei, Chloroplasts, Mitochondria, Ribosomes.
   b. Nuclei, Mitochondria, Chloroplasts, Ribosomes.
   c. Chloroplasts, Mitochondria, Nuclei, Ribosomes.
   d. Chloroplasts, Nuclei, Mitochondria, Ribosomes.

6. Compound epithelial tissue is classified into [   ]
   a. Squamous & Cuboidal epithelium   b. Stratified & Transitional epithelium
   c. Columnar & Ciliated epithelium   d. Stratified & Peudostratified epithelium

7. In which of these tissues do fat cells dominate? [   ]
   a. Skeletal tissue   b. Adipose tissue   c. Areolar tissue   d. Dentine tissue

8. Areolar tissue is found around [   ]
   a. All organs of the body   b. The lungs
   c. The alimentary canal   d. The skeletal system

9. The walls of arteries and alveoli of lungs are made up of [   ]
   a. adipose tissue   b. Yellow elastic tissue
   c. Cartilage   d. Areolar tissue

10. Harversian canals are found in [   ]
    a. Liver   b. Brain   c. Bones   d. Heart muscle

11. Scwann cells are seen in [   ]
    a. Cartilage   b. Nervous tissue
    c. Loose connective tissue   d. Blood

12. *Nodes of Ranvier* are found in [   ]
    a. Sensory neuron   b. Motor neuron   c. Both a & b   d. Neither a nor b

13. Haustoria are found in [   ]
    a. *Taenia*   b. *Fasciola*   c. *Schistosoma*   d. None of these

14. Nematoblasts [   ]
    a. Adhesive cells along the suckers of a tapeworm's scolex
    b. Filtering cells in the oral region of filter feeding mussel
    c. Stinging cell present along the surface of tentacles
    d. Sucking cells in the mouth parts of a butter fly

15. Bile helps in   [   ]
    a. Stimulation of intestinal wall to secrete acids
    b. Digestion of proteins
    c. Improve absorption of water and salts along the gut wall
    d. Emulsification of fats

16. Lamina propria is found   [   ]
    a. In the innermost layer of the gut   b. Within the liver
    c. Within the heart muscle   d. Within the pancreas

17. Pancreas are characterized by the presence of   [   ]
    a. *Islets of Langerhans*   b. Brush border cells
    c. Auerbach's plexus   d. Meissner's plexus

18. Match the following   [   ]
    A. Antherosclerosis   i. Nervous system disorder
    B. Alzheimer's disease   ii. Cancer
    C. Lymphosarcoma   iii. Cardiovascular disease
    D. Kwashiorkar disease   iv. Protein deficiency

    a. A-iii, B-iv, C-ii, D-I
    b. A-iii, B-i, C-ii, D-iv
    c. A-ii, B-iv, C-i, D-iii
    d. A-i, B-ii, C-iii, D-iv

19. The hormone rennin is secreted in   [   ]
    a. Brain   b. Duodenum   c. Kidney   d. Testis

20. The site of action of Luteinising Hormone (LH)   [   ]
    a. Most tissues   b. Ovary and testis   c. Thyroid gland   d. Mammary glands

21. Loop of Henle is found in   [   ]
    a. Neuron   b. Nephron   c. Osteon   d. Hepatocyte

22. Urine of a fish contains   [   ]
    a. Ammonia   b. Uric acid   c. Skatol   d. Indole

23. Which of these does NOT come under connective tissue?   [   ]
    a. White collagen fibres   b. Yellow elastic fibres
    c. Reticular fibres   d. Pseudostratified epithelial tissue

24. Tough bonds of fibrous connective tissue which holds bones together at the joints   [   ]
    a. White elastic fibres   b. Ligaments
    c. Cartilage   d. Tendons

25. Bone is derived from   [   ]
    a. Ectoderm   b. Mesoderm   c. Endoderm   d. Both a & c

26. The three types of cartilage in mammals   [   ]
    a. Ectodermal, endodermal and mesodermal cartilage
    b. Hard, Semisoft and Soft cartilage
    c. Hyaline, Elastic and Fibrous cartilage
    d. Organ, Tissue and Bone cartilage

27. Ribs and vertebrae are   [   ]
    a. Spongy bones   b. Compact bones   c. Cartilages   d. Hollow bones

28. Chondrology is   [   ]
    a. Study of cartilage
    b. Study of hormones
    c. Study of skeletal system
    d. Study of fluid tissues

29. In a normal blood sample, the number of   [   ]
    a. Neutrophils > Eosinophils > Lymphocytes > Basophils
    b. Neutrophils > Lymphocytes > Eosinophils > Basophils
    c. Lymphocytes > Neutrophils > Basophils >Eosinophils
    d. Neutrophils > Lymphocytes > Basophils > Eosinophils

30. Plasma is   [   ]
    a. (Blood) – (Cells)
    b. (Blood) – (RBCs)
    c. (Blood) – (WBCs)
    d. (Blood) – (Platelets)

31. Serum is   [   ]
    a. (Blood) – (Fibrin)
    b. (Blood) – (Cells)
    c. (Blood) – (Cells + Fibrin)
    d. (Blood) – (Plasma)

32. Which of these are voluntary muscle fibres?   [   ]
    a. Cardiac muscle fibres
    b. Striped muscle fibres
    c. Unstriped muscle fibres
    d. b & c

33. In a neuron, synaptic knobs are present at   [   ]
    a. The tips of dendrites
    b. The tips of axons
    c. a & b
    d. All along the myelin sheath

34. Small, motile, phagocytic neuroglial cells   [   ]
    a. Mesoglia   b. Chondroblasts   c. Adipocytes   d. Microglia

35. Tightly packed parallel bundles of collagen fibres   [   ]
    a. Smooth muscles   b. Striated muscles   c. Tendons   d. Ligaments

36. Sarcoplasm   [   ]
    a. Wall of blood vessels
    b. Transparent covering on the brain
    c. The cytoplasm of a muscle cell
    d. Transparent covering on the heart

37. Match the following   [   ]

    A. Annelida   i. *Pheretima posthuma*
    B. Arthropoda   ii. *Ascaris lumbricoides*
    C. Porifera   iii. *Periplaneta americana*
    D. Nematoda   iv. *Scypha*

    a. A-i, B-iv, C-iii, D-ii
    b. A-iii, B-i, C-ii, D-iv
    c. A-i, B-iii, C-iv, D-ii
    d. A-ii, B-iv, C-iii, D-i

38. Which of these classes does not belong to phylum Annelida   [   ]
    a. Oligochaeta   b. Polychaeta   c. Hexapoda   d. Hirudinea

39. S shaped structures present around the segments of earthworm are called   [   ]
    a. Cilia   b. Seta   c. Chaeta   d. b or c.

40. Gizzard is usually described as   [   ]
    a. The grinding mill
    b. The locomotory muscle
    c. Oesophagal heart
    d. Testes sacs

41. Which of these statement is CORRECT about earthworm?   [   ]
     a. Plasma of the blood is red due to dissolved hemoglobin
     b. Blood corpuscles are colorless and nucleated
     c. Both a & b are correct
     d. Both a & b are incorrect

42. The excretory organ of the earthworm   [   ]
     a. Nephridium   b. Nephron   c. Glomerulus   d. Rectal papillae

43. Which of these is NOT a male reproductive organ?   [   ]
     a. Seminal vesicles   b. Spermatheca   c. Vasa diferentia   d. Prostate glands

44. In Unani Medicine, earthworms are used to treat   [   ]
     a. Gout   b. Certain cancers   c. Conjunctivitis   d. Skin eruptions

45. The cockroach shows *cusorial locomotion*   [   ]
     a. It runs on the ground with the help of legs
     b. It flies over short distances with the help of wings
     c. It crawls by means of muscular contraction and relaxation
     d. a & b

46. Drones are   [   ]
     a. Worker bees which feed the female egg laying bee
     b.   Male bees born in brood chambers from unfertilized eggs
     c. The female egg laying bees
     d. Mulberry silkworms

47. Honey bee   [   ]
     a. *Bombyx mori*   b. *Apis indica*
     c. *Periplaneta Americana*   d. *Cimex hemipterus*

48. *Musca nebula* is   [   ]
     a. Indian house fly   b. Indian bedbug
     c. Indian head louse   d. Lac insect

49. Which of these is NOT found within the *Anapheles* mosquito infected with *Plasmodium*? [   ]
     a. Zygote   b. Ookinet   c. Trophozoite   d. Sporozoite

50. Schizonts are found in   [   ]
     a. Liver cells   b. RBCs   c. Both a & b   d. Neither a nor b

51. Strobila   [   ]
     a. Head region of the tapeworm
     b. Neck region of the tapeworm
     c. Body region of the tapeworm
     d. Muscular structure present on the head region of the tapeworm

52. Each segment of the tapeworm is called   [   ]
     a. Collumella   b. Proglottid   c. Rostellum   d. Acetabula

53. The female *Wuchereria* is shorter than the male worm   [   ]
     a. TRUE   b. FALSE

54. Plant like animals   [   ]
     a. Cnidarians   b. Poriferans   c. Molluscs   d. Chitons

55. Which of these is NOT a Cnidarian? [ ]
 a. *Hydra* b. *Aurelia* c. *Adamsia* d. *Fasciola*

56. Sponges are mostly marine except [ ]
 a. *Scypha* b. *Hyalonema* c. *Spongilla* d. *Leucosolinia*

57. Portuguese man of war [ ]
 a. *Obelia* b. *Aurelia* c. *Physalia* d. Rhizostoma

58. Which of these is NOT a flat worm? [ ]
 a. *Schistosoma* b. *Palaemon* c. *Echinococcus* d. *Taenia*

59. Horshoe crab [ ]
 a. *Heterometru* b. *Triathrus* c. *Limulus* d. *Scolopendra*

60. Pick out the WRONG pair [ ]
 a. *Pediculus*-Head louse b. *Palaemon*-Fresh water prawn
 c. *Ancylostoma*-Pinworm d. *Fasciola*-Liver fluke

61. Bioluminiscent animals [ ]
 a. Emit light
 b. Emit light and generate heat (infrared rays)
 c. Generate heat
 d. Can emit light or heat depending on the organisms requirement

62. Animals which cannot maintain constant body temp [ ]
 a. Homeothermic b. Aestivating animals
 c. Poikilothermic d. Hibernating animals

63. Summer sleep [ ]
 a. Hibernation b. Aestivation c. Homeostasis d. Thermal brood

64. Allen's Rule & Bergmann's Rule are related to [ ]
 a. Effects of temp on animals b. Effects of pressure on animals
 c. Effects of humidity on animals d. Effects of light on animals

65. Pick out the FALSE statement [ ]
 a. Pyramid of biomass in a pond is inverted
 b. Pyramid of numbers in a lake ecosystem is upright
 c. Pyramid of energy is inverted in a tree ecosystem
 d. Pyramid of numbers is inverted in a food chain involving parasites

***

Answer Key and Validation

Zoology

1. a	12. c	23. d	34. d	45. d	56. c
2. c	13. d	24. b	35. c	46. b	57. c
3. c	14. c	25. b	36. c	47. b	58. b
4. b	15. d	26. c	37. c	48. a	59. c
5. a	16. a	27. a	38. c	49. c	60. c
6. b	17. a	28. a	39. d	50. c	61. a
7. b	18. b	29. b	40. a	51. c	62. c
8. a	19. c	30. a	41. c	52. b	63. b
9. b	20. b	31. c	42. a	53. b	64. a
10. c	21. b	32. b	43. c	54. b	65. c
11. b	22. a	33. b	44. a	55. d	

13. *Haustoria are found in the parasitic plant Cuscuta (Dodder)*

60. * *Common name for Ancylostoma is Hookworm. Pinworm is Enterobius*

65 *Pyramids of energy are always upright*

19
Biochemistry

Retrace your subject...

Principles of Biochemistry: Water, Proteins, Carbohydrates, Lipids, Enzymes.

Biochemistry constitutes the study of structural and functional components of living organisms. It is a focused subdivision of Science that forms a **bridge between Biology and Chemistry**. Biochemistry scrutinizes life processes such as respiration and synthesis in terms of energy economics and interrelationships within the living system.

The **scope** for the biochemist is varied and multidirectional. The in-depth knowledge of Biochemistry has lead to the understanding of **diseases**, their diagnosis, treatment and even prevention to a large extent. The influence of Biochemistry on industrial **fermentative processes** and **food manufacturing units** should not be underestimated, as often these industries are monitored purely on the basis of the biochemical changes that occur during product formation. The Canon of Biochemistry is also intricately utilized in **pharmaceutical industries** to safeguard the quality of drugs and vaccines being manufactured. On the **fundamental research** front, Biochemistry constitutes the backbone of discovering the contents of complex innate matter by the help of standard and refined procedures such as electrophoresis, various chromatography and centrifugation techniques. **Modern Biochemistry** is a more recent approach that has gained rapid recognition and compliments owing to its vast applicability to advanced sciences. Several disciplines such as Molecular Biology, Enzymology, Physiology, Endocrinology and Clinical Biochemistry are said to have established on the groundwork of biochemical values. The authority of Biochemistry on **Microbiology** results in understanding the structural and functional features of microorganisms so that they can be cultivated in a better manner for enhanced utility. Undoubtedly, the microbiologist must know the fundamental principles *of Biochemistry* which begin with the study of proteins, Carbohydrates, fats, enzymes, vitamins and nucleic acids.

It is beyond the scope of this book to present the theoretical data under each heading. However, this chapter includes an extensive coverage of exercises based on the fundamental facts of **water, proteins, carbohydrates and enzymes**. The learner is requested to refer to standard textbooks of Biochemistry for theoretical hardcore facts. Nevertheless, the answers for the following exercises are provided with due explanations wherever felt necessary so as to clear any doubts that may arise. The knowledge of Biochemistry to this extent as is given in this chapter is more than sufficient for a competitive exam of Microbiology.

Note: **The exercises related to Metabolism, Nucleic acids (DNA, RNA, Mutations, etc) and Vitamins are not included in this Chapter as these will appear in chapters 7, 8 & 20 respectively.**

Exercises

1. The term Biochemistry was first introduced by _______   [   ]
   a. Carl Neuberg　　　b. Theophrastus　　　c. Karl Scheele　　　d. Emil Fisher

2. Match the columns   [   ]
   A. Justus Von Liebig　　　　　　i. Lab synthesis of urea
   B. Hermann Emil Fisher　　　　　ii. Father of agricultural chemistry
   C. Fredrich Wohler　　　　　　　iii. Composition of air
   D. Antoine Lavoisier　　　　　　iv. Structural Biochemistry

   a. A-iii, B-i, C-iv, D-ii　　　　　　b. A-ii, B-iv, C-i, D-iii
   c. A-iv, B-iii, C-ii, D-i　　　　　　d. A-i, B-ii, C-iv, D-iii

3. Pick out the odd biomolecule   [   ]
   a. Lactose　　　　　b. Maltose　　　　　c. Sucrose　　　　　d. Galactose

4. Which of these is a polysaccharide?   [   ]
   a. Pectin　　　　　b. Inulin　　　　　c. Chitin　　　　　d. All

5. Pick out the FALSE statement about pure water   [   ]
   a. Melting point = 0^0C　　　　　　b. Boiling point = 100^0C
   c. Heat capacity = $100 cal/g$　　　　d. Surface tension = 72.8

6. Freezing point for sea water   [   ]
   a. 0^0C
   b. 2^0C
   c. -5^0C
   d. no definite freezing point can be designated for sea water

7. Which of these properties of water is due to hydrogen bonding   [   ]
   a. resist change in temp　　　　　b. resist change of state
   c. is cohesive　　　　　　　　　　d all

8. Ice floats on water because   [   ]
   a. Ice is less dense than liquid water
   b. Ice exhibits a different polarity
   c. Ice is at a lower temp than liquid water
   d. Water exhibits pressure over ice.

9. The H-O-H bond angle in water is 104.5^0 which is 5^0 less than the bond angle of a perfect
   tetrahedron (109.5^0) because of   [   ]
   a. Van der Waal's forces pull H atoms closer
   b. Non bonding orbitals of oxygen atoms slightly compress the orbitals shared by H
   c. Equal sharing of electrons between H and O
   d. Electrostatic attraction of neighboring water molecules

10. What is FALSE regarding the structure of water molecule   [   ]
    a. O is more electronegative than H
    b. Two electric dipoles exist in the water molecule
    c. The O atom bears a partial negative charge (δ^-) and the H atom bears a partial
       positive charge (δ^+)
    d. Electrostatic attraction between H atoms of different water molecules constitute
       the hydrogen bonds

11. Water easily dissolves biomolecules which are   [   ]
    a. Polar   b. Non-polar   c. Amphipathic   d. a & c

12. Waxes do not dissolve in water because   [   ]
    a. Water is polar solvent and waxes are non-polar
    b. Water is non-polar solvent and waxes are polar
    c. Water is non-polar solvent and waxes are non-polar
    d. Water is polar solvent and waxes are polar

13. The correct order of increasing bond dissociation energy required is   [   ]
    a. C-H, Non covalent bonds, C=O, C≡C
    b. C≡C, C=O, C-H, Non covalent bonds
    c. C-H, C=O, C≡C, Non covalent bonds
    d. Non covalent bonds, C-H, C=O, C≡C

14. What is common about: H bonds, Ionic bonds, Hydrophobic bonds and Vander Waal's forces.   [   ]
    a. All are repulsive interaction
    b. All are non-covalent interactions
    c. All are covalent interactions
    d. All are strong interactions of attraction

15. The most abundant intracellular macromolecules   [   ]
    a. Carbohydrates   b. Vitamins   c. Proteins   d. Enzymes

16. None of the proteins of *E.coli* is identical with any of the human proteins   [   ]
    a. True   b. False

17. Insulin   [   ]
    a. A sulphur containing protein   b. A polysaccharide
    c. A phospholipids   d. A lipoprotein

18. In any amino acid, the carboxyl group constitutes the ____   [   ]
    a. First carbon atom   b. Second carbon atom
    c. Last carbon atom   d. Any carbon atom linked to the amino group

19. In an amino acid, the α carbon is the_________   [   ]
    a. First carbon atom
    b. Last carbon atom
    c. Second carbon atom to which the amino group is attached.
    d. Any carbon to which the carboxyl group is attached.

20. Pick out the FALSE statement   [   ]
    a. The α carbon atom in almost all amino acids is asymmetric
    b. Generally amino acids occur in two optically active forms
    c. Almost all amino acids found in proteins belong to L series
    d. Those amino acids having amino group to the left are D forms

21. Which of these amino acids have two asymmetric carbon atoms?   [   ]
    a. Threonine   b. Isoleucine   c. Both   d. None of these

22. Match the following   [   ]

    A. Tyrosine   i.   Hexagonal plates
    B. Cystine   ii. Slender needles
    C. Alanine   iii. Bitter
    D. Arginine   iv. Sweet

    a. A-ii, B-i, C-iv, D-iii   b. A-ii, B-iii, C-iv, D-i
    c. A-iv, B-iii, C-ii, D-i   d. A-i, B-ii, C-iv, D-iii

23. Pick out the FALSE statement   [   ]
    a. Human serum albumin with 585amino acids, contains only one tryptophan
    b. All pulses characteristically lack sulphur containing amino acid methionine
    c. Amino acids are easily soluble in benzene and ether
    d. Cereals characteristically lack lysine.

24. Which of these is NOT an aromatic amino acid?   [   ]
    a. Tyrosine   b. Proline   c. Tryptophan   d. Phenylalanine

25. What is FALSE about tryptophan?   [   ]
    a. It is the most complex amino acids found in proteins
    b. It is heterocyclic and derivative of indole
    c. It is the only amino acid which is completely destroyed by acid hydrolysis.
    d. It was the first amino acid to be discovered.

26. Synthesis of proteins from amino acids is   [   ]
    a. A multiple decarboxylation process   b. A multiple dehydrogenation process
    c. A multiple dehydration process   d. A multiple deamination process

27. Which is the correct order of bonds arranged in terms of decreasing bond strengths?   [   ]

    a. Covalent bonds, hydrogen bonds & ionic bonds
    b. Covalent bonds, ionic bonds & hydrogen bonds.
    c. Hydrogen bonds, covalent bonds & ionic bonds.
    d. Covalent bonds, ionic bonds & hydrogen bonds

28. The secondary structure of a protein is mostly   [   ]
    a. Helical   b. Globular   c. Linear   d. Supercoiled

29. What is FALSE about collagen?   [   ]
    a. The basic collagen monomer is a double helix.
    b. It is the most abundant protein of mammals
    c. Nearly every third residue of collagen is glycine
    d. It contains 4-hydroxyproline, which is rarely found elsewhere

30. The spatial arrangement of amino acids that are far apart in linear polypeptide
    chain is brought about by   [   ]
    a. Primary structure of the protein
    b. Secondary structure of the protein
    c. Tertiary structure of the protein.
    d. Quarternary structure of the protein

31. Interaction on one site of the protein affecting another distantly located site on
    the same molecule is called   [   ]
    a. Isosteric interaction   b. Allosteric interaction
    c. Asteric interaction   d. Autosteric interaction.

32. Which of these is a fibrillar protein?   [   ]
    a. Collagen   b. Keratin   c. Elastin   d. All.

33. Which is the Wrongly matched pair?   [   ]
    a. Conjugated proteins – Lipoproteins and phosphoproteins
    b. Glycoproteins – serum globulins, egg albumin
    c. Mucoproteins – Hemoglobin
    d. Chromoproteins – cytochromes and flavoproteins

34. Which of these is a Correct pair?   [   ]
    a. Transport protein – Hemoglobin   b. Storage protein – Collagen
    c. Structural protein – Casein   d. Contractile protein – Keratin

35. Which of these is a defence protein   [   ]
    a. Immunoglobulin   b. Thrombin
    c. Fibrinogen   d. All

36. Most proteins, upon hydrolysis with conc HCl at $100\text{-}110^0$C for 6h to 20h   [   ]
    a. Yeild amino acids in the form of their hydrochlorides.
    b. Undergo intramolecular dehydration
    c. Undergo deamidation
    d. Yeild amino acids in their natural forms

37. Which of the following statement is Wrong?   [   ]
    a. Amino acids react with alkalies to form salts
    b. Amino acids react with alcohols to form volatile esters.
    c. Amino acids react with formaldehyde to form highly soluble complexes
    d. Amino acids react with benzaldehyde to form Schiff's bases.

38. Which of these amino acids tests does NOT produce a red color?   [   ]
    a. Ninhydrin test   b. Sakaguchy test   c. Pauly test   d. Nitroprusside test

39. Plants are comparatively many times richer in carbohydrates than animals   [   ]
    a. Always True
    b. Always False
    c. True in some cases

40. Which of these biomolecules are richer in animals than in plant   [   ]
    a. Lipids   b. Proteins   c. Both a and b   d. Water

41. Monosaccharides are   [   ]
    a. Simple sugars which possess free aldehyde or ketone group
    b. Simple sugars which possess free aldehyde and ketone group
    c. Simple sugars which possess free aldehyde group
    d. Simple sugars which possess free keto group

42. The most abundant monosaccharide in nature   [   ]
    a. 1 L-Glucose   b. D-Glucose   c. L-Fructose   d. Sucrose

43. Which of these is a disaccharide   [   ]
    a. Sucrose   b. Maltose   c. Lactose   d. All

44. Trehalose is a   [   ]
    a. Homo reducing disaccharide
    b. Homo non-reducing disaccharide
    c. Hetero non-reducing disaccharide
    d. Hetero reducing disaccharide

45. An oligosaccharide is a compound sugar which on hydrolysis yields   [   ]
    a. 2 to 10 molecules of the same monosachharide
    b. 2 to 10 molecules of different monosachharides
    c. 2 to 10 molecules of the same or different disaccharides
    d. 2 to 10 molecules of the same or different monosachharides

46. The general formula for disaccharides is   [   ]
    a. $C_n(H_2O)_n$ b. $C_n(H_2O)_{n-1}$ c. $C_n(H_2O)_{n-2}$ d. $C_n(H_2O)_{n-3}$

47. Which of these is a heteropolysachharide?   [   ]
    a. Hyaluronic acid   b. Inulin   c. Pectin   d. Chitin

48. Which is the wrong pair?   [   ]
    a. Threose – Tetrose   b. Glycerose – Triose
    c. Arabinose – Pentose   d. Lyxose - Hexose

49. Glycosides are formed when sugars react with   [   ]
    a. Alcohols   b. Aldehydes   c. Ketones   d. Water

50. Reducing sugars   [   ]
    a. Carbohydrates with free aldehyde/free ketone group
    b. Exhibit mutarotation
    c. Form osazones with phenyl hydrazine
    d. All of these

51. Which of these is NOT a reducing sugar?   [   ]
    a. Lactose   b. Maltose   c. Fructose   d. Sucrose

52. The Barfoed's reagent is employed to distinguish   [   ]
    a. Sugars from proteins
    b. Monosaccharides from Disaccharides
    c. Reducing sugars from non reducing sugars
    d. Fermentation reactions

53. Sucrose is found in   [   ]
    a. Pineapple   b. Carrot   c. Maple   d. All

54. Upon hydrolysis sucrose yields   [   ]
    a. Glucose + Glucose   b. Glucose + Fructose
    c. Fructose + Fructose   d. Glucose + Galactose

55. The sugar with a higher *sweetness index* than sucrose   [   ]
    a. Galactose   b. Raffinose   c. D-Fructose   d. Glycol

56. Lactose is made up of   [   ]
    a. Glucose + Glucos   b. Glucose + Fructose
    c. Fructose + Fructose   d. Glucose + Galactose

57. Polysaccharides are   [   ]
    a. Polymeric anhydrides of simple sugars
    b. Polymeric hydrides of simple sugars
    c. Polymeric carboxylates of simple sugars
    d. Polymeric decarboxylates of simple sugars

58. At room temperatures, starch is insoluble in   [   ]
    a. Water   b. Ether   c. Alcohol   d. All

59. Xylan is the most common representative of [ ]
    a. Homopolysaccharide
    b. Heteropolysaccharide
    c. Hemicellulose
    d. Saccharoprotein

60. The heteropolysaccharide comprising of D-glucuronic acid + N-acetyl,

    D-glucosamine [ ]
    a. Chitin
    b. Hyalurnic acid
    c. Chondritin
    d. Keratin

61. Which of these yields only fructose units on complete hydrolysis? [ ]
    a. Starch
    b. Cellulose
    c. Pectin
    d. inulin

62. Fats combined with proteins (lipoproteins) are important components of [ ]
    a. Cell membranes
    b. Mitochondria
    c. Both a & b
    d. Ribosomes

63. Chemically, fats are defined as [ ]

    a. Esters of glycerol and fatty acids
    b. Ethers of glycerol and fatty acids

    c. a or b
    d. None of these

64. The general formula for saturated fatty acids is [ ]

    a. $C_nH_{2n}COOH$
    b. $C_nH_{2n+1}COOH$
    c. $C_nH_{2n-1}COOH$
    d. $C_nH_{2n-2}COOH$

65. Wool fat is [ ]

    a. Palmitic acid
    b. Capric acid
    c. Lauric acid
    d. Cerotic acid

66. Natural source of butyric acid [ ]

    a. Coconut oil
    b. Palm oil
    c. Butter
    d. Groundnut oil

67. Pick out the Wrong pair [ ]

    a. Monoethenoid acid (one double bond) – Oleic acid

    b. Diethenoid acid (two double bond) – Linoleic acid

    c. Triethenoid acid (three double bond) – Stearic acid

    d. Tetraethenoid (four double bond) – Arachidonic acid

68. Match the pairs [ ]

    A. Octanoic, (C_8) saturated   i . Caprylic acid
    B. Tetradecanoic (C_{14}), Saturated   ii. Myristic acid
    C. Hexadecanoic (C_{16}), Saturated   iii. Palmitic acid
    D. `Eicosanoic (C_{20}), Saturated   iv. Arachidic acid

    a. All the pairs are correctly matched
    b. The pairs are exactly reversely matched from bottom to top
    c. Pairs in B and C are interchanged
    d. Pairs in A and D are interchanged

69. Pick out the odd fatty acid [ ]
    a. Vaccenic Acid
    b. Arachidonic acid
    c. Lauric acid
    d. Oleic acid

70. Which of these is a component of human fat? [ ]

    a. Butyric acid
    b. Lauric acid
    c. Oleic acid
    d. Linolenic acid

71. Soybean oil is rich in [ ]

    a. Linoleic acid
    b. Linolenic acid
    c. Caproic acid
    d. Stearic acid

72. Pick out the Wrong statement   [   ]
    a. Animal fat is richer in saturated fatty acids, while plant fat is richer in unsaturated fatty acids.
    b. Animal fat is solid at room temp whereas plant fat is liquid at room temp
    c. Oxidative rancidity is rarely observed in animal fat while it is frequent in plant fat
    d. Animal fat is stored in liver and bone marrow while plant fat is stored in Seeds and fruits.

73. The most abundant membrane lipids   [   ]

    a. Phospholipids   b. Glycolipids   c. Steroids   d. Sulfolipids

74. Arrange in increasing order of cholesterol content   [   ]
    a. Brain, Milk, Chicken, Egg yolk
    b. Milk, Egg yolk, Brain, Chicken
    c. Milk, Chicken, Egg yolk, Brain
    d. Chicken, Milk, Egg yolk, Brain

75. Which of these has highest cholesterol content?   [   ]

    a. Butter   b. Cheese   c. Cream   d. Condensed milk

76. Ergosterol has been found to be present in   [   ]

    a. Algae   b. Yeast   c. Pork   d. Cow milk

77. Which of these are considered derived lipids?   [   ]
    a. Steroids
    b. Steroids and Terpenes
    c. Steroids, Terpenes & Carotenoids
    d. Steroids, Terpenes, Carotenoids & Sphingolipids

78. Lecithins, Cephalins and sphingomyelins are   [   ]

    a. Phospholipids   b. Glycolipids   c. Carotenoides   d. Homolipids

79. Lipids when treated with lipases yield   [   ]
    a. Fatty acids   b. Glycerol
    c. Fatty acids and glycerol   d. None of these

80. Saponification is   [   ]
    a. Hydrogenation of fats by alkali
    b. Hydrolysis of fats by alkali
    c. Dehydration of fats by alkali
    d. Oxidation of fats by alkali

81. Antioxidants in vegetable oils   [   ]
    a. Prevent oxidative rancidity
    b. Enhance oxidative rancidity
    c. Prevent dehydration reactions
    d. Enhance dehydration reactions

82. A measure of degree of unsaturation of fatty acids in the fat is given by   [   ]
    a. Acid value   b. Saponification number
    c. Iodine value (Koettstorfer number)   d. Reichert-Meissl number

83. Which of these measures the quantity of short chain fatty acids in the fat molecule   [   ]
    a. Acetyl value   b. Saponification number
    c. Iodine value (Koettstorfer number)   d. Reichert-Meissl number

84. Which of these is NOT a health friendly fat?   [   ]
    a. Coconut oil   b. Groundnut oil   c. Sunflower oil   d. Safflower oil

85. The advantage of reactions brought about by biocatalysts (enzymes) over
    chemically catalysed reactions   [   ]
    a. Higher reaction rates
    b. Milder reaction conditions (temp/pressure/pH)
    c. Greater reaction specificity
    d. All

86. The term enzyme was coined in 1878 by   [   ]
    a. Fredrich Wilhelm Kühne   b. Edward Buchner
    c. John Northrop   d. Moses Kunitz

87. Emil Fisher is best known for   [   ]
    a. Cell free fermentation reaction   b. Enzymatic action of pepsin
    c. Lock-and-key hypothesis   d. 'Vital force' concept.

88. Identify the TRUE statement(s)   [   ]
    a. Enzymes are absolutely stereospecific
    b. Enzymes vary considerably in their degree of geometric specificity
    c. both a and b are true
    d. The amount of enzyme in the cell is independent of its rate of synthesis and rate of degradation

89. Enzymes need cofactors to catalyse   [   ]
    a. Oxidation-reduction reactions   b. acid-base reactions
    c. Both   d. None

90. An apoenzyme is   [   ]
    a. (Active holoenzyme) – (Protein component)
    b. (Active holoenzyme) – (Non-protein component)
    c. (Active holoenzyme) + (Protein component)
    d. (Active holoenzyme) + (Non-protein component)

91. A prosthetic group   [   ]
    a. The cofactors which are strongly associated with the protein part of the enzyme
        through covalent bonds
    b. The cofactors which are transiently associated with the given enzyme molecule.
    c. The chemically changed coenzyme after participation in the reaction
    d. The coenzyme molecule before participation in the reaction.

92. Coenzyme A mediates   [   ]
    a. Carboxylation reactions   b. Amino group transfer reactions
    c. Acyl transfer reactions   d. Alkylation reactions

93. Michaelis-Menten equation   [   ]
    a. $v_0 = V_{max}[S] / \{K_M + [S]\}$ b. $v_0 = \{K_M + [S]\} / V_{max}[S]$
    c. $V_{max}[S] = v_0 / \{K_M + [S]\}$ d. $V_{max}[S] = [S] / \{K_M + [S]\}$

94. A competitive inhibitor   [   ]
    a. Competes directly with the normal substrate for an enzyme binding site.
    b. Competes by changing substrate configuration so that it can no longer be bound by the
        enzyme
    c. Competes by increasing the product conc levels at the site of the reaction
    d. Competes by creating unfavorable reaction conditions in terms of temp/pH/pressure
        so that the enzyme cannot bind to the substrate.

95. When the inhibitor binds directly to the enzyme-substrate complex instead of
binding to the free enzyme, it is [   ]
    a. Competitive inhibitor
    b. Uncompetitive inhibitor.
    c. Mixed inhibitor
    d. Bound inhibitor

96. What is FALSE about the enzyme lysozyme? [   ]
    a. Found in hen egg white
    b. Hydrolyse NAG-NAM linkage in bacterial cell walls.
    c. Hydrolyses the poly NAG components of fungal cell wall chitin
    d. It is a highly complex molecule with several interlinked polypeptide chains

97. Zymogens are [   ]
    a. Large inactive precursors of proteolytic enzymes
    b. Active protein subunits of large enzyme molecules
    c. Inactive parts of complex enzyme molecules
    d. A common term for all saccharolytic enzymes

98. Which of these amino acids tests produces blue color? [   ]

    a. Folin's test   b. Ehrlich test   c. Both   d. None

Answer Key and Validation

Biochemistry

1. a	18. a	35. d	52. b	69. c	86. a
2. b	19. c	36. a	53. d	70. c	87. c
3. d	20. d	37. c	54. b	71. a	88. c
4. d	21. c	38. a	55. c	72. d	89. a
5. c	22. a	39. a	56. d	73. a	90. b
6. d	23. c	40. c	57. a	74. c	91. a
7. d	24. b	41. a	58. d	75. a	92. c
8. a	25. d	42. b	59. c	76. b	93. a
9. b	26. c	43. d	60. b	77. d	94. a
10. d	27. b	44. a	61. d	78. a	95. b
11. a	28. a	45. d	62. c	79. c	96. d
12. a	29. a	46. b	63. a	80. b	97. a
13. d	30. c	47. a	64. b	81. a	98. c
14. b	31. b	48. d	65. d	82. c	
15. c	32. d	49. a	66. c	83. d	
16. a	33. c	50. d	67. c	84. a	
17. a	34. a	51. d	68. a	85. d	

23. *Amino acids are soluble in polar solvents such as water and ethanol, but insoluble in non-polar solvents such as benzene and ether*

25. *The first amino acid to be discovered was Asparagine in 1806, and was first isolated from the plant Asparagus.*

26. *The C atom of carboxyl group of one amino acid is linked to the nitrogen atom of the amino group of the adjacent amino acid to form a peptide bond, that involves removal of H_2O.*

29. *The basic collagen monomer is a triple helix*

48. *Lyxose is a pentose sugar.*

67. *Linolenic acid is the representative of trerthanoid acid. Stearic acid is a saturated fatty acid with*

69. *18 C atoms Lauric acid is a saturated fatty acid whereas the others are unsaturated fatty acids.*

72. *Oxidative rancidity is rarely observed in plant fat while it is frequent in animal fat*

77. *The hydolysis products of simple and compiound lipids are called derived lipids. Stroids, Terpenes and Carotenoids are derived lipids. Sphingolipids are a type of compound lipids in which sphingosine is the main alcohol component.*

84. *Coconut oil is saturated, and hence is not considered health friendly.*

96. *Lysoszyme is a small protein with single polypeptide chain of just 129 amino acid residues.*

20

Life Sciences

Retrace your subject…

Forms of life, Modes of nutrition, micro & macro nutrients, vitamins, modes of reproduction, eutrophication, evolution,…

Life Sciences condenses the core matter of Biology in the light of fundamental and applied aspects. Over the past decade, modern sciences have assumed an almost alternative position to linear science. Microbiology, Biochemistry and Biotechnology are now reputable and focused subjects in the scientific pitch. As a consequence, many universal facts of Biology in relation to social, applied and ethical aspects have been shelved from the minds of contemporary students. **Hence there is a need to educate students of biological sciences with fundamental principles of life, whether or not they form the crux of their specialized subject**. All such specifics can only be assorted under the common umbrella of Life Sciences.

Life sciences is the complex subject of the living organisms, including their structure, function, growth, origin, evolution, and distribution. It is a perfect amalgamation of all biological sciences, particularly Biochemistry, Biomedicine, Biotechnology, Cytology, Genetics, Marine biology, Microbiology, Molecular Biology, Botany and Zoology. Life science explores the techniques, procedures and equipment aspects of biology. It even elucidates the ultimate physiochemical nature of organisms under one headline.

This Chapter covers exercises on those biological topics that may not find appropriate place in any other Chapters offered in this book. Many of these exercises may not be of direct use to the graduate who is attempting a competitive test on Microbiology. However, knowledge of such principles gives the requisite proximity to Biology, and is supportive in attempting certain tricky and analytical exercises with no difficulty.

It is out of the scope of this book to provide theoretical notes on all the aspects of Life Sciences. Nonetheless, the reader will find the explanations to the answers wherever necessary so as to elucidate uncertainties.

Exercises

1. Which of these functions of water occurs both in plants and animals?   [   ]
   a. Translocation
   b. Osmosis and turgidity
   c. Reagent in hydrolysis
   d. Lubricaton

2. Synthesis of NAD and NADP requires   [   ]
   a. Pentose sugars
   b. Hexose sugars
   c. Both a & b
   d. Neither a nor b

3. Which of these is characteristic of plants but not of animals?   [   ]
   a. Plasmodesmata
   b. Plasmodesmata and cll wall
   c. Plasmodesmata, cell wall and lamella
   d. None

4. Tonoplast   [   ]
   a. Small unidentifiable structures in living cells
   b. Membrane surrounding vacuoles
   c. Cementing layer between two cells
   d. Minute opening that connects cytoplasm of neighboring cells.

5. The purpose of tissue in higher forms of life   [   ]
   a. To maintain continuity of life
   b. To offer division of labour
   c. To enhance mechanism of speciation
   d. To offer homostatis

6. Functional unit of a specialized organ   [   ]
   a. Tissue
   b. System
   c. Cell
   d. Entire organism

7. Which of these is a macronutrient for life   [   ]
   a. Zn
   b. P
   c. Fe
   d. Mg^{2+}

8. Which of these is not a macronutrient   [   ]
   a. Mg^{2+}
   b. SO_4^{-2}
   c. Ca^{2+}
   d. Mo

9. Which of these associated with Vit B_{12} [   ]
   a. Cu
   b. Co
   c. Ca^{2+}
   d. Cl^-

10. Fe does not play a significant role in   [   ]
    a. Respiration
    b. Photosynthesis
    c. Malanin production
    d. Chlorophyll synthesis

11. Match the column   [   ]
    A. K^+
    B. SO_4^{2-}
    C. Cl^-
    D. Mg^{2+}

    (i) Coenzyme A
    (ii) Cofactor
    (iii) Membrane functions
    (iv) Gastric juice

    a. A-iv, B-i, C-iii, D-ii
    b. A-ii, B-i, C-iii, D-iv
    c. A-iii, B-i, C-iv, D-ii
    d. A-i, B-ii, C-iv, D-iii

12. Which is a wrong pair   [   ]
    a. Iodine – Thyroxime
    b. Fluorine – Component of tooth enamel
    c. Manganese – Bone development
    d. Copper – RBC development

13. Which of these require Boron for normal cell division   [   ]
    a. Plants   b. Animals   c. Protozoa   d. Bacteria

14. Saprophytes show   [   ]
    a. Holozoic nutrition   b. Predation
    c. Extracellular digestion   d. Intracellular digestion

15. Nematoblasts are used in   [   ]
    a. Excretion   b. Ingestion   c. Digestion   d. All

16. Which of these shows biting & chewing moth parts   [   ]
    a. Hydra   b. Earthworm   c. Locust   d. Butterfly

17. The mouth parts of Aphids (greenfly) are   [   ]
    a. Fluid feeding type   b. Piercing & sucking type
    c. Detritus feeding   d. Biting & chewing type

18. In humans, completion of digestion and absorption of food is accomplished by   [   ]
    a. Duodenum   b. Ileum   c. Colon   d. Caecum

19. Stimulation of salivation is supported by   [   ]
    a. Eye receptors   b. Olfactory receptors
    c. Both   d. None

20. Pushing of food through the gut by muscularis externa   [   ]
    a. Pinocytosis   b. Peristalsis   c. Pericycle   d. Probosis

21. Composition of a balanced diet varies with   [   ]
    a. Age and activity   b. Body size
    c. Temp of the external environment   d. All

22. What is FALSE about dietary fibre   [   ]
    a. Complex mixture of indigestible compounds
    b. Helps to reduce blood cholesterol and bowel cancer
    c. Helps to absorb vitamins
    d. Consists of polysaccharides such as cellulose

23. Fat soluble vitamin   [   ]
    a. B_{12} b. C   c. K   d. H

24. Water soluble vitamin   [   ]
    a. D   b. A   c. B_2 d. E

25. Which of these is tocophrol   [   ]
    a. Vit E   b. Vit B_3 c. Vit A   d. Vit H

26. Pellagra is th deficiency disease of   [   ]
    a. Vit B_3 b. Vit B_1 c. Vit D   d. Vit C

27. Match the columns   [   ]
    A. Retenol   i. Vit A
    B. Calciferol   ii. Vit M
    C. Phylloquinone   iii. Vit D
    D. Folic acid   iv. Vit K

    a. A-i, B-iii, C-ii, D-iv
    b. A-i, B-iii, C-iv, D-ii
    c. A-iii, B-iv, C-i, D-ii
    d. A-ii, B-i, C-iii, D-iv

28. Bacteria in intestine synthesise   [   ]
    a. Vit D   b. Vit B_2 c. Vit K   d. Vit E

29. A lake is said to be dystrophic when   [   ]
    a. It has bo nutrients   b. It is very poor in nutrient content
    c. It is very rich in nutrients   d. None of these

30. Discharge of phosphates into lakes from sewage works   [   ]
    a. Causes oligotrophism   b. Causes eutrophism
    c. Does not affect the lake   d. Decreases phytoplanktons

31. Which of these can cause eutrophication   [   ]
    a. Use of nitrogen fertilizers   b. Use of biopesticides
    c. Use of chemical pesticides   d. All of these

32. Eutrophication is   [   ]
    a. The process of nutrient enrichment in water bodies
    b. The process of nutrient depletion in water bodies
    c. The process of progress in aquatic life in water bodies
    d. The process of decrease in aquatic life in water bodies

33. Dissolved Oxygen (DO) of a water body   [   ]
    a. Increases with increase in organic matter
    b. Decreases with increase in organic matter
    c. Is unaffected with increase in organic matter

34. What is FALSE about eutrophication   [   ]
    a. Decrease in plant algal and animal biomass
    b. Anoxic conditions may develop
    c. Turbidity of the lake incrases
    d. Shortens the life span of the lake

35. Desertification is caused due to   [   ]
    a. Overgrazing by live stock   b. Deforestation
    c. Over cultivation   d. All of these

36. Horizontally growing underground stem that serves for vegetative reproduction is called   [   ]
    a. Stolon   b. Rhizome   c. Rhizoid   d. Collumella

37. Which of these shows swollen tap roots?   [   ]
    a. *Daucus* (carrot)   b. *Brassica rapa* (Turnip)
    c. *Raphanus satirus* (Radish)   d. All

38. In human reproductive system, cowpers glands are seen in   [   ]
    a. Male   b. Female   c. Both   d. None

39. The physiological and psychological sensations associated with sexual climax in males and females is called   [   ]
    a. Capacitation   b. Copulation   c. Orgasm   d. Ejaculation

40. *Dahlia* is characterized by   [   ]
    a. Stem tubers   b. Root tubers   c. Fleshy leaves   d. Dehiscent fruits

41. Stalk of a flower is called   [   ]
    a. Pedicel   b. Perianth   c. Receptacle   d. Nectaris

42. Very high level of drinking alcohol in pregnancy can cause Fetal Alcohol Syndrome
    (FAD). It is characterized by [   ]
    a. Flat face
    b. Microcephaly (small head/brain)
    c. Hyperactivity and poor conc. power
    d. All

43. The theory that says life arrived on this planet from elsewhere [   ]
    a. Theory of special creation
    b. Theory of steady state
    c. Theory of cosmozoan
    d. Theory od spontaneous generation

44. Mesohippus, Merychippus and Pilohippus are [   ]
    a. Evolutionary history of the monkey
    b. Evolutionary history of man
    c. Evolutionary history of the snake
    d. Evolutionary history of the horse

45. Theory of Natural selection of inherited characteristics [   ]
    a. Lamarckism
    b. Neo-Darwinism
    c. Mendellism
    d. Morghanism

46. First successful gene therapy was for [   ]
    a. AIDS
    b. SCID
    c. Smut disease of maize
    d. Diabetes mellitus

47. First cloned mammal produced from a single cell [   ]
    a. Rat
    b. Sheep
    c. Rabit
    d. Frog

48. First product made by genetically engineered bacteria [   ]
    a. Testosterone
    b. Monoclonal Ab
    c. Insulin
    d. Estrogen

49. Which of these is genetic disorder? [   ]
    a. Phenylketonuria
    b. Huntington's chorea
    c. Turner's syndrome
    d. All of these

50. Which of these is NOT a genetic disorder? [   ]
    a. Down's syndrome
    b. Sickle cell anaemia
    c. Wool sorter's disease
    d. Hemophilia

51. Klinfelter's syndrome is characterized by [   ]
    a. Sterile female
    b. Feminised male
    c. Reduced intelligence
    d. Thick mucus along liver and lungs

Answer Key and Validation

Life Sciences

1. c	10. c	19. c	28. c	37. d	46. b
2. a	11. c	20. b	29. c	38. a	47. b
3. c	12. d	21. d	30. b	39. c	48. c
4. b	13. a	22. c	31. a	40. b	49. d
5. b	14. c	23. c	32. a	41. a	50. c
6. a	15. b	24. c	33. b	42. d	51. b
7. d	16. c	25. a	34. a	43. c	
8. d	17. b	26. a	35. d	44. d	
9. b	18. b	27. b	36. b	45. b	

21

Current Trends in Modern Microbiology

Retrace your subject...
Biodeterioration, Biofouling, Bioremediation, Human genome project, Bioinformatics, Advanced techniques: PCR, Scanning tunneling microscopy...

Biodeterioration

Undesirable changes brought about in any material by the action of microorganisms. Modern trends in this area include detailed studies of the biodeterioration of Paper, wood, paints, Stones, textile, leather, tar roads, and such other material. Humidity, temp, pH of the substance and ecological factors influence biodeterioration

Material	Biodeteriogen
Walls and Painted surfaces	Liches, Mossess, liverworts, and a variety of fungi
Stone Monuments	Fungi – Cause colored patches, black-brown patinas, exfoliation and powdering. Actinobacteria – Cause white efflorescence, white grey powering and exfoliation Cyanobacteria – Produce colored sheets and scars Lichens – Cause crusts, ultrafine cracks and patches. Sulfur oxidizing bacteria – Attack sandstone monuments *Staphylococcus and Micrococcus* – Attack Limestone monuments
Marble	Fungi: *Alternaria, Aspergillus nidulans, Aspergillus niger, Fusarium, Curvularia verrugulosa, Humicola grisea, Macrophoma, Penicillium notatum*
Marble, limestone or sandstone	Algae: *Dermococcus, Pleurococcus, Navicula*
Wood	Brown rot fungus – *Postia placenta*, Attack on cellulose fibres – *Aspergillus , Trichoderma viridae*, other harmful fungi – *Gloeophyllum trabeum, L. olivascens , Daedalea berkeleyi,*
Lignin	*Galactomyces geotrichum* and *Myrothecium verrucaria*
Paper	The filamentous iron bacterium *Sphaerotilus natans*, Slime forming bacteria - *Aerobacter aerogens and Bacillus* Slime depositing fungi - *Mucor, Penicillium, Fusarium, and Trichoderma.* Yeast - *Torula, Rhodotorula*
Textile (cotton/woolen clothing)	*Alternaria, Cladosporium herbarum, and Fusarium ,*(especially *F. moniliforme, Phoma* and related forms, *Leptosphaeria* and related forms, and *Pullularia pullulans, Aspergillus, Penicillium, Stachybotrys, Chaetomium,* Bacteria- Mixed culture activity of aerobic gram positive, *Clostridium felsineum,* etc.
Preventive Measures	Use of biocidal chemicals such as Zn compounds, Mg compounds, Copper compounds, esters and aldehydes, diluted phenolic compounds, etc.

Biofouling	
Definition	Build-up of organisms on moist or wet material, causing undesirable results. Accordingly, *Microfouling* is the accumulation of microorganisms and *Macrofouling* refers to the attachment of larger organisms, such as seaweed, mussels, polychaete worms and others.
Wells and Pipelines	The insides of groundwater wells suffer from reduced flow rates due to biological accumulation. The outer and inner surfaces of ocean laying pipes are also subject to biofouling that retards the seawater flow through the pipe. Ordinary pipelines carrying hard water from or into tanks develop a flora inside called *scaling*.
Ships hulls	Ship hulls are constantly in touch with water and hence subject to biofouling. This causes reduced performance and increased fuel requirements of the vessel.
Other Examples	Is also found in all circumstances where water is in contact with the material. Example: Spiral wound membrane filters in Reverse Osmosis Unit of water treatment, membrane bioreactors and Cooling water cycles of large industrial equipment and power stations. Certain oil pipelines that are water entrained.
Epibiosis	Biofouling that occurs on the surfaces of living aquatic plants or animals is called epibiosis.
Prevention & Control	<ul><li>Use of copper bottomed ships instead of wood hulls. However, copper is not safe for all marine life and may disturb the ecosystem.</li><li>Coating of highly diluted tributyltins as antifouling agent on sbmarine surfaces. However, it is poisonous to certain shellfishes</li><li>Deliberate removal of the *scale* by backwash of pipes or treatment with citric acid to kill microflora.</li></ul>

Bioremediation	
Definition	Treatment of a polluted environment using live microrganisms. when green plants or their enzymes are used for the purpose, the process is called *Phytoremediation*. Bioremediation was first proposed by George M. Robinson. He spent his spare time experimenting with dirty jars and various mixes of microbes.
Major areas of application	Degradation of chlorinated hydrocarbons in soil by adding certain bacteria. Decomposition of crude oil by bacteria inorder to clean-up oil spills in the sea or ocean. *Deinococcus radiodurans* (a radioresistant bacterium) has been genetically modified to consume and digest toluene and ionic mercury from highly radioactive nuclear waste.
In situ bioremediation *ex situ* bioremediation	Treating the unwanted material at the site of pollution itself. Removal of the contaminated material to be treated elsewhere.
Bioventing	A kind of insitu bioremediation technology that employs naturally occurring microbes to degrade certain organic substances adsorbed to soils. Bioventing is effective in cleaning up jet fuels. Kerosense, diesel fuel and gasolene. The technique involves induction of airflow to provide oxygen for the degradation.
Boaugmentation & Biostimulation	Addition of genetically engineered bacteria (or natural strains) in the polluted area for rapid treatment. Bioaugmentation is commonly used in municipal wastewater in activated sludge bioreactors. *Biostimulation* is a closely related term that refers to addition of nutrients in the polluted area so that the biodegrading organisms multiply rapidly in that area. In a way, Biostimulation can be enhanced by Bioaugmentation
Rhizofiltration	A form of phytoremediation that involves cleaning up contaminants in water through the root system of cetrain plants. The plants are put in contact with the contaminated liquid. They absorb contaminants through their root systems and may or may not transport them up into the stems and/or leaves. The plants continue to absorb contaminants until they are harvested. Rhizofiltration is used for treatment in aquatic environments, while phytoextraction deals with soil bioremediation.

Biostimulation	Involves the modification of the environment to stimulate existing bacteria capable of bioremediation. This can be done by addition of various forms of rate limiting nutrients and electron acceptors, such as phosphorus, nitrogen, oxygen, or carbon (e.g. in the form of molasses). Additives are usually added to the subsurface through injection wells, although injection well technology for biostimulation purposes is still emerging. Removal of the contaminated material is also an option, albeit an expensive one.
Limitations	All contaminants cannot be treated by bioremediation. For example, heavy metals such as cadmium and lead are not readily degraded by organisms. Moreover, most of the bioremediation processes in practice are at a slow pace and require further research to hasten them. The process of degradation of the pollutants must be understood thoroughly with respect to the biochemical pathways and regulatory networks to accelerate the degradation.
Phytoremediation	Natural plants or transgenic plants are able to bioaccumulate hazardous pollutants such as cadmium and lead. These plants containing the toxic substances are then harvested for removal.
Mycoremediation	Paul Stamets introduced the term to the use of fungal mycelia including mushrooms in bioremediation. The major areas of application include treatment of nerve gases and sarin pollution, using fungi. Cultivation of Oyster mushrooms in soils contaminated with diesel oils is effective in the clean-up.

Human Genome Project (HGP)	
Aim	To decipher the genetic make-up of the human species To determine the exact sequence of chemical base pairs that make up the entire DNA
Challenge	HGP is a magnanimous task initially expected to take 15 years! It is a strenuous task of preparing the data of 3million base pairs with minimum error The finished data is equivalent to writing 200 Books each of 1000 pages.! May take about 9.5yrs to read the entire data (Reading 10bases/sec)
Historical	HGP began in 1990 initially headed by D Watson and later taken over by F Collins April 2003 – the essentially complete Human Genome was announced May 2006 – the sequence of the last chromosome was published in the Journal Nature
Consortium	Foundation of the $3 billion project was laid by the US Dept of Energy and US National Institutes of Health The international consortium for HGP comprised of Geneticists from China, France, Germany, Japan & UK Meanwhile the HGP was simultaneously conducted by a Private Company Celera Genomics
Patents	March 2000 – President Clinton Announced that the genome sequence would NOT be patented The international consortium decided that the Human Genome shall be made freely available to all researchers without patenting. Celera Genomics refused to deposit its data in the unrestricted database However, Celera incorporated the Public Data into their own.
Database	A Free Database of The US National Center for Biotechnology information housed the sequences of known & hypothetical Genes and Proteins Other organizations (Univ. of California, Santa Cruz and Ensemble), present additional data and annotation Computer programs have been developed to analyze the data as the data is too massive to interpret as it is.
GenBank	GenBank receives sequences from labs throughout the world from more than 100,000 distinct organisms GenBank continues to grow at an exponential rate, doubling every 18months. It contains over 65billion nucleotide bases in more than 61million sequences
Limitations (Gaps in the genome	The HGP did not study the entire DNA found in the human cells..! The improved draft of the Human Genome shows approx. 92% of the sequence. Some heterochromatic areas remain un-sequenced still.

sequence)	The remaining part of the genome is difficult to be sequenced using current technology. The central regions of each chromosome, known as centromeres are highly repetitive DNA sequences that are difficult to sequence. The centromeres are several base pairs long that are entirely un-sequenced The ends of the chromosomes called telomeres are also highly repetitive and are largely left un-sequenced
The method	Hierarchical Shotgun Approach was used in the HGP The genome was broken into smaller pieces (150,000bp) called BACs (Bacterial Artificial Chromosomes) Each of these pieces was sequenced separately and then assembled. Celera used a riskier and faster technique called whole genome shotgun sequencing.
Errata	The data provided by the HGP may NOT be 100% correct The data published by HGP does not represent the exact sequence of each and every individual. The current effort in identifying differences involves Single Nucleotide Polymorphisms (SNPs). Transposons or jumping genes can cause mutations and change the amount of DNA in the genome. There are also repeat content of human genome: Simple Sequence Repeats, Interspersed repeats, segmental duplications.
ELSI (Ethical, Legal & Social Issues)	Who owns the genetic information and who has access to it? To what extent do genes determine behavior? What is the relation between Genes & Race? Such queries about HGP have created enough excitement in the scientific community.
	The knowledge of human genome is now as necessary as the knowledge of human anatomy Human Genome Code will provide new avenues for advances in Medicine & Biotechnology New therapeutic procedures Genetic tests to show predisposition to a variety of illnesses…
The Domain of Bioinformatics	Identifying boundaries between Genes and other features in a raw DNA sequence. Expert biologist may be the best annotator, but very slow. Computer programs are increasingly used to meet the demand of genome sequencing projects. All these have given rise to a need based branch of modern science – Bioinformatics.

Bioinformatics	
Historical	The term *Bioinformatics* was coined by Paulien Hogeweg in 1978 for the study of informatic processes in biological systems. The primary goal of bioinformatics is to facilitate easy methods to study, analyze and understand voluminous biological data such as the data of complete chromosomal make up of an organism.
Designation.	Bioinformatics is the application of information technology to the field of molecular biology. It has developed rapidly to gain a distinct position among the branches of Biology.
Scope & Realm	The creation and advancement of databases Algorithms, computational and statistical techniques, Solving formal and practical problems of data analysis.
Applications	Mapping and analyzing DNA and protein sequences, Creating and viewing 3-D models of protein structures. Measuring biodiversity Analysis of gene expression, regulation, protein expression
Major Research Areas	Sequence alignment, gene finding, genome assembly, protein structure alignment, protein structure prediction, prediction of gene expression and protein-protein interactions, and the modeling of evolution.

Table *Contd…*

| Terminology | **Algorithm**: A set of rules for calculating or problem solving carried out by a computer program
Pascal: A programming language designed to teach computing. It is largely superseded by C language which is very widely used bioinformatics software.
Annotation: the process of locating and marking genes or other imp sequences, in a given raw sequence data.
WWW: World Wide Web - An internet based system for information exchange using http protocol.
URL: Uniform Resource Locator – A text address for a web page.
Browser: An application that enables display of formatted web pages by processing and interpretation of documents written in *HTML* [Hypertext Markup Language]. The most widely used browsers are Netscape Navigator and Internet Explorer.
BLAST: Basic Local Alignment Search Tool – A program for searching sequence database similarity.
FASTA: A sequence alignment algorithm (similar to BLAST) that provides quick search of sequence databases. BLAST and FASTA are many times faster than dynamic programming but less accurate relatively.
Phylogenetics: The branch of science that deals with determining the evolutionary relationships among organisms.
Cladogram: The figurative expression of evolutionary relationship between organisms presented in the form of a series of bifurcations (two branches) is called a cladogram. The study and principles underlying the construction of cladograms is referred to as *cladistics.*
Ontology: Study to describe the relationships between objects or other entities within a given area of interest. It is used in artificial intelligence and information science.
Microarray: A tiny chip that contains thousands of immobilized molecules in a definite pattern. Example: A DNA microarray is used for genotyping & expression analysis A Protein microarray is used for expression analysis and detecting protein interactions. |

Modernized Techniques	
Polymerase Chain Reaction (PCR)	Developed by Kary Mullis (1984), PCR is a DNA amplification technique widely used in Molecular Biology particularly in medical and biological research labs.
Steps Involved 1. **Denaturation** at 94°C [1min process]	The double strand melts open to single stranded DNA, all enzymatic reactions stop.
2. **Annealing** at 54°C [45sec process]	Ionic bonds are constantly formed between the single stranded primer and the single stranded template. The polymerase attaches and starts copying the template.
3. **Extension** at 72°C [2min process]	The bases complementary to the template are coupled to the primer on the 3' side by action of the polymerase that adds dNTP's from 5' to 3', reading the template from 3' to 5' side. Steps 1, 2 and 3 are repeated for 30-40 cycles.
Applications	DNA cloning for sequencing, DNA-based phylogeny, functional analysis of genes, diagnosis of hereditary diseases and in detection and diagnosis of several infectious diseases.
Scanning Tunneling Microscopy (STM)	Developed by Gerd Binnig and Heinrich Rohrer in 1981. (STM) is a technique for viewing surfaces at the atomic level. It has a lateral resolution of 0.1nm and works by probing the density of states of a material using tunneling current. The STM can be used in ordinary air or liquid atmospheres and also in conditions of ultra high vacuum. Its functional temp ranges from near zero Kelvin to a few hundred degrees Celsius. The STM is based on the principle of quantum tunneling. When a conducting tip is brought very near to a metallic or semi conducting surface, a bias between the two can allow electrons to tunnel through the vacuum between them. Variations in current as the probe passes over the surface are translated into an image. STM however requires extremely clean surfaces and sharp tips.

Exercises

1. The Ames test is used to _________ [ ]
   a. Screen chemicals for potential carcinogenicity
   b. Assay micro-quantities of drugs
   c. Diagnose enteric fevers
   d. Check the validity of a chemical to be used in culture media

2. An example of biodeterioration [ ]
   a. Curing of tobacco leaves
   b. Steroid transformation
   c. Fungal rot of paper pulp
   d. Idli batter fermentation

3. The advantage of using microbial biopolymers. [ ]
   a. Production is possible with in expensive substrate
   b. They are biodegradable and eco-friendly
   c. They are usable
   d. They are more miscible

4. The most widely used microbially produced biosurfactant are [ ]
   a. Xanthans
   b. Pullulans
   c. Dextrans
   d. Glycolipids

5. The Organism used to kill Japanese beetle [ ]
   a. Bacillus thurigiensis
   b. Nuclear Polyhedrons virus (NPVs)
   c. Cytoplasmic Polyhedrosis Viruses (CPVs)
   d. Bacillues popilliae

6. An example of engineered bioremediation. [ ]
   a. Addition of xanthan gum to enhance oil recovery
   b. Addition of cyclodextrins to reduce bitterness of Pharmaceuticals
   c. Pumping of oxygen to enhance oil spill cleanup process.
   d. All

7. The *kerosene fungus* isolated from airplane fuel tanks [ ]
   a. Leptosporium phytae
   b. Cladosporium resinae
   c. Plamerochaete chrysosporium
   d. Aureobasidium pullulans.

8. *Bacillus thuringiensis* is found to be active against pests belonging to [ ]
   a. Lepidoptera
   b. iptera
   c. Coleoptera
   d. All

9. Bio sensors may be used for [ ]
   a. Measurement of ions and molecules
   b. monitoring pollutants
   c. Direction of Flavor compounds in foods
   d. All

10. Use of microbes extract metals from ores gave rise to __________ [ ]
    a. Micro mine technology
    b. nanotechnology
    c. Biohydrometallurgy
    d. Biotransformation

11. Copper has been extracted from low grade ores using [ ]
    a. Saccharomycopsis
    b. Phanerochaete chrysosporium
    c. Pachysolen tannophilus
    d. Thiobacillus ferrooxidans

12. Which of these bacterium was found to ferment sugar to alcohol twice as fast as yeast can? [   ]
    a. Zymomonas mobilis
    b. Propiombacterium shermanii
    c. Pseudomonas denitrificans
    d. Clostridium acetobutylium

13. Binding this to the β Lactam ring of antibiotics can prevent damage by β Lactamases [   ]
    a. Aspartic acid   b. Nitric acid   c. Clavulanic acid   d. Lactic acid

14. Industrial Production of lactase employs [   ]
    a. Saccharomycopsis
    b. Trichoderma
    c. Kluyveromyces
    d. Saccharomyces

15. Amino acid that can be Synthesized by microbes [   ]
    a. Phenylalanine   b. aspartic acid   c. tryptophan   d. All

16. Pick out the FALSE statement [   ]
    a. Recently, *Candida* species has been employed to make protein from paper pulp wastes.
    b. *Rhizopus nigricans* has a key role in certain steroid biotransformation reactions.
    c. Microbial leaching of metal from low grade ores is presently more expensive than chemical extraction processes.
    d. Riboflavin is produced from using *Ashbya gossypii*

17. Recently developed *Atomic Force Microscopes* are useful in [   ]
    a. Mapping plasmids within a cell
    b. Studying interaction between proteins within a cell
    c. Locating restriction enzymes within a cell
    d. All

18. Scanning tunneling microscope is a modification of [   ]
    a. Scanning probe microscope
    b. Scanning electron microscope
    c. Transmission electron microscope
    d. None of these

19. Molecular chaperones occur in __________ [   ]
    a. Procaryotes only
    b. Eucaryotes only
    c. both procaryotes & eucaryotes
    d. artificially synthesized proteins

20. Carl Woese's domain concept places slime molds in ________ [   ]
    a. Plantae
    b. Eucarya
    c. Archea
    d. Eubactria

21. Grouping organisms together based on their phenotypic characters is called [   ]
    a. Biological classification
    b. Phenetic classification
    c. Jaceard classification
    d. Phyletic classification

22. What is/are the characteristicfeature(s) of Sphingobacteria? [   ]
    a. They all have sphingolipids in their cell walls
    b. They all show gliding motility
    c. They are all non–photosynthetic, gram negative eubacteria
    d. all

23. The Largest and most diverse group of bacteria is ________ [   ]
    a. Methanobacteria
    b. Lactic acid bacteria
    c. Proteobacteria
    d. Halobacteria

24. In Bio-informatics GOLD indicates [   ]
    a. Gene Omnibus Laplacian Dichroism
    b. Gateway of Laser Desorption
    c. Genome Online Database
    d. Garnier – osguthorpe Language Database

25. www refers to   [   ]
    a. world wide web   b. weighted wrong word
    c. world watershed word   d. web world wide

26. Netscape Navigator refers to   [   ]
    a. Parts of text that are linked to other sites
    b. A domain name server
    c. A browser Programme for working
    d. A Transmission contract Protocol.

27. The Uniform resource locator that identifies the protocol for communication over www,

    has a format beginning with   [   ]
    a. http://   b. //http: c. http\\:   d. http:\\

28. Generation of random fragments that are sequenced to provide entire genome coverage in   [   ]
    a. clone contig sequencing   b. shotgun sequencing
    c. single pass sequencing   d. Multiple sequencing

29. RNA sequencing is most easily determined by   [   ]
    a. NMR   b. Spectroscopy
    c. MS   d. Deducing data from corresponding DNA sequence

30. Which of these is employed fro determination of protein structure   [   ]
    a. X-Ray crystallography   b. NMR spectroscopy
    c. X-Ray fibre diffraction   d. All of these

31. The file formats FASTA & GDE are widely used for   [   ]
    a. Storing RNA sequences   b. Understanding multi subunit complexes
    c. Studying protein-protein interactions   d. Representing nucleic acid protein sequences

32. Which of these is a data retrieval tool   [   ]

    a. Entrez   b. BLAST   c. GOLD   d) All

33. Study of Similarities and differences between species to trace their common ancestor is   [   ]
    a. Evolution   b. Substitution score matrices
    c. Phylogenetics   d. Proteomics

34. Ontology is the study of   [   ]
    a. Relationship between entities within a given area of interest
    b. Dynamic programming algorithms
    c. Database type
    d. Sequence similarity searches

35. The figure made from series of paired bifurcations from every ancestral node is called   [   ]
    a. Hidden Markov Model (HMM)   b. Phylogenetics tree
    c Cladogram   d. Histogram

36. A set of computer based rules for calculating is called   [   ]
    a. Alignment   b. Algorithm
    c. Annotations   d. Archives

37. The miniature device made from immobilization of thousands of different molecules is called   [   ]
    a. Enantiomers   b. Middleware   c. Matrix   d. microarray

38. Which of these is a UNIX like operating system [   ]
    a. FASTA   b. LINUX   c. MALDI   d. OMIM

39. Which of these is a Procedural Programming Language designed for teaching computing [   ]
    a. PSI-BLAST   b. QSAR   c. Pascal   d. SRS

40. PERL is [   ]
    a. A binary code interpreted by a computer's processor
    b. Functional site in DNA sequence that identifies a gene
    c. Serial analysis of gene expression
    d. Scripting language used in Bioinformatics for analysis of sequence data.

41. In the PCR technique, target DNA sequence in denatured by heating the Reaction mixture to [   ]
    a. 65°C   b. 72°C   c. 80°C   d. 95°C

42. *Taq polymerase* is the enzyme of choice in PCR due to [   ]
    a. Its heat stability   b. Its rapid reaction rate
    c. Its specificity   d. Error proof function

43. In PCR technique, maintenance of the system at 65°C encourages [   ]
    a. Denaturing   b. Annealing   c. Extension   d. None of these

44. The discovery of Bacterial Artificial chromosomes (BACs) is the outcome of [   ]
    a. Preparation of monoclonal antibodies   b. PCR technique
    c. Southern blotting   d. Human Genome Project (HGP)

45. The Biotech company Celera Genomics gained popularity due to [   ]
    a. Preparation of monoclonal antibodies   b. PCR technique
    c. Southern blotting   d. Human Genome Project (HGP)

46. The HGP largely employed [   ]
    a. Hierarcheal shotgun sequencing   b. Bacterial conjugation experiments
    c. PCR technique   d. NMR spectroscopy

Answer Key and Validation

Current Trends in Modern Microbiology

1. a	9. b	17. d	25. a	33. c	41. d
2. c	10. c	18. a	26. c	34. a	42. a
3. a	11. d	19. c	27. a	35. c	43. b
4. d	12. a	20. b	28. b	36. b	44. d
5. d	13. c	21. c	29. d	37. d	45. d
6. c	14. b	22. d	30. d	38. b	46. a
7. b	15. c	23. c	31. d	39. c	
8. d	16. c	24. c	32. a	40. d	

18. *Scanning probe microscopes are a new class of microscopes used to measure surface features by moving sharp probes over object surfaces. Atomic force microscope and scanning tunneling microscope are two kinds of scanning probe microscopes.*

22. *Best examples of Sphingo bacteria are Beggiatoa, Thiothrix, Cytophaga, Leucothrix etc. They are non–fruiting, gliding, non–photosynthetic bacteria.*

23. *Proteobacteria are the entire group of purple photosynthetic bacteria which are further divided into several sub groups.*

26. *A Program used to exchange information over internett is called a browser. The most widely used browsers are Internet Explorer & Netscape navigator.*

40. *PERL = Practical Extraction & Reporting Language*

43. *After denaturation at 95°C, the reaction mixture is cooled to 65°C where the oligonucleotide primers anneal to their complementary sequences on the single stranded templates.*

Grand Test I

Introduction

The Grand Test 1 is intended to check your fundamental principles of Microbiology and allied subjects. The conventional graduate of Microbiology may find this test very easy and simple. However, such a test is essential to examine your fundamental grip on the subject at the surface. Also, it is recommended that **graduates of non-microbiology origin should attempt the Grand Test I first**, as the other two tests are more challenging and need further in-depth subject knowledge.

After having taken the test, you may assess your performance and realize where you stand on the competitive front based on the **Anaylsis-Sheet** provided.

1. Father of modern Microbiology [ ]
   a. Robert Koch
   c. Anton van Leeuwenhoek
   b. Louis Pasteur
   d. Paul Ehrlich

2. Freeze etching is a technique employed in [ ]
   a. Dark field microscopy
   c. Transmission electron microscopy
   b. Fluorescence microscopy
   d. Scanning electron microscopy

3. Periplasmic space is the gap between the [ ]
   a. capsule and cell membrane
   c. peptidoglycan and outer membrane
   b. cell membrane and cell wall
   d. lipid layers of cell membrane

4. Secretary vesicles are seen on [ ]
   a. Golgi apparatus   b. Chloroplasts   c. Mitochondria   d. Endoplasmic reticulum

5. Coenzyme is [ ]
   a. Protein part of an enzyme
   b. Non-protein part of an enzyme
   c. Inorganic ion that improves enzyme performance
   d. Any molecule that regulates rate of enzyme reaction

6. What is FALSE about photosynthesis [ ]
   a. End products are usually O_2 and glucose
   b. Dark reactions are light independent
   c. Energy is produced in both light and dark reactions
   d. Light is the source of energy

7. Chocolate blood agar is [ ]
   a. a selective medium
   c. a transport medium
   b. an enriched medium
   d. an enrichment medium

8. Ames test is carried out to [ ]
   a. diagnose carcinoma in humans
   c. establish mutagenesis of a substance
   b. detect cancer causing viruses
   d. distinguish cancer cells from normal cells

9. Conjugation was first discovered by   [   ]
   a. Lederberg & Tatum
   b. Joshua & Lederberg
   c. Jacob & Wollman
   d. Barbara McClintock

10. Father of taxonomy   [   ]
    a. Bergey   b. Aristotle   c. Linnaeus   d. Darwin

11. Polyacrylamide gel electrophoresis (PAGE)   [   ]
    a. Separates proteins on the basis of their ionic charges
    b. Separates proteins on the basis of their mass
    c. Separates proteins on the basis of their molecular size
    d. Separates proteins on the basis of their affinity to standard molecules

12. What is TRUE about *Orthomyxoviridae*?   [   ]
    a. Enveloped visuses
    b. Genome is segmented in eight pieces
    c. They are RNA viruses with minus sense strand
    d. a, b and c TRUE

13. The hyphae of *Aspergillus niger* are   [   ]
    a. Aseptate and multinucleate
    b. Septate and uninucleate
    c. Septate and multinucleate
    d. Aseptate and uninucleate

14. Typhus fever is caused by   [   ]
    a. *Rickettsia*   b. *Rochalimia*   c. *Coxiella*   d. *Francisella*

15. The organisms used to determine phenol coefficient   [   ]
    a. *E.coli* and *Bacillus*
    b. λ *phage* and T phage
    c. *S.aureus* and *Salmonella typhi*
    d. *Aspergillus* and *Mucor*

16. One of the best techniques of preserving pure cultures   [   ]
    a. Chick embryo inoculation
    b. Lyophilization
    c. Mineral oil overlay on agar slants
    d. Refrigeration

17. UV light kills microorganisms effectively at a wavelength of   [   ]
    a. 70nm   b. 90nm   c. 200nm   d. 300nm

18. Kirby-Bauer method is used to   [   ]
    a. Identify pathogenic strains of *Salmonellae*
    b. Establish toxigenicity of *Vibrio cholerae*
    c. Study relation between two Ags or Abs.
    d. Detect antibiotic sensitivity pattern for an organism.

19. Mebendazole and Piperazine   [   ]
    a. Anti viral drugs
    b. Anti fungal drugs
    c. Anti helminthic drugs
    d. Anti allergens

20. Severe itching sensation at the anal region is due to   [   ]
    a. *Trypanosoma gondii*
    b. *Enterobius vermicularis*
    c. *Entamoeba histolytica*
    d. All

21. Result IMV$_i$C test for *E.coli*   [   ]
    a. + + - -   b. - - + +   c. + - + -   d. - + - +

22. Tetanus toxin is   [   ]
    a. a neurotoxin   b. an endotoxin   c. an enterotoxin   d. a cytotoxin

23. Hospital acquired infections are called   [   ]
    a. Zoonotic   b. Chronic   c. Transient   d. None of these

24. What prevails when a large proportion of a population is immune to a disease   [ ]
    - a. Innate immunity
    - b. Herd immunity
    - c. Humoral immunity
    - d. acquired immunity

25. Which of these is a lymphatic organ?   [ ]
    - a. Thymus gland
    - b. Spleen
    - c. Both a & b
    - d. None of these

26. The dual nature of immune system   [ ]
    - a. Active and Passive immunity
    - b. Innate and Acquired immunity
    - c. Humoral and Cell mediated immunity
    - d. Natural and Artificial immunity

27. Match the following   [ ]

    | A. Bacterial skin disease | i. Tinea |
    | B. Viral skin disease | ii. Warts |
    | C. Fungal skin disease | iii. Madura foot |
    | D. Actinomycetes skin disease | iv. Acne |

    - a. A-iii, B-ii, C-i, D-iv
    - b. A-iv, B-ii, C-i, D-iii
    - c. A-i, B-iv, C-iii, D-ii
    - d. A-iii, B-ii, C-iv, D-i

28. Tilak samplers are used to collect   [ ]
    - a. Soil
    - b. Air
    - c. Urine
    - d. Feces

29. Humus is best described as   [ ]
    - a. Inorganic soil component
    - b. Subsurface earth material
    - c. Nonliving organic matter in soil
    - d. Microflora of soil

30. Which of these tests is NOT employed for examination of water?   [ ]
    - a. MPN
    - b. MFT
    - c. MBRT
    - d. APC

31. Bacterial soft rot of foods is due to   [ ]
    - a. *Fusarium spp.*
    - b. *Erwinia carotovora*
    - c. *Monilia fructicola*
    - d. All.

32. Patulin – A toxic apple cider contaminant is produced by   [ ]
    - a. *Clostridium*
    - b. *Penicillium*
    - c. *Aspergillus*
    - d. *Pseudomonas*

33. Which of these is a food borne pathogen   [ ]
    - a. *Bacillus cereus*
    - b. *Campylobacter jejuni*
    - c. *Vibrio parahaemolyticus*
    - d. All

34. Flat sour spoilage is observed in   [ ]
    - a. Raw market vegetables
    - b. Milk pouches
    - c. Canned foods
    - d. Slaughter house meat

35. Ripening of camembert cheese is by   [ ]
    - a. *Brevibacterium*
    - b. *Penicillium*
    - c. *Streptococcus*
    - d. *Propionibacterium*

36. Baird Parker Agar (BPA) is used in the selective isolation of   [ ]
    - a. *Staphylococcus aureus*
    - b. *Streptococcus pyogenes*
    - c. *Corynebacterium diphtheriae*
    - d. *Salmonella spp.*

37. The prefixes *pico* and *nano* represent   [ ]
    - a. 10^{-10} and 10^{-9}
    - b. 10^{-12} and 10^{-9}
    - c. 10^{-11} and 10^{-12}
    - d. 10^{-9} and 10^{-8}

38. Poliovirus is placed in the family   [ ]
    - a. Caliciviridae
    - b. Paramyxoviridae
    - c. Picornaviridae
    - d. Arenaviridae

39. The word *myxo* is used to indicate   [ ]
    a. Dry and scaly
    b. Slimy and mucoidal
    c. Tumerous
    d. Branched and filamentous

40. The reaction sequence that begins with conversion of pyruvic acid to AcetylCoA   [ ]
    a. Embden Mayerhoff Parnas Pethway
    b. Calvin Benson Cycle
    c. Entner Doudoroff Pathway
    d. Kreb's cycle

41. A protein aggregate that forms the viral body   [ ]
    a. Capsule
    b. Carbuncle
    c. Capsid
    d. Cascade

42. An enzyme that breaks down hydrogen peroxide to water and molecular oxygen   [ ]
    a. Catalase
    b. Coagulase
    c. Amylase
    d. Kinase

43. A tumor of the jaw caused by Epstein Barr virus   [ ]
    a. Chaga's disease
    b. Brill-Zinsser disease
    c. Burkitt's disease
    d. Cat scratch disease

44. A deep wound infection caused by *Clostridium*   [ ]
    a. Furuncle
    b. Gangrene
    c. Gingivitis
    d. Erysepalas

45. The prion mediated disease due to the consumption of raw human brain by humans (cannibalism)   [ ]
    a. Kuru
    b. Leionnaire's disease
    c. Hansen's disease
    d. Hide Porter's disease

46. Test for the water purity   [ ]
    a. ONPG & MUG test
    b. Ames Test
    c. Mantoux test
    d. Shick test

47. Rheumatoid factor   [ ]
    a. IgG
    b. IgE
    c. IdM
    d. IgD

48. Axenic culture   [ ]
    a. Contains organisms of same genus, species and strain
    b. Contains organisms of same genus & species, but different strains
    c. Contains organisms of same genus but different species
    d. Contains organisms of different genera

49. HIV infects and destroys   [ ]
    a. RBCs
    b. T suppressor cells
    c. T helper cells
    d. All of these

50. Tertiary structure of protein   [ ]
    a. Single polypeptide chain of 50 or more amino acids
    b. More than 75% of polypeptide chain is α helical
    c. Non-uniform folding of α helical chain to form a compact shape
    d. Separate polypeptide chains held together

51. Which of these is ABSENT in xylem tissue?   [ ]
    a. Tracheids
    b. Sclereids
    c. Parenchyma
    d. Vessel elements

52. The cells in the cartilage tissue are called   [ ]
    a. Squames
    b. Chondrocytes
    c. Goblets
    d. Osteoblasts

53. Which of these are accessory pigments?   [ ]
    a. Chlorophylls
    b. Carotenoids
    c. Fluoresceins
    d. Pyocyanins

54. CO_2 fixation occurs twice - First in mesophylls and then in bundle sheeth cells, in case of   [ ]
    a. C_3 Plants
    b. C_4 plants
    c. Both a & b
    d. None of these

55. Holozoic nutrition is characterized   [   ]
    a. Presence of alimentary canal
    b. Intracellular digestion
    c. Predation
    d. Saprophytism

56. The sea anemone- hermit crab association symbolizes   [   ]
    a. Mutualism   b. Parasitism   c. Commensalism   d. Predation

57. Cell replacement and regeneration is the function of   [   ]
    a. Mitotically dividing cells
    b. Meiotically dividing cells
    c. Both a & b
    d. Nervous system

58. The laws of inheritance were given by   [   ]
    a. August Weismann
    b. Charles Darwin
    c. Gregor Mendel
    d. James Hutton

59. The condition in which urine turn black upon exposure to air   [   ]
    a. Albinism   b. Alkaptonuria   c. Phenylketonuria   d. Diabetes mellitus

60. Skeletal muscle pigment   [   ]
    a. Myoglobin   b. Hemoglobin   c. Leg Hemoglobin   d. Melanin

61. Transpiration occurs from   [   ]
    a. Stomata   b. Cuticle   c. Lenticels   d. All

62. Rhizobia in root nodules are   [   ]
    a. Gram positive cocci
    b. Gram negative rods
    c. Gram negative pleomorphic forms
    d. Gram positive short rod to coccoidal forms

63. Pesticide poisoning has shown most devastating effect on   [   ]
    a. Birds   b. Humus   c. Wild animals   d. Earthworms

64. Unidirectional, unbranched flow of energy is depicted in   [   ]
    a. Food chains
    b. Food webs
    c. Ecological pyramids
    d. Standard deviation curves

65. The Dutch merchant who called microorganisms as *animalcules*   [   ]
    a. Anton van Leeuwenhoek
    b. Francesco Redi
    c. Elie Metchnikoff
    d. Alexander Fleming

66. Water forms thin layers because   [   ]
    a. It is polar in nature
    b. It has high surface tension
    c. It has high specific heat
    d. It is a miscible fluid.

67. Steroids may be categorized as   [   ]
    a. Proteins   b. Carbohydrates   c. Lipids   d. Nucleic acids

68. Match the following   [   ]
    A. Negative staining   i. *Treponema*
    B. ZNCF staining   ii. Fungi
    C. Fontana's staining   iii. *Mycobacterium*
    D. Lactophenol cotton blue   iv. Capsules

    a. A-i, B-iv, C-iii, D-ii
    b. A-iv, B-ii, C-iii, D-i
    c. A-iv, B-iii, C-i, D-ii
    d. A-iii, B-i, C-ii, D-iv

69. Which of these is present in animal cells but absent in plant cells?   [   ]
    a. Centrioles   b. Microvilli   c. Lysosmes   d. All

70. Which of these is based on molecular kinetic energy?　　　　　　[ ]
    a. Simple diffusion　　　　　　　b. Facilitated diffusion
    c. Active transport　　　　　　　d. Osmosis

71. Photoheterotrophs　　　　　　　　　　　　　　　　　　[ ]
    a. Green sulfur bacteria　　　　　b. Green non-sulfur bacteria
    c. Algae　　　　　　　　　　　　d. All

72. Number of ATP molecules that result during conversion of pyruvate to AcetylCoA　[ ]
    a. 2　　　　　b. 6　　　　　c. 18　　　　　d. 38

73. Use of microorganisms to treat polluted areas　　　　　　　[ ]

    a. Bioleaching　　　b. Biodeterioration　c. Bioremediation　d. Biosensor

74. What is FALSE about Bioluminiscence?　　　　　　　　　[ ]
    a. *Achomobacter* is a bioluminescent bacterium
    b. ATP is not generated during bioluminescence
    c. Luciferase catalyses an oxygen independent reaction to emit light
    d. A carrier molecule derived from $VitB_2$ is used in the process of emitting light

75. Direct Microscopic Count (DMC) of bacteria is widely used in　[ ]
    a. Sewage treatment plant　　　b. Dairy industry
    c. Paint industry　　　　　　　d. Pathology lab

76. The core of bacterial endospore consisits of　　　　　　　[ ]
    a. Ca ions　　　　　　　　　　b. Dipicolinic acid
    c. Dipicolinic acid and Ca ions　d. None of these

77. What is FALSE about rRNA?　　　　　　　　　　　　　[ ]
    a. Combines with specific proteins to form ribosomes
    b. Serves as a site of protein synthesis
    c. Has a typical clover leaf shape
    d. Associated enzymes function in controlling protein synthesis.

78. Deletion or insertion of one or more bases results in　　　　[ ]
    a. Point mutations　　　　　　　b. Frame shift mutations
    c. Phenotypic variation　　　　　d. None of these

79. The result of *Hfr* x F⁻ conjugation　　　　　　　　　　[ ]
    a. F⁺ cell
    b. *Hfr* cell
    c. F⁻ cell with variable quantity of chromosomal DNA
    d. None of these

80. Aerating device in the fermentor is called　　　　　　　　[ ]

    a. Impeller　　　b. Baffles　　　c. Sparger　　　d. None of these

81. Yeast cell wall is different from mold cell walls in:　　　　[ ]
    a. Phosphomannoproteins are present in the yeast cell wall
    b. βGlucans are present in the yeast cell wall
    c. Both a & c
    d. Chitin is totally absent in yeast cell wall

82. Key enzyme in Entner Doudoroff pathway　　　　　　　　[ ]
    a. Pyruvate dehydrogenase　　　b. KDPG aldolase
    c. Aconitase　　　　　　　　　　d. Pyruvate kinase.

83. Source of energy for nitrifying bacteria [   ]
    a. HNO_2 & NH_3 b. HNO_3 c. H_2S d. S

84. Viruses were referred to as *contagium vivum fluidum* b [   ]
    a. Dimitri Ivanowsky   b. Martinus Beijerinck
    c. Twort & d'Herelle   d. Stanley

85. Urease positive, H_2S positive and gelatin liquefaction positive [   ]
    a. *Pseudomonas*   b. *Klebsiella*   c. *Salmonella*   d. *Proteus*

86. Scrub typhus is caused by [   ]
    a. *Rickettsia tsutsugamushi*   b. *R. rickettsii*
    c. *R. Quintana*   d. *R. prowazeki*

87. Fever (pyrexia) [   ]
    a. A pathological condition   b. A first line of defense
    c. Specific host defense   d. Second line of defense

88. Roundworm infestation is associated with activation of [   ]
    a. IgG   b. IgE   c. IgM   d. IgA

89. Which if these is prevented by administration of purified toxoid [   ]
    a. Tetanus   b. AIDS   c. Rabies   d. Poliomyelitis

90. Treatment of milk through Ultra High Temp (UHT) employs [   ]
    a. 71.7^0C for 15seconds   b. 87.7^0C for 3seconds
    c. 137.8^0C for 2seconds   d. 148.9^0C for 2seconds

91. Idli fementation is brought about by [   ]
    a. *Leuconostoc mesenteroides*   b. *Streptococcus faecalis*
    c. *Pediococcus cerevisiae*   d. All of these

92. Gram negative bacteria appear [   ]
    a. Pink or Red   b. Purple   c. Blue   d. Pitch Black

93. Heterocysts are seen in [   ]
    a. Animal cells   b. Cyanobacteria
    c. Helminth parasites   d. Protozoa

94. The correct order of basal rings of bacterial flagella, from top to bottom [   ]
    a. M. S, P and L rings   b. S, P, M and L rings
    c. P, S, L and M rings   d. L, P, S and M rings

95. The three dimensional structure of DNA was first studied through [   ]
    a. Transmission electron microscopy   b. X Ray crystallography
    c. Fluorescence labeling   d. Scanning tunneling microscopy

96. According to the Indian Patents Act 1970, which of these food stuffs have been granted patent? [   ]
    a. Idli & Dosa   b. Vada & Pickle
    c. Badam Halwa & Lemon pickle rice   d. All of these

97. Adansonian taxonomy is based on [   ]
    a. Linnaeus system of nomenclature   b. Numerical arrangement
    c. Intuitive categorization   d. Phage type & serotype characters

98. Heterofermentative lactic acid bacterium [   ]
    a. *Lactobacillus delbruckii*   b. *Streptococcus lactis*
    c. *Leuconostoc mesenteroides*d   d. *Lactobacillus bulgaricus*

99. What is FALSE about allosteric enzymes?   [   ]
    a. They are also called regulatory enzymes
    b. They have multiple subunits and multiple binding sites
    c. They have multiple inhibitor or activator binding sites
    d. Binding of the first substrate automatically inhibits binding of the second substrate.

100. Pick out the WRONG pair of mutualism   [   ]
    a. *Azolla-Nostoc*
    b. Cycads-*Nostoc*
    c. Termites-Citrobacters
    d. *Anthoceros-Anabena*

Answer Sheet – Grand Test I

1.	2.	3.	4.	5.	6.	7.	8.	9.	10.
11.	12.	13.	14.	15.	16.	17.	18.	19.	20.
21.	22.	23.	24.	25.	26.	27.	28.	29.	30.
31.	32.	33.	34.	35.	36.	37.	38.	39.	40.
41.	42.	43.	44.	45.	46.	47.	48.	49.	50.
51.	52.	53.	54.	55.	56.	57.	58.	59.	60.
61.	62.	63.	64.	65.	66.	67.	68.	69.	70.
71.	72.	73.	74.	75.	76.	77.	78.	79.	80.
81.	82.	83.	84.	85.	86.	87.	88.	89.	90.
91.	92.	93.	94.	95.	96.	97.	98.	99.	100.

Solution to Grand Test I

1. b	2. c	3. b	4. a	5. b	6. c	7. b	8. c	9. b	10. c
11. c	12. d	13. c	14. a	15. c	16. b	17. c	18. d	19. c	20. b
21. a	22. a	23. d	24. b	25. c	26. c	27. b	28. b	29. c	30. c
31. b	32. b	33. d	34. c	35. b	36. a	37. b	38. c	39. b	40. d
41. b	42. a	43. c	44. b	45. a	46. a	47. c	48. a	49. c	50. c
51. b	52. b	53. b	54. b	55. a	56. a	57. a	58. c	59. b	60. a
61. d	62. c	63. a	64. a	65. a	66. b	67. c	68. c	69. d	70. a
71. b	72. b	73. c	74. c	75. b	76. c	77. c	78. b	79. c	80. c
81. c	82. b	83. a	84. b	85. d	86. a	87. d	88. b	89. a	90. c
91. d	92. a	93. b	94. d	95. c	96. d	97. b	98. c	99. d	100. a

Assessment Sheet

Your Score	Observation	
95 -100	Very Good:	Your fundamentals are adequately clear. You can definitely proceed to attempt the Grand Tests II and III.
80 – 95	Good:	Your subject interest is fairly good. But may be some more reading is needed to improve on your fundamentals.
60 – 80	Satisfactory:	You have a good knowledge on the basic facts. Nevertheless, it is suggested you should read much more and make your self thorough with the subject.
<60	Not satisfactory:	As Grand Test I is supposed to be the easiest paper of the three, it is clear that your fundamentals of Microbiology need to be definitely reviewed. You must read a lot and catch up with your basics., before you proceed with Grand Test II

B

Grand Test II

If you score an A in GT 1 you are in a position to take GT 2. The exercises in are more challenging. The knowledge based questions are based on slightly inner details of the topics instead of mere fundamental principles. The options given for choosing the correct answer are interlinked and may have very close relations. A few logic based questions are also incorporated in GT 2.

GT 2 will check your in-depth subject knowledge and test your analytical ability to select the most appropriate answer for a problem

Note: Some of the following exercises may have more than one correct option. However, you are required to select *the most appropriate* option as the correct one.

1. The relatively new species of bacteria *Jenthinobacterium* and *Sphingobacterium* have been identified as   [   ]
   a. Psychrophiles
   b. Mangrove bacteria
   c. thermophiles
   d. Non-legume root nodulating symbionts

2. Orientation of bacilli in parallel rows   [   ]
   a. Stallectite arrangement
   b. hizoidal arrangement
   c. Pallisade arrangement
   d. Chinese letter configuration

3. Immunity acquired by injecting presynthesised antibody   [   ]
   a. Natural active immunity
   b. Natural passive immunity
   c. Artificial passive immunity
   d. Artificial active immunity

4. Auxotroph   [   ]
   a. Uses CO_2 as sole source of carbon
   b. bears an additional nutrient requirement
   c. Is independent of a particular chosen nutrient for survival
   d. Always requires cyanocobalamine as growth factor

5. Morbidity is defined as   [   ]
   a. The course of pathogenesis of a disease.
   b. Death toll due to a disease
   c. The number of times a disease has caused epidemics
   d. The number of cases of a disease in a given population

6. What is FALSE about the bacterium *Sulfolobus*   [   ]
   a. Produces sulfuric acid
   b. Optimum growth temp is $25 - 37^0C$
   c. Growth is strictly aerobic
   d. Highly irregular lobed cocci

7. Irregular non-sporing Gram positive rods   [   ]
   a. *Arthrobacter*
   b. *Stigmatella*
   c. *Chondromyces*
   d. *Cystobacter*

8. Viruses that replicate in neutrophils   [   ]
   a. Cytomegaloviruses
   b. Influenza virus
   c. Measles virus
   d. Polio virus

9. The oriental fermented food Tempeh is the outcome of [ ]
   a. Fungal fermentation
   b. Fungal followed by bacterial fermentation
   c. bacterial fermentation
   d. Yeast fermentation

10. The most exterior region of lipoploysaccharide is [ ]
   a. Lipid A region   b. Fatty acid region   c. O Ag region   d. Polysaccharide region

11. Which of these cannot be grown in a cell free culture media [ ]
   a. *Mycobacterium tuberculosis*
   b. *Clostridium perfringens*
   c. *Mycobacterium leprae*
   d. All

12. Photosynthetic nitrogen fixing bacterium [ ]
   a. *Gloeocapsa*
   b. *bacillus polymyxa*
   c. *Klebsiella pneumoniae*
   d. *Acetobacter vinelandii*

13. In alginate immobilization of enzymes [ ]
   a. The yield efficiency decreases with decrease in bead size
   b. The yield efficiency is independent of the bead size
   c. The yield efficiency increases with decrease in bead size

14. Match the plant biomass with its application [ ]
   A. *Jatropha curcus*   i. Whole oil substitute
   B. *Jojoba* plantation   ii. H_2 production
   C. *Prosopis juliflora*   iii. Diesel substitute
   D. BGA   iv. Fuel wood

   a. A-iii, B-i, C-iv, D-ii
   b. A-iii, B-ii, C-iv, D-i
   c. A-ii, B-iii, C-i, D-iv
   d. A-iv, B-iii, C-i, D-ii

15. *Thiobacillus ferrooxidans* gets energy for growth from the oxidation of [ ]
   a. Iron   b. Sulfur   c. Iron or sulfur   d. Phosphorus

16. Whooping cough is the common name of [ ]
   a. Influenza   b. Pneumonia   c. Pertussis   d. Tuberculosis

17. Wool sorter's disease is the common name for [ ]
   a. Pulmonary anthrax
   b. Meningitis
   c. Purpural fever
   d. Sinusitis

18. Facilitated diffusion [ ]
   a. Protein mediated process
   b. Requires biologically derived energy
   c. a & b
   d. Not protein mediated and does not require biologically derived energy

19. The most abundant donor of electrons to the mitochondrion [ ]
   a. FADH   b. NADH   c. $FMNH_2$ d. CoQ

20. The germ tube test is a rapid and reliable test to identify [ ]
   a. *Apergillus spp.*
   b. *Vibrio cholerae*
   c. *Candida albicans*
   d. *Mycobacterium tuberculosis*

21. Lyme disease is associated with [ ]
   a. *Crystispira*   b. *Leptorspira*   c. *Treponema*   d. *Borrelia*

22. F. Loeffler & P. Frosch   [   ]
    a. Proposed hemagglutinin test
    b. Tissue culture technique
    c. Plaque assay of polio virus
    d. First demonstrated filterable animal viruses

23. Production of cDNA's is the outcome of the study of   [   ]
    a. Pox viruses   b. Retroviruses   c. Reoviruses   d. Myxoviruses

24. Gram negative bacteria that can induce formation of self-proliferating galls   [   ]
    a. *Agrobacterium spp*
    b. *Fusarium spp*
    c. *Nocardia spp*
    d. *Azotobacter spp*

25. Pseudomurein is a cell wall polymer present in members of   [   ]
    a. Actinobacteria   b. Archaeobacteria   c. Eubacteria   d. *Pseudomonas spp.*

26. The fundamental law of population genetics that provides the basis for studying
    Mendelian Populations   [   ]
    a. Maxam-Gilbert Law
    b. Fischer's Law
    c. Hardy-Weinberg Law
    d. Griffith's Law

27. A tower fermentor finds special advantage in the production of   [   ]
    a. Beer   b. Lactic acid   c. Citric acid   d. Ethanol

28. A live-in association in which one of the partner receives benefit from the other while
    harming the latter is called   [   ]
    a. Mutualism   b. Commensalism   c. Parasitism   d. Predation

29. The consequence[s] of HF [hydrogen fluoride] pollution in air   [   ]
    a. Loss of weight in cattle
    b. Severe damage to young leaves of many plants
    c. Abnormal bones and teeth in many animals
    d. All of these

30. Speciation implies   [   ]
    a. Drawing the evolutionay history of a species
    b. Differenciation of individuals of the same species
    c. Evolution of new variety from a pre-existing species.
    d. Calculating the carrying capacity for a species in a given ecosystem

31. A chemical which when added to a fermentation broth, gets directly incorporated
    into the product   [   ]
    a. Growth factor   b. Precursor   c. Chelator   d. Inducer

32. A ribozyme is a molecule of RNA that has the ability to catalyze transesterification
    in RNA molecules   [   ]
    a. TRUE   b. FALSE

33. Burn abscesses are typically caused by   [   ]
    a. *Pseudomonas*   b. *Brucella*   c. *Treponema*   d. *E.coli*

34. Clamydospores   [   ]
    a. Asexual algal spores
    b. Asexual fungal spores
    c. Sexual algal spores
    d. Sexual fungal spores

35. Pepsin and Trypsin belong to the class of   [   ]
    a. Hydrolases   b. Lyases   c. Isomerases   d. Ligases

36. These come under the group of Terpenoid compounds   [   ]
    a. Carotenes   b. Xanthophils   c. Both   d. None

37. When both shorter and longer wavelengths of light are given simultaneously, the rate of
photosynthesis is higher than that of the sum of the individual rates   [   ]
    a. Hill Reaction
    b. Blackman's Law
    c. Warburg effect
    d. Emerson's Enhancement Effect

38. Scwann cells are found in   [   ]
    a. Epithelial tissue
    b. The nervous tissue
    c. Bone
    d. Blood

39. Pyrenoids are found in   [   ]
    a. *Rhizopus*   b. *Entamoeba*   c. *Spyrogyra*   d. *Enveloped viruses*

40. Cellular homeostasis is maintained with the help of   [   ]
    a. Microtubules
    b. Lysozomes
    c. Transition vesicles
    d. Plasma membrane

41. Hospital acquired lobar pneumonia in old age   [   ]
    a. *E.coli*
    b. *Mycobacterium* species
    c. *Klebsiella* species
    d. *Pseudomonas* species

42. Pentameric immunoglobulin   [   ]
    a. IgM   b. IgE   c. IgD   d. IgA

43. Biodiversity focuses on   [   ]
    a. Population dynamics in an ecosystem
    b. Spectrum of life on earth
    c. Ecological climax
    d. All of these

44. Which of these concepts formulates Mendel's law of independent assortment   [   ]
    a. Concept of alleles
    b. Development of each character is controlled by a gene
    c. All chromosomes of a species are identical with each other
    d. Equal contribution of the two parents to the development of characters of the hybrid [F_1]

45. *Trypanosoma*   [   ]
    a. Parasitic flagellate protozo
    b. Symbiont in many lichens
    c. PO_4 solubilizing mycorrhiza
    d. Poisonous mushrooms

46. Swineherd's disease   [   ]
    a. Borreliosis
    b. *Bacillus anthracis*
    c. Leptospirosis
    d. *Haemophilus parasuis*

47. Absence of visible agglutination due to Ab excess   [   ]
    a. Prozone   b. Prezone   c. Postzone   d. Zonal equilibrium

48. Nucleolus within the nuclei of eukaryotes is associated with   [   ]
    a. Metabolic regulation
    b. Synthesis of rRNA
    c. Synthesis of nuclear membrane
    d. Site for translation

49. The most interchangeable term for Bioinformatics   [   ]
    a. Computational Biology
    b. Microarrays
    c. Expression Profiling
    d. Molecular systematics

50. The DNA sequences of hundreds of organisms have been decoded after the
successful sequencing of   [   ]
    a. *Nucella lapillus*
    b. *E.coli*
    c. *Haemophilus influenzae*
    d. Phage Øx174

51. Biofouling occurring on the surfaces of living marine organisms may be referred to as   [ ]
    a. Microfouling   b. Epibiosis   c. Commensalism   d. Bioremediation

52. Extraction of specific metals from their ores through the use of microorganisms   [ ]
    a. Bioleaching   b. Microleaching   c. Bioremediation   d. Bioextraction

53. Which of these is NOT associated with microbial pathogens   [ ]
    a. Disseminated Intracellular coagulation   b. Madhura foot
    c. Arthrosclerosis   d. Scalded Skin Syndrome

54. The energy yielded in ED pathway is   [ ]
    a. Better than that of glycolysis   b. Less than that of glycolysis
    c. Same as that of glycolysis   d. Same as that of Kreb's cycle

55. Phenylketonuria is the inability to convert   [ ]
    a. Tyrosine to diphenylamine   b. Diphenylamine to tyrosine
    c. Phenyalanine to tyrosine   d. Tyrosine to Phenyalanine

56. Golgi apparatus was discovered by   [ ]
    a. Camillo Gogi   b. Koshland   c. Fischer   d. Singer & Nicolson

57. Triploblastic bilateral symmetry, un-segmented acoelomate, presence of mouth but no anus are characterisitic of   [ ]
    a. Nematoda   b. Annelida   c. Platyhelminthes   d. Arthropoda

58. Phylum Filicinophyta comprises of   [ ]
    a. Ferns   b. Mosses   c. Liverworts   d. All

59. Polysaccharides are   [ ]
    a. non-sweet   b. non-reducing   c. non-crystalline   d. All

60. Cloning of plants by tissue culture   [ ]
    a. Layering   b. Grafting   c. Explantation   d. Micropropagation

61. Biodiesels are   [ ]
    a. Short chain alkyl ethers   b. Short chain alkyl esters
    c. Long chain fatty acids   d. Acyclic fatty acids

62. Dietary supplements containing therapeutically beneficial bacteria   [ ]
    a. Bakery foods   b. Probiotics   c. SCP   d. Fermented foods

63. Large scale study of structures and functions of proteins   [ ]
    a. Enzymology   b. Proteinomics   c. Proteomics   d. Genomics

64. MIC is determined for   [ ]
    a. Antibiotics   b. Xenobiotics   c. Cosmetics   d. Fermented foods

65. The most reliable method to determine counts of bacteria in drinking water   [ ]
    a. Pour plate method   b. Spread plate method
    c. Membrane filter technique   d. Streak plate technique

66. The best way to sterilize plastic Petri dishes, catethers and rubber goods   [ ]
    a. UV [260nm for 3h]   b. Steam without pressure [$< 100^0$C]
    c. Hot air oven [75^0C for 3h]   d. Ethylene oxide gas

67. Effective and safe ophthalmic solution   [ ]
    a. Parabens   b. Chlorhexidine   c. Formalin   d. Benzoates

68. Which of these is commonly referred to as death angel   [ ]
    a. Salmonella - Infectious bacterium   b. Vibrio cholerae – Toxic bacterium
    c. HIV - AIDS Virus   d. Amanita – poisonous mushroom

69. *Phytophthera infestans* is associated with   [   ]
    a. Irish potato blight tragedy   b. Pacific oil spill tragedy
    c. Gujarat SARS tragedy   d. Punjab maize smut tragedy

70. Trepanosomes are characterized by   [   ]
    a. Undulating membrane   b. Multinuclear cyst
    c. Axoneme   d. Chromatoidal bodies

71. What is FALSE about Diatoms   [   ]
    a. Producers of free oxygen   b. Involved in Vit A and Vit D production
    c. Useful in Filters and abrasives   d. Agents of paralytic shellfish poisoning

72. Evolutionary transition forms between algae and protozoa   [   ]
    a. Chlorophyta [green algae]   b. Euglenophyta, [Euglenoids]
    c. Rhodophyta [Red algae]   d. Pyrrophyta [Dinoflagellates]

73. Interferons are produced by   [   ]
    a. Host cell infected by virus   b. Cells surrounding the infected cell
    c. Host immune cells   d. Tissue macrophages

74. Virtually no antibiotics are therapeutically effective against any viral diseases   [   ]
    a. TRUE   b. FALSE

75. Lymphokine production is   [   ]
    a. Humoral immunity   b. Cell mediated immunity
    c. Artificial immunity   d. None of these

76. Common causative agent of sore throat   [   ]
    a. *Klebsiella*   b. *Streptococcus*   c. *Staphylococcus*   d. *Pseudomonas*

77. One step growth curve is described for   [   ]
    a. Synchronus culture   b. Continuous culture
    c. Bacteriophages   d. Batch cultures

78. Production of GLycocalyx is enhanced by incorporating   [   ]
    a. Sugar in the medium   b. Extra proteins in the medium
    c. Calcium compound in the medium   d. Fatty acids in the medium

79. Inactive viral vaccine against Polio   [   ]
    a. NMR vaccine   b. DPT vaccine   c. Sabin vaccine   d. Salk vaccine

80. Cell free alcoholic fermentation was first understood by   [   ]
    a. Holley   b. Buchner   c. Schwann   d. Kary Mullis

81. Biostatistics finds application in the field of   [   ]
    a. Health service research   b. Biological sequence analysis
    c. Ecology and Population Genetics   d. All

82. The cell walls of Halobacteria are rich in   [   ]
    a. Polysaccharides   b. Glycoproteins
    c. Glycocalyx   d. Layers of Peptidoglycan

83. Replica plate technique requires   [   ]
    a. Sterile velveteen cloth   b. Sterile cotton cloth
    c. Sterile Millipore membrane   d. A replication machine

84. Targeting unique sequences of an organism chromosomes is the main function of   [   ]
    a. PCR Technique   b. DNA Probe technology
    c. Monoclonal Ab technology   d. Biosensor technology

85. *Bacillus popillae* is a biopesticide against   [   ]
    a. Aphids   b. Caterpillars   c. Japanese beetle   d. Rust mites

86. Microbe mediated Biotechnology on wheat is mainly to   [   ]
    a. Increase crop yield
    b. Obtain herbicide resistant varieties
    c. Reduce foliage and improve inflorescence
    d. Decrease gluten content

87. *Hfr* cell is the outcome of   [   ]
    a. F^- x F^+ cells
    b. Fusion of F plasmid in the bacterial chromosome
    c. F^- x F^- cells
    d. F^+ x F^+ cells

88. Hops are employed in the production of   [   ]
    a. Champagne   b. Wine   c. Beer   d. Whiskey

89. Which of these is quantitative immunological assay   [   ]
    a. RIA   b. SRID   c. Oudine method   d. a & b

90. Use of chemical herbicides causes   [   ]
    a. Retardation of photosynthetic rates in some plants
    b. Late ripening of fruits in some plants
    c. Non-chloropylous leaves in some plants
    d. All

91. Acid rains can be prevented by   [   ]
    a. Fitting automobiles with catalytic converters
    b. Use of low sulfur coal
    c. Both a & b
    d. Neither a nor b

92. Enveloped viruses acquire the envelop   [   ]
    a. When they are released from host cell
    b. when they are released from the cell membrane or nuclear membrane or ER
    c. completely through viral gene expression
    d. After they are released from the host cell

93. Spill of broth media on the work table should be cleaned by using   [   ]
    a. sponge   b. cotton   c. muslin cloth   d. paper towel

94. Which of these is recommended for preserving fresh fruits and vegetables [   ]
    a. Blanching before freezing   b. Blanching after freezing
    c. Avoid blanching   d. Peeling before blanching

95. Which of these is an opportunistic pathogen   [   ]
    a. *Mycobacterium tuberculosis*   b. *E.coli*
    c. *Bacillus anthrasis*   d. All

96. In chicken, waste and reproductive material empty in   [   ]
    a. Cloaca   b. Thymus   c. Spleen   d. Bursa of fabricus

97. The *chromatogram* in paper chromatography refers to   [   ]
    a. The glass container in which the run takes place
    b. the solution poured into the glass chamber
    c. The paper on which the substances are spot
    d. The solution sprayed on the paper

98. Whitlow occurs due to　　　　　　　　　　　　　　　　　　[　]
    a. *Neisseria gonorrhoeae*　　　　　b. *Lymphogranuloma venereum*
    c. Herpes simplex virus　　　　　　　d. *Staphylococccus aureus*

99. *Serratia, Thiobacillus, Micrococcus* are known for　　　　[　]
    a. Nitrification　　b. Denitrification　　c. Nitrosification　　d. Ammonification

100. A graft between individuals of different species　　　　　[　]
     a. Biograft　　　　b. Heterograft　　　c. Xenograft　　　d. Allograft

Answer Sheet – Grand Test 2

1.	2.	3.	4.	5.	6.	7.	8.	9.	10.
11.	12.	13.	14.	15.	16.	17.	18.	19.	20.
21.	22.	23.	24.	25.	26.	27.	28.	29.	30.
31.	32.	33.	34.	35.	36.	37.	38.	39.	40.
41.	42.	43.	44.	45.	46.	47.	48.	49.	50.
51.	52.	53.	54.	55.	56.	57.	58.	59.	60.
61.	62.	63.	64.	65.	66.	67.	68.	69.	70.
71.	72.	73.	74.	75.	76.	77.	78.	79.	80.
81.	82.	83.	84.	85.	86.	87.	88.	89.	90.
91.	92.	93.	94.	95.	96.	97.	98.	99.	100.

Assessment Sheet

Your Score	Observation
95 -100	**Very Good**: You have an excellent hold on the subject matter. You can definitely proceed to attempt the Grand Test 3.
80 – 95	**Good**: Your subject interest is fairly good. However, further practice and reading is required to excel.
60 – 80	**Satisfactory**: You have a good knowledge on the indepth facts. Nevertheless, you may have to check with your illustrations, examples and logical reasoning.
<60	**Not satisfactory:** You must read a lot and catch up with your basics, before you proceed with Grand Test 3. It is suggested you should read much more and make your self thorough with the individual topics.

Solution to Grand Test 2

1. a	2. c	3. c	4. b	5. d	6. b	7. a	8. b	9. b	10. c
11. c	12. a	13. c	14. a	15. c	16. c	17. a	18. a	19. b	20. c
21. d	22. d	23. b	24. a	25. b	26. c	27. a	28. c	29. d	30. c
31. b	32. a	33. a	34. b	35. a	36. c	37. d	38. b	39. c	40. d
41. c	42. a	43. b	44. b	45. a	46. c	47. a	48. b	49. a	50. d
51. b	52. a	53. c	54. b	55. c	56. a	57. c	58. a	59. d	60. d
61. b	62. b	63. c	64. a	65. c	66. d	67. b	68. d	69. a	70. a
71. d	72. b	73. a	74. a	75. b	76. b	77. c	78. a	79. d	80. b
81. d	82. b	83. a	84. b	85. c	86. b	87. b	88. c	89. d	90. a
91. c	92. b	93. d	94. a	95. b	96. d	97. c	98. c	99. b	100. c

6   *Sulfolobus is a thermophile with optimum growth temp between 78-88^{0}C.*

7.   *b, c and d are the gliding Myxobacteria*

12.   *Gloeocapsa is a cyanobacterium that fixes nitrogen anaerobically*

20   *Isolates of C.albicans produce hyphal outgrowth called germ tubes when incubated in serum suspension at 37^{0}C within 3h.*

23.   *The discovery of the enzyme reverse transcriptase in retroviruses provided an essential tool to produce complementary DNA's [cDNA]*

27.   *In a tower fermentor, the wort is introduced into the base of the tower and passes through a porous plug of yeast. The wort rises through the tower and conveniently leaves the fermentor via a yeast separation zone. Thus the major advantage of tower process is reduction of wort residence time from one week to 4-8h*

38.   *Scwann cells are a type of glial cells of the nervous tissue that surround axons*

39.   *Pyrenoids are special proteneicious structures that store starch in several algae*

40.   *The selectively permeable nature of plasma membrane helps in regulating the movement of material in and out of the cell thus maining homeostasis.*

53.   *Arthrosclerosis starts with the deposition of yellow fatty streaks containing a high portion of cholesterol in the inner coat of arteries.*

71.   *Dinoflagellates are responsible for paralytic shellfish poisoning. Diatoms are not known to exhibit this pathogenesis.*

78.   *Glycocalyx is particularly enhanced around bacterial cells on addition of sucrose in the medium*

79.   *Sabin vaccine is live attenuated vaccine against polio.*

98.   *Whitlow is a herpetic lesion on the finger that helps in its spread to genital or oral areas.*

C

Grand Test III

If you have scored A Grade in GT 2, you are in a position to take GT 3. The exercises in GT 3 are largely challenging. They have been framed to examine your logical and analytical thinking.

Thorough subject knowledge, compounded with genuine intelligence is required to crack most of these puzzles. Along with, there are also some through and through knowledge based questions that demand an in-depth know-how of Microbiology.

Note: Some of the following exercises may have more than one correct options. However, you are required to select 'the most appropriate' option as the correct one.

1. Generation of an F'cell is the result of   [   ]
   a. Aberrant reversion of F factor   b. Aberrant integration of F factor
   c. Interrupted mating of F^- and F^+ cells   d. a & c

2. DNA polymerization requires   [   ]
   a. DNA polymerase + 4deoxyribonucleotide triphosphates
   b. DNA polymerase + 4deoxyribonucleotide triphosphates + DNA template
   c. DNA polymerase + 4deoxyribonucleotide triphosphates + DNA template + Mg^{++}
   d. DNA polymerase + 4deoxyribonucleotide triphosphates + DNA template + Mg^{++} + a primer for initiation

3. The oriental fermented food that contains significant amount of mold biomass in the finished product   [   ]
   a. Natto   b. Tempeh   c. Poi   d. Kimichi

4. Which of these foods does not contain fermented Soya beans   [   ]
   a. Miso   b. Tofu   c. Sushi   d. Tempeh

5. Ochratoxin is a mycotoxin from   [   ]
   a. *Apergillus*   b. *Fusarium*   c. *Penicillin*   d. *Agaricus*

6. *Bacillus anthracis* and sheep prophylaxis is the original work of   [   ]
   a. Pasteur   b. Jenner   c. Koch   d. Chinese & Turks

7. Release of histamine from injured tissue causes   [   ]
   a. Elevated temp   b. Pain
   c. Accumulation of WBCs   d. All

8. Margination and Diapedesis are a part of   [   ]
   a. Complement fixation   b. Phagocytosis
   c. Humoral immune response   d. Delayed hypersensitivity

9. Burkitt's Lymphoma is caused by   [   ]
   a. *Actinomyces israelii*   b. *Amanita phalloides*
   c. *Epstein-Barr virus*   d. SV 40

10. The *ras* protein is [ ]
    a. An oncogene protein
    b. Polymerase repressor protein
    c. Trans membrane protein
    d. Cytokine

11. Tobacco Mosaic virus is transmitted by [ ]
    a. Mechanical inoculation
    b. Insect vectors
    c. Pollination
    d. soil-root-soil

12. *Curvularia lunata* is involved in [ ]
    a. Bioleaching
    b. Camphor degradation
    c. Degradation of petroleum
    d. Steroid biotransformation

13. Procaryotic ribosomes are less dense than eukaryotic ribosomes [ ]
    a. TRUE
    b. FALSE

14. Cytoplasmic streaming is absent in [ ]
    a. *E. coli*
    b. *Giardia*
    c. Epithelial cells
    d. *Chlorella*

15. Molecular chaperons play a role in [ ]
    a. Endospore formation
    b. Spore germination
    c. Protein folding
    d. Plasmid replication

16. Heat shock proteins are related to [ ]
    a. Lyophilization
    b. Chemotaxis
    c. Bacterial endotoxins
    d. Molecular chaperons

17. Transduction can occur most readily between [ ]
    a. Gram negative and Gram positive bacteria
    b. Bacteria of the same genus and different species
    c. Distantly related genera of bacteria
    d. Procaryotic and eukaryotic cells

18. Microbiologically safest drinking water is water stored in vessels made of [ ]
    a. Copper
    b. Silver
    c. Steel
    d. Glass

19. Which of these is involved in symbiotic nitrogen fixation [ ]
    a. *Mesorhizobium*
    b. *Sinorhizobium*
    c. *Bradyrhizobium*
    d. All

20. *Leg* Hemoglobin is absent in [ ]
    a. Cyanobacterial symbiotic system
    b. Frankia
    c. Parasponia
    d. a, b and c

21. Small circular infectious ssRNAs [ ]
    a. Virions
    b. Viroids
    c. Prions
    d. Sat viruses

22. Which of these have a relatively longer generation time [ ]
    a. *Yersinia*
    b. *Mycobacterium*
    c. *Klebsiella*
    d. *Salmonella*

23. Growth of microaerotolerant bacteria in a long tube of nutrient agar is [ ]
    a. Only on the surface of the agar
    b. All over the tube except the surface
    c. Just under the surface
    d. Only in the bottom most region of the tube

24. Mineral oil overlay on agar slants to preserve bacterial cultures [ ]
    a. Prevents dessication
    b. Minimises entry of O_2
    c. Removes water from the cells so that they become inactive
    d. a & b only.

25. Artistic patterns of bacterial colonies are obtained in [   ]
    a. Enriched media
    b. Undernourished media
    c. Semi solid media [0.5% agar]
    d. Gelatin based media

26. Pick out the FALSE statement [   ]
    a. Coxsackie virus is spread by sexual contact
    b. Dengue fever shares the same mosquito vector as yellow fever
    c. Hanta virus causes fatal respiratory disese
    d. Respiratory syncytial virus is associated with infant bronchiolitis

27. Hepatitis D cannot replicate independently without the presence of [   ]
    a. Hepatitis A   b. Hepatitis B   c. Hepatitis C   d. Hepatitis E

28. Peptic ulceration is caused by [   ]
    a. *Clostridium difficle*
    b. *Staphylococcus aureus*
    c. *Listeria monocytogens*
    d. *Helicobacter pylori*

29. A 24y woman with a firm, oval and round and painless sore on the lower lip
    must be subjected to [   ]
    a. VDRL test   b. Tri-dot test   c. Shick test   d. Mantoux test

30. A 12y old boy complains of nausea, mild diarrhea and slight abdominal pain after 8h of
    consuming ladoo bought from a roadside vender. The suspected pathogen could be [   ]
    a. *Clostridium botulinum*
    b. *Shigella dysenteriae*
    c. *Staphylococcus aureus*
    d. *Brucella*

31. Staining for metachromatic granules helps in the diagnosis of [   ]
    a. Tetanus   b. Diphtheria   c. Pertussis   d. All

32. The suggested test for diagnosis of a case of fever, mild diarrhea, vomiting and nausea [   ]
    a. Stool culture   b. Widal test   c. Dick test   d. Kanagawa test

33. Presence of encapsulated yeast in pathological specimen is suggestive of [   ]
    a. *Candida*
    b. *Helminthosporium*
    c. *Cryptococcus*
    d. *Cunninghamella*

34. Action of pepsin and papain enzymes proved helpful in explaining [   ]
    a. Enzyme kinetics
    b. Mechanism of action of bacterial enterotoxins
    c. Structure of a typical Immunoglobulin
    d. Mechanism of complement fixation

35. The medium used to demonstrate ammonifiers in soil [   ]
    a. Peptone water   b. Nitraite broth   c. Nitrite broth   d. Mannitol broth

36. Monoclonal antibodies specific for human antigens are generally raised in [   ]
    a. Mice   b. rabbits   c. Horses   d. Chimpanzees

37. The roots of more than 80% of land plants are found to be symbiotically associated with [   ]
    a. *Rhizobium*   b. Fungi   c. Viruses   d. Algae

38. The air that flows into laminar flow bench is [   ]
    a. Always in parallel beams
    b. In circular microcurrents
    c. In large cyclonic currents
    d. In random directional flow

39. Uv treatment is best to disinfect [   ]
    a. Animal fodder   b. Test tubes   c. Water   d. Syrups and tablets

40. Pascalization [ ]
    a. Application of high pressure in food processing
    b. Use of high temp to process dairy products
    c. Radiation treatment to onions and potatoes
    d. Microwave radiation of bakery products

41. Soft potatoes with or without fluids ooze may be due to [ ]
    a. *Trichothesium roseum*   b. *Xanthomonas compestris*
    c. *Erwinia caratovora*   d. *Botrytis allii*

42. *Campylobacter jejuni* was earlier known as [ ]
    a. *Vibrio jejuni*   b. *Akinetobacter jejuni*
    c. *Bacillus jejuni*   d. *Pseudomonas jejuni*

43. The main site for synthesis of cytoplasmic proteins in eukaryotic cell [ ]
    a. Mitochondria   b. Golgi complex
    c. Endoplasmic Reticulum   d. Nucleus

44. Secondary lysozome fusion is a mechanism for a cell to degrade its own cytosolic proteins. [ ]
    a. TRUE   b. FALSE

45. An objective lens that claims to have a numerical aperture 1.25 gives best resolution when the sample is viewed in [ ]
    a. Air   b. Oil   c. Vaccum   d. Water

46. In metric equivalent, *nano* corresponds to [ ]
    a. One hundredth   b. One Thousandth   c. One millionth   d. One billionth

47. In clinical diagnostics clean catch method refers to collection of this sample [ ]
    a. Seminal fluid   b. Stool   c. Urine   d. Sputum

48. Which of these finds use in diagnosis of viral infections [ ]
    a. Phage typing   b. Selective enrichment
    c. Enzyme immunoassay   d. Gram's staining

49. Bacteria gain resistance to heat and desiccation due to [ ]
    a. Capsule   b. Envelope   c. Endospore   d. Pili

50. Prusiner won the Nobel Prize for the study on [ ]
    a. Prions   b. Discontinuous genes
    c. Identification of the Dengue fever virus   d. Anaphylaxis

51. The magnetite present in the magnetosmes of magnetotactic bacteria enable them to [ ]
    a. Generate magnetic fields   b. Attract iron from the soil
    c. Respond to magnetic fields   d. None of these

52. The angler fish in the deep ocean depths is symbiotically associated with [ ]
    a. Bioluminiscent bacteria   b. *Halobacterium*
    c. Iron Sulfur bacteria   d. Barophilic bacteria

53. Identification of carcinogens can be made by the [ ]
    a. Fluctuation Test   b. Holley's Test   c. Ames Test   d. Replica plating

54. Extracellular digestion of foods [ ]
    a. Fungi   b. Protozoa   c. Viruses   d. All

55. *Mucor* produces [ ]
    a. Zygospores   b. Sporangiospores   c. Both   d. None

56. The fungi form a plasmodium body [ ]
    a. Slime molds   b. Sac fungi   c. Club fungi   d. All

57. Elementary bodies [EBs] are the infectious forms of [   ]
    a. *Rickettsia*   b. Chlamydia   c. Archaea   d. Antinobacteria

58. Anoxygenic photosynthesis is seen in [   ]
    a. Purple sulfur bacteria
    b. Green sulfur bacteria
    c. Purple non-sulfur baceria
    d. All

59. Zoogloeal mass is formed in [   ]
    a. Activated sludge process
    b. Trickling filter bed
    c. Anaerobic digestion of sludge
    d. Oxidation ponds [lagoons]

60. BOD value will be highest in [   ]
    a. Spring water
    b. Lake water
    c. Effluent from a dairy plant
    d. Sewage water

61. In routine analysis of drinking water sulfite reducing anaerobes are detected using [   ]
    a. Bismuth sulfite agar
    b. Endo agar
    c. Differential reinforced clostridial agar
    d. TCBS agar

62. Pick the FALSE statement about Epstein Barr virus [EBV] [   ]
    a. Also called Human Hepes virus 4
    b. Commonly cause inctious mononucleosis
    c. Transmitted by Contaminated meat
    d. It is a dsDNA virus

63. Brucellosis is also called [   ]
    a. Undulant fever
    b. Malta fever
    c. Mediterranian fever
    d. All.

64. Unidirectional transfer of genetic material is seen in [   ]
    a. Prokayotes   b. Eukaryotes   c. Both   d. None

65. Gram negative aerobic rods [   ]
    a. Sarcinae
    b. Methanobacteriaceae
    c. Rickettsiaceae
    d. Legionellaceae

66. Fever, headache, chest pain, nausea and vomiting are symptoms of [   ]
    a. Botulism   b. Meningitis   c. Q Fever   d. Peurperal fever

67. Pick out the WRONG pair [   ]
    a. Citrus canker – *Xanthomonas*
    b. Black rust of wheat – *Puccinia*
    c. Blast of rice – *Ustilago*
    d. Powdery mildew – *Erysiphae*

68. Which of these is FALSE statement with respect to killed vaccines [   ]
    a. They must be administered by intravascular injection
    b. They are more effective against systemic viruses than against viruses that replicate in local mucosal sites
    c. They are more stable under environmental conditions when compared to attenuated vaccines
    d. Relatively smaller amounts of Ag is required for injection before induction of immunity

69. The process in which the plant fibres [jute, hemp etc] are separated by action of microorganisms [   ]
    a. Brining   b. Worting   c. Retting   d. Pitching

70. Glycerin can be produced fermentatively by the action of [   ]
    a. *Bacillus spp*
    b. *Saccharomyces spp*
    c. *Ashbya spp*
    d. *Claviceps spp.*

71. Contributions to soil microbiology [   ]
    a. Lister & Fleming
    b. Redi & Spallanzani
    c. Pasteur & Koch
    d. Beijerinck & Winogradsky

72. Dixon & Jolly are associated with [ ]
    a. Active absorption
    b. Cohesion of water and transpiration pull
    c. Turgur pressure
    d. Process of Guttation

73. Phytoplankton in water grows in abundance in the [ ]
    a. Limnetic zone   b. Profundal zone   c. Benthic zone   d. Aphotic zone

74. The functional role of an organism in a given ecosystem is described as [ ]
    a. Ecological pyramid
    b. Ecological niche
    c. Batesian mimicry
    d. Ecotype

75. Which of these association can survive in snowcapped mountaneous regions [ ]
    a. Plant root – Fungus
    b. Crab- Sea anemone
    c. Cyanobacteria – Fungi
    d. Shark – Remora fish

76. Proto co-operation is involved in [ ]
    a. Predation   b. Mutualism   c. Commensalism   d. Ammensalism

77. The spectrum of life at the level of genetic, generic and ecological characteristics is referred to as [ ]
    a. Biome   b. Biodiversity   c. Ecosystem   d. Biomass

78. What is the most desired feature for formation of a biofilm [ ]
    a. Flagellation
    b. Capsulation
    c. Sporulation
    d. Aggregation and co-aggregation

79. Yeast is inoculated in a sucrose solution and incubated in a sealed can. The possible pathway for metabolism will be [ ]
    a. TCA   b. ETC   c. Acetone-butanol   d. None of these

80. In the preparation of nutrient agar, when is the correct time to check and adjust the pH [ ]
    a. Before addition of agar
    b. After addition of agar
    c. After digesting the agar in a boiling water bath.
    d. Any time; it does not make much difference

81. Pigmented aquatic thallophytes [ ]
    a. Microalgae   b. Microfungi   c. a & b   d. Niether a nor b

82. Flagella are absent in [ ]
    a. Algae   b. Protozoa   c. Cyanobacteria   d. All

83. Which of these show sterol dependence for growth [ ]
    a. Rickettsia & Mycoplasma
    b. Mycoplasma and Protozoa
    c. Mycoplasma, Rickettsia & Protozoa
    d. Rickettsia & Archaea

84. Which of these are the smallest by size [ ]
    a. Eubacteria   b. Protozoa   c. Rickettsia   d. Mycoplasmas

85. Pick the WRONG pair of parasite and its target system [ ]
    a. *Enterobius* – Intestine
    b. *Brugia* – Lymphatic system
    c. *Necator* –Lungs
    d. *Loa loa* – Subcutaneous tissue

86. Pick the WRONG pair of parasite and its common name [ ]
    a. *Trichuris trichura* – Whipworm
    b. *Ascaris* – Tapeworm
    c. *Ancylostoma* – Threadworm
    d. *Taenia*-Hookworm

87. Arrange the steps A, B, C and D in the correct sequence for Gram's staining   [   ]
    A. Stain with saffronin   B. Flood Gram's Iodine
    C. Treat with alcohol   D. Stain with Crystal violet
    a. D, C, B, A   b. D, B, C, A   c. A, B, C, D   d. B, D, A, C

88. Arrange the steps A, B, C & D in the correct sequence for sewage treatment   [   ]
    A. Sludge composting   B. Coarse straining to remove solid wastes
    C. Activated sludge process   D. Sludge press
    a. C, D, A, B   b. B, A, D, C   c. A, B, C, D   d. B, C, D, A

89. Arrange the steps A, B, C & D in the correct sequence for Manufacture of Beer   [   ]
    A. Boiling the wort   B. Malting   C. Mashing   D. Adding Hops
    a. A, D, C, B   b. C, B, D, A   c. B, C, A, D   d. B, D, A, C

90. Arrange the steps A, B, C & D in the correct order of adding reagents for Staining VAM fungi   [   ]
    A. Glycerin   B. Lactophenol cotton blue
    C. KOH + HCl   D. Tryphan blue
    a. C, D, A, B   b. A, D, C, B   c. B, C, D, A   d. A, B, C, D

91. Arrange the steps A, B &C in the correct sequence in the PCR technique   [   ]
    A. Denaturation   B. Annealing   C. Polymerization
    a. B, C, A   b. A, B, C   c. A, C, B   d. C, A, B

92. Arrange the steps A, B, C & D in the correct sequence for T_4 Phage replication cycle   [   ]
    A. Adsorption   B. Synthesis & Assembly
    C. Lysis & Release   D. Penetration
    a. A, D, B, C   b. A, C, B, D   c. B, C, D, A   d. A, B, C, D

93. Which of these is the *Venus flower basket*   [   ]
    a. *Euplectella*   b. *Hyalonema*   c. *Chalina*   d. *Euspongia*

94. Which of these is a horse disease?   [   ]
    a. Scrapie   b. Glanders
    c. Lumpy skin disease   d. Ectromelia

95. Which DNA contains *hydroxymethyl cytosine* in place of usual cytosine.   [   ]
    a. Actinobacteria   b. Mycoplasmas   c. Archaeobacteria   d. Bacteriophage

96. The Exon Valdez oil spill tragedy was handled by   [   ]
    a. Bioremediaion   b. Biofouling   c. Biostimulation   d. Bioaugmentation

97. Dendrology is the study of   [   ]
    a. Ecological calamities   b. Plant amphibians
    c. Trees   d. Nerve cells

98. The actual coding region of every prokaryotic gene begins with   [   ]
    a. 3'TAC5'   b. 5'AUG3'   c. 5'TATAAT3'   d. 5'AGGA3'

99. The private company Celera gained international fame due to its work on   [   ]
    a. Scanning tunneling microscopy   b. PCR technology
    c. Human Genome Project   d. Shotgun method

100. BLAST and FASTA are   [   ]
    a. Text addresses for web pages
    b. The figurative expressions of evolutionary relationship between organisms
    c. Tiny chips that contains thousands of immobilized molecules in a definite pattern.
    d. Sequence alignment algorithms

Answer Sheet – Grand Test 3

1.	2.	3.	4.	5.	6.	7.	8.	9.	10.
11.	12.	13.	14.	15.	16.	17.	18.	19.	20.
21.	22.	23.	24.	25.	26.	27.	28.	29.	30.
31.	32.	33.	34.	35.	36.	37.	38.	39.	40.
41.	42.	43.	44.	45.	46.	47.	48.	49.	50.
51.	52.	53.	54.	55.	56.	57.	58.	59.	60.
61.	62.	63.	64.	65.	66.	67.	68.	69.	70.
71.	72.	73.	74.	75.	76.	77.	78.	79.	80.
81.	82.	83.	84.	85.	86.	87.	88.	89.	90.
91.	92.	93.	94.	95.	96.	97.	98.	99.	100.

Solution to Grand Test 3

1. a	2. d	3. b	4. c	5. a	6. a	7. d	8. b	9. c	10. a
11. a	12. d	13. a	14. a	15. c	16. d	17. b	18. a	19. d	20. d
21. b	22. b	23. c	24. d	25. b	26. a	27. b	28. d	29. a	30. c
31. b	32. b	33. c	34. c	35. a	36. a	37. b	38. a	39. c	40. a
41. c	42. a	43. c	44. b	45. b	46. d	47. c	48. c	49. c	50. a
51. c	52. a	53. c	54. a	55. c	56. a	57. b	58. d	59. b	60. c
61. c	62. c	63. d	64. a	65. d	66. c	67. c	68. d	69. c	70. b
71. d	72. b	73. a	74. b	75. c	76. b	77. b	78. d	79. d	80. a
81. a	82. c	83. b	84. d	85. c	86. a	87. b	88. d	89. c	90. a
91. c	92. a	93. a	94. b	95. d	96. a	97. c	98. a	99. c	100. d

4   *Sushi is a fermented fish product.*

5.   *Ochratoxins are released by Aspergillus species. These are known to cause renal tumors and spread through coffee, wheat flour, bread and corn*

13   *Ribosomes of prokaryotes are often called 70S ribosomes while those eukaryotes is 80S ribosomes.*

14   *Prokaryotes lack cytoplasmic streaming*

16.   *Heat shock proteins are chaperons that protect the cells from thermal damage and stress.*

26.   *Coxackie virus is spread by oro-fecal route*

29.   *The primary stage of syphilis is characteristic of a painless sore on the lips, genitals or rectum called chancre. Syphilis is diagnosed by the VDRL test.*

30.   *Symptoms of Staphylococcal food poisoning appear within 8 to 12h of consumption of contaminated food that mainly includes sweets and meat.*

32.   *The most suspected infection in this case should be typhoid or paratyphoid fever as per the symptoms. Hence Widal test should be the first choice for diagnosis.*

42.   *Campylobacter jejuni are non-sporeforming, oxidase positive, gram negative, curved rods that shared features of Vibrio. However, based on their biochemical, serological and G+C ratio, they were re-classified as Campylobacter.*

46. *$1nm = 0.000\ 000\ 0001m = 10^{-9}m$*

52. *The Angler fish harbours symbiotic bioluminescent bacteria in its long 'lure' that lights up the dark ocean underwater and attracts prey.*

55. *Zygospores and Sporangiospores are the sexual and asexual spores of the fungus Mucor*

67. *Blast of rice is caused by Pyricularia oryzae. Ustilago is associated with smut diseases*

75. *Cyanobacteria – fungus = Lichens that can survive the adverse condition in the snowcapped mountainous region*

78. *Aggregation is important in the development of microcolonies that lead to eventually biofilm architecture*

81. *Microalgae are pigmented while the microfungi are colorless*

85. *Necator is an intestinal parasite*

86. *Ascaris = Round worm, Ancylostoma = Hookworm, Taenia = Tapeworm*

Pictorial Quiz

1. Match the pictures of the Microbiologists with their contribution

A

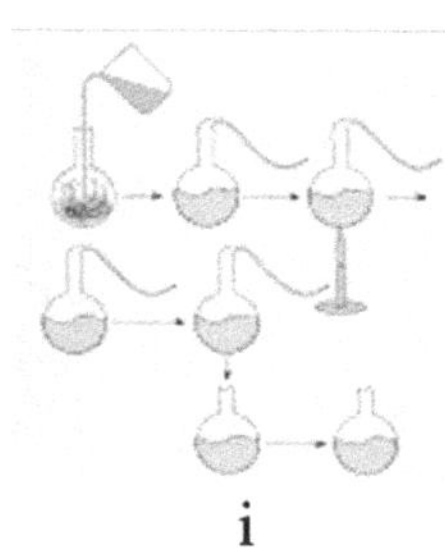

i

B

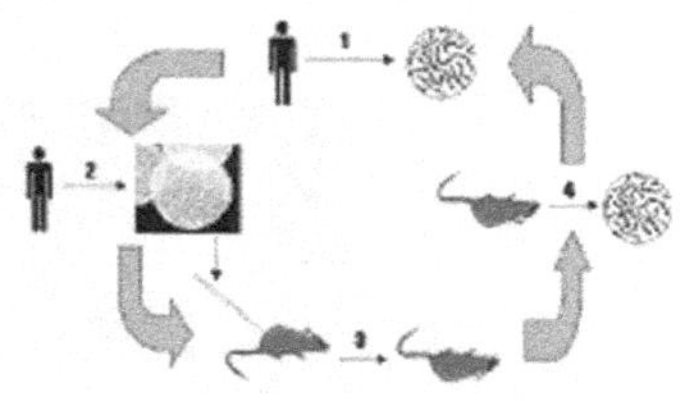

ii

C

iii

D

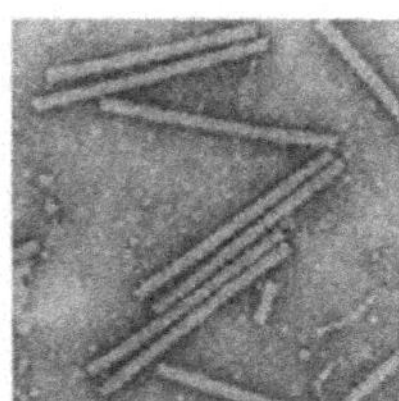

iv

2. The following painting depicts insult to the original thoughts of a physician. Who was that physician. Select from the three pictures given below the painting.

A

B

C

3. Match the flagellation with the corresponding pictures

A.	Peritrichous	i.		
B.	Lophtrichous	ii.	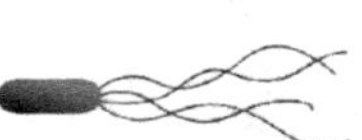	
C.	Monotrichous	iii.	 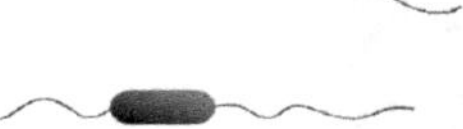	
D.	Amphitrichous	iv.	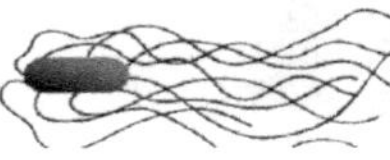	

4. Label the pictures A to J as per the morphological features Choose from: Coccobacilli, Bacilli, Streptococci, Tetrads, Slender rods, Filamentous rods, Staphylococci, Spirilla and Diplococci.

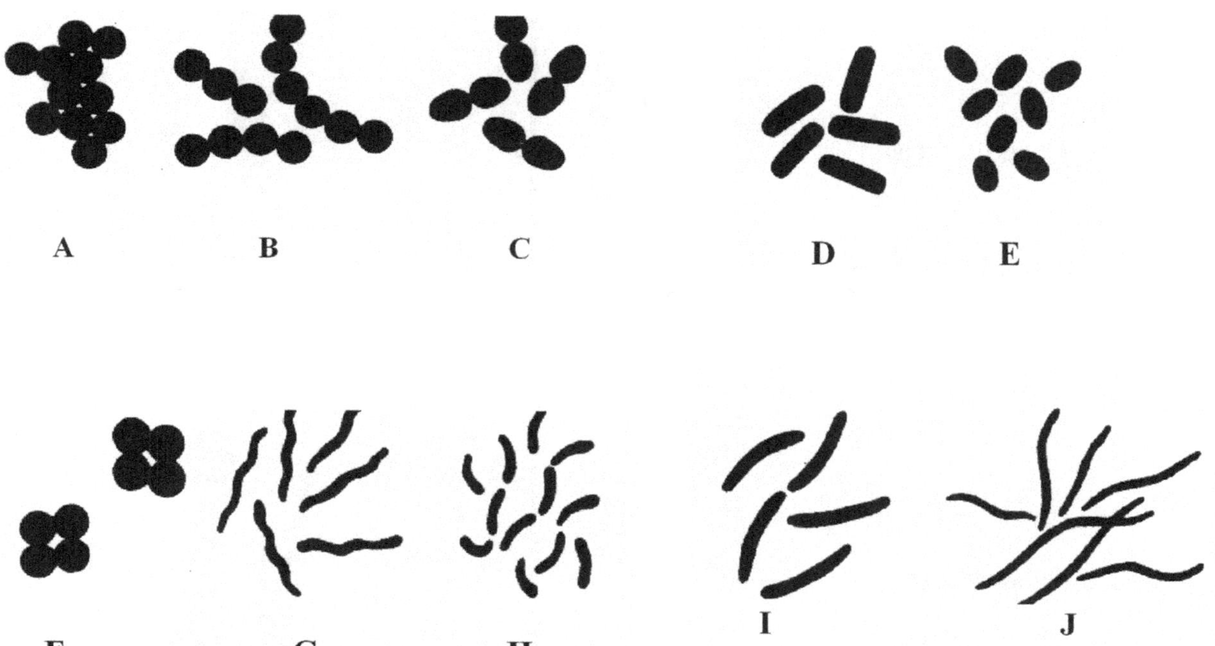

A   B   C   D   E

F   G   H   I   J

5. Identify the type of hemolysis exhibited

    a.   Alpha

    B.   Beta

    C.   Gamma

    d.   Delta

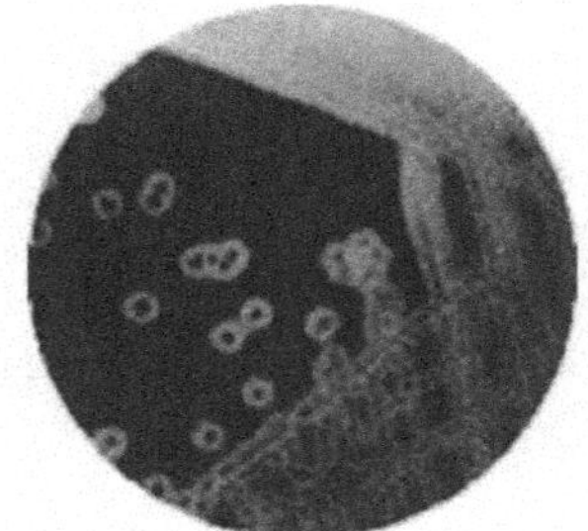

6. Match the Biochemical tests with their correct pictures

A. Coagulase test

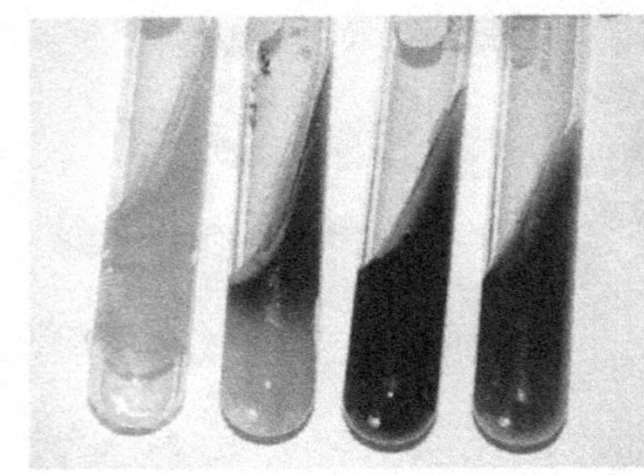

i

B. Catalase test

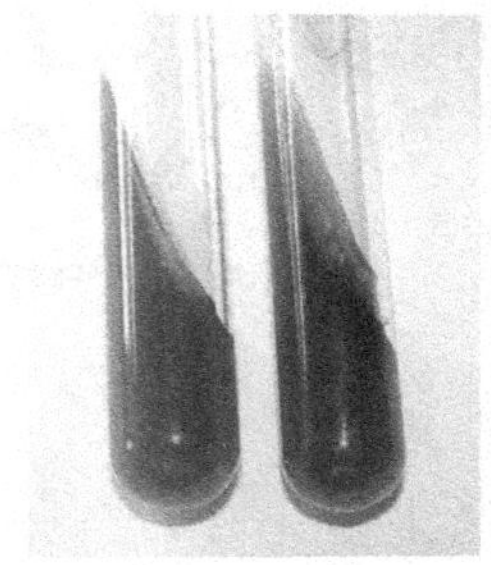

ii

C. Citrate utilization

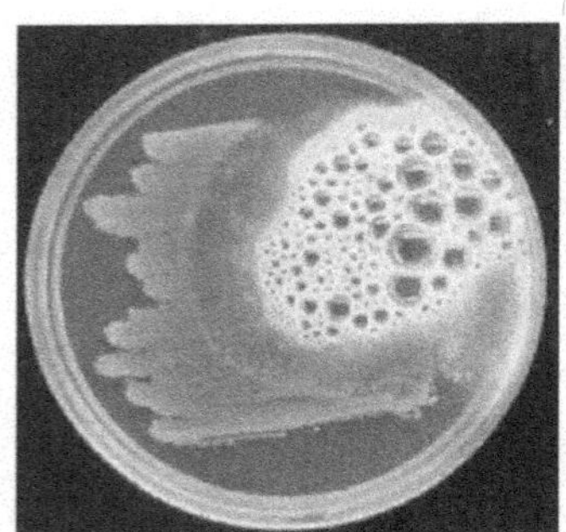

iii

D. Triple Sugar Iron (TSI)

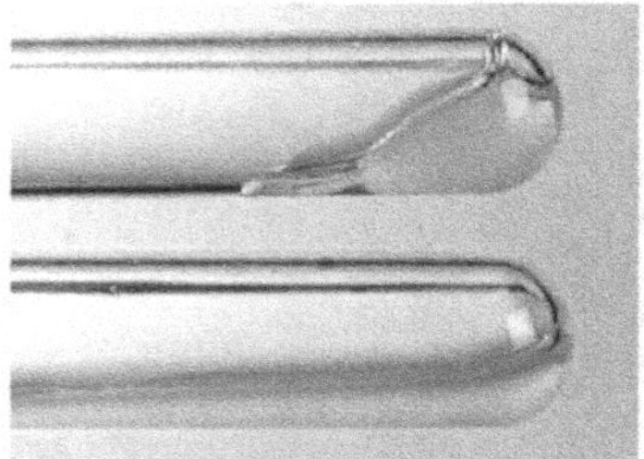

iv

7. Identify the test carried out

    a. Sugar fermentation test

    b. Indole test

    c. MR test

    d. VP test

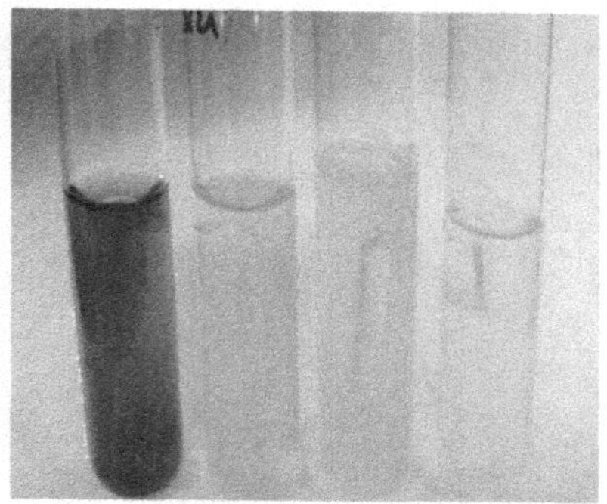

8. The colonies in the picture are showing

    a. Phosphotase activity

    b. DNAse activity

    c. Lecithinase activity

    d. Amylase activity

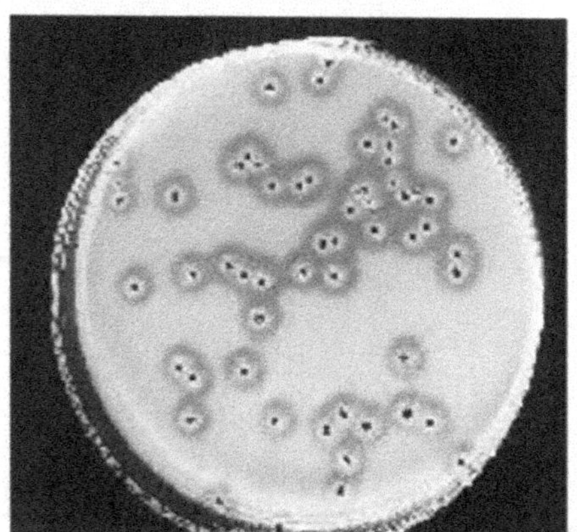

9. Identify the culture medium used

    a. Mac Conkey agar

    b. EMB agar

    c. Salmonella Shigella agar

    d. TCBS agar

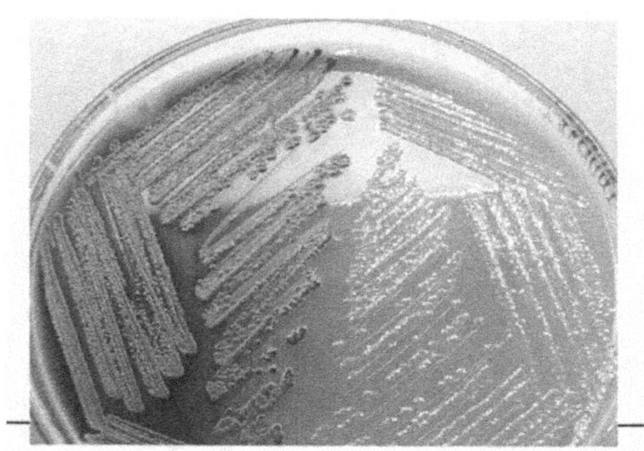

10. Based on the growth, choose the most probable medium

    a. Brain Heart

    b. Muller Hinton agar

    c. Nutrient agar

    d. Potato Dextrose Agar

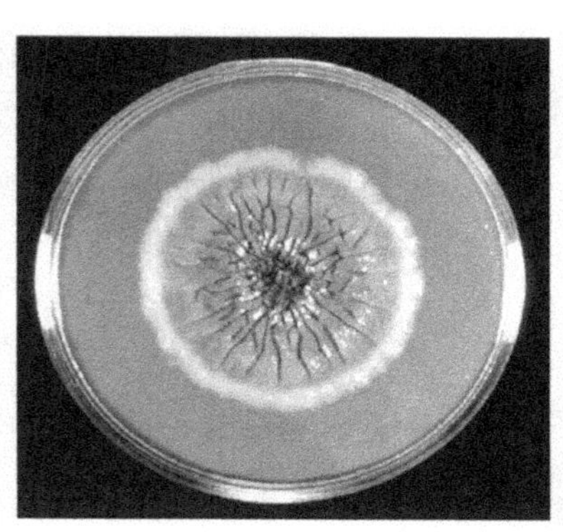

11. Match the dairy products with their correct pictures

i

ii

iii

iv

A. Swiss Chese

B. Kefir Butter

C. Cottage Cheese

D. Cheddar Cheese

12. Identify this oriental fermented food

    a. Miso

    b. Ang-khak

    c. Tempeh

    d. poi

13. Which of these is primarily a lactic fermented product?

A   B   C

14. Identify the oriental fermented food

    a. Tofu Ju

    b. Sausage

    c. Natto

    d. Dill Pickle

15. This Indian fermented food is the outcome of fermentation by

    a. *Leuconostoc and Pediococcus*

    b. *Glucanobacter and Acetobacter*

    c. *Monascus perpureus*

    d. *Aspergillus oryzae*

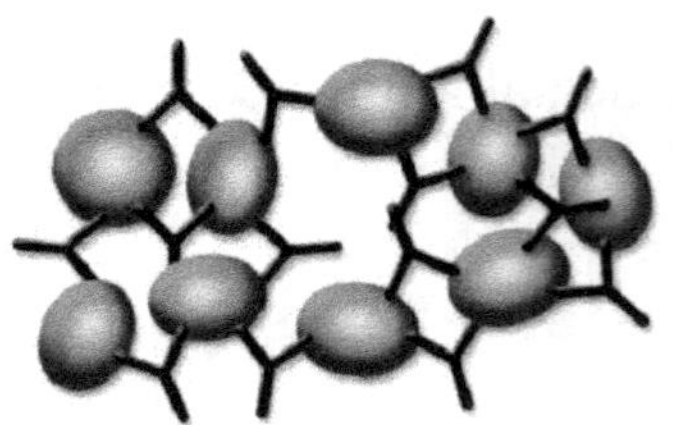

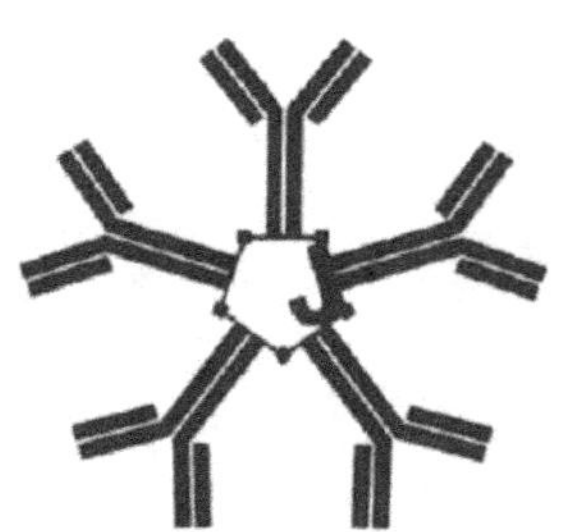

16. Identify the immunological reaction
    a. Precipitation
    b. Agglutination
    c. Flocculation

17. The picture shows
    a. Secretory Antibody
    b. Transplacental Antibody
    c. Antibody against parasites
    d. Millionaire Antibody

18. Which of these techniques is quantitative?

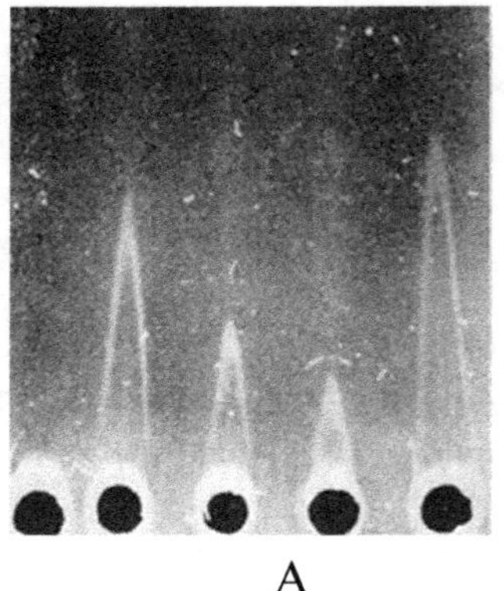

A

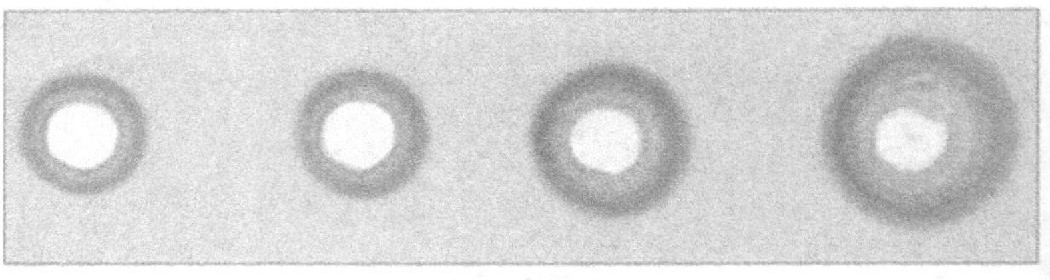

B

a. A
b. B
c. Neither A nor B.
d. Both A & B

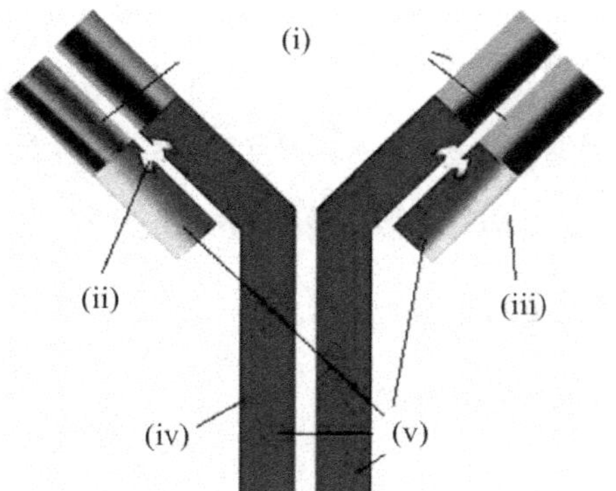

19. Which of the parts should be labeled as *Variable Regions*
    a. i
    b. ii
    c. iii
    d. iv

20. Match the machinery with its correct picture

a. Impeller

i

b. Baffles

ii

c. Distillation apparatus

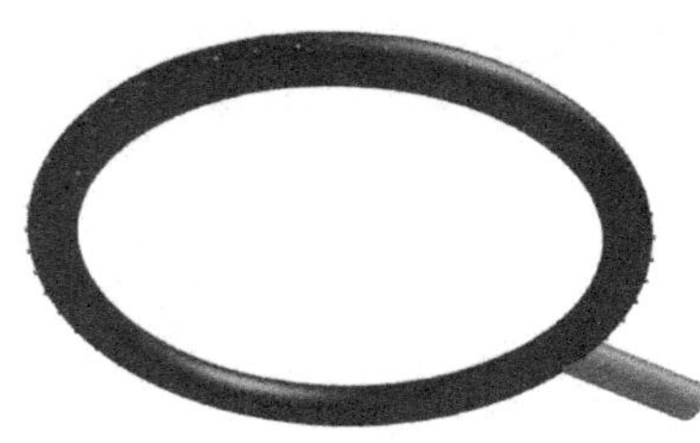

iii

d. Sparger

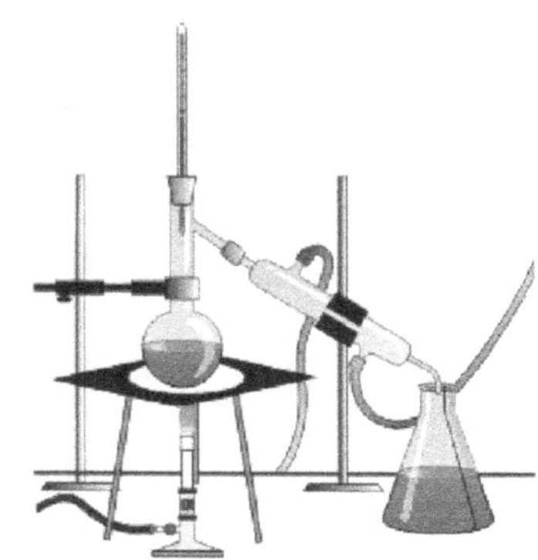

iv

iv

21. Match the disease with its correct picture

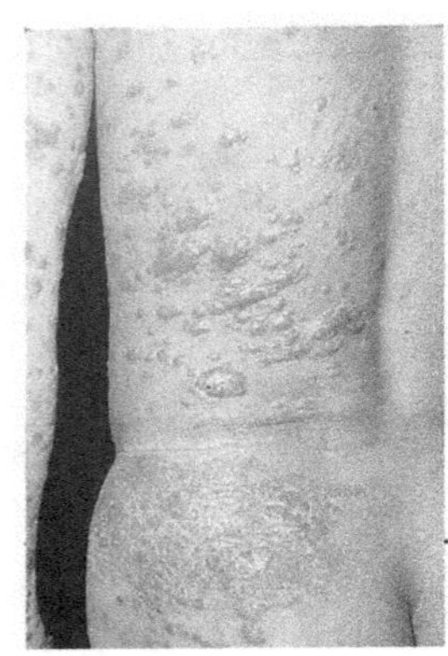

a. Syphilis

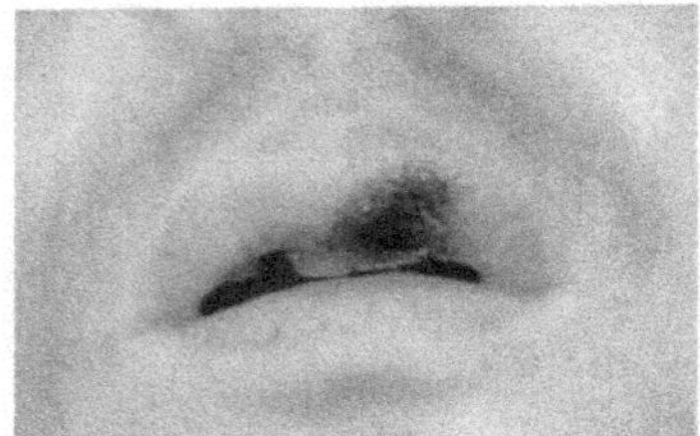

b. Anthrax

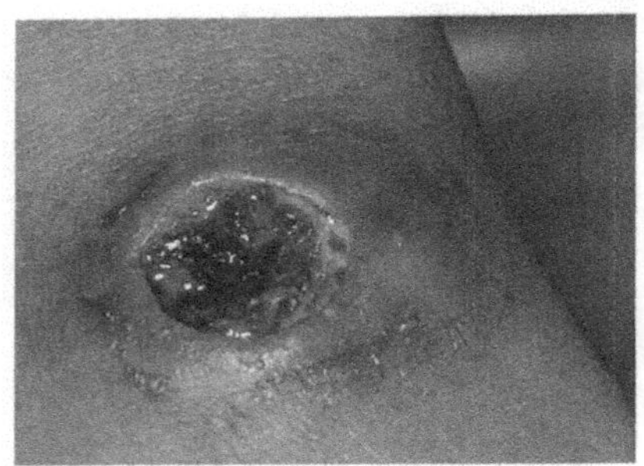

c. Candidiasis

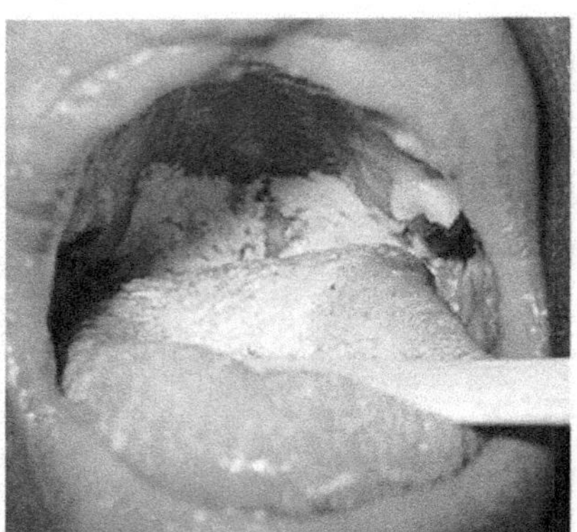

d. Lepromatous leprosy

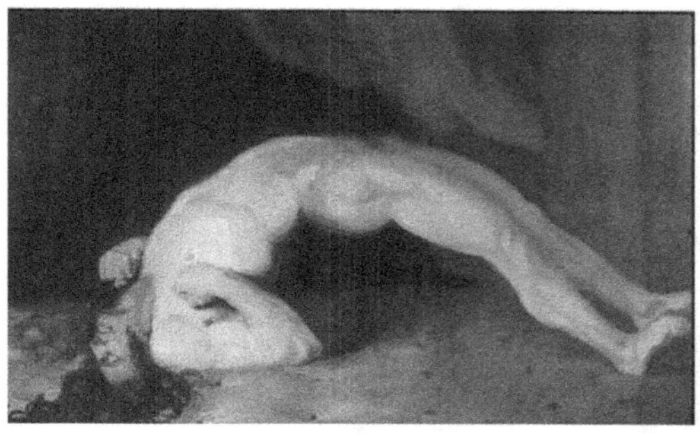

22. What disease is depicted in the painting?

    a. Polio

    b. Tetanus

    c. Break bone fever

    d. Kuru

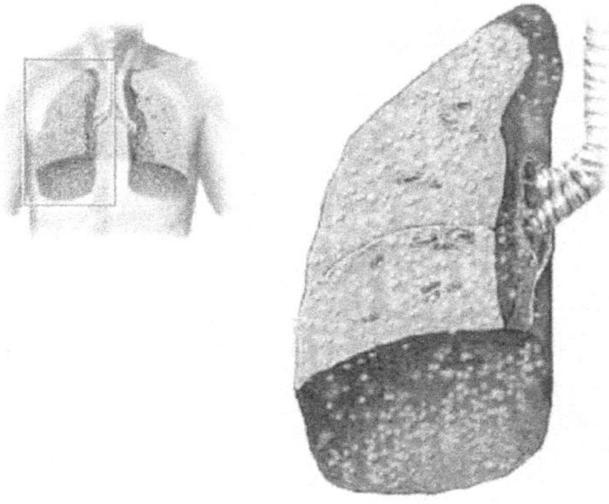

23. The picture shows

    a. Lung abscesses by *Pseudomonas*

    b. Granulomas by *Mycobacteria*

    c. Pustules by pneumococci

    d. Lesions by Corynebacterium

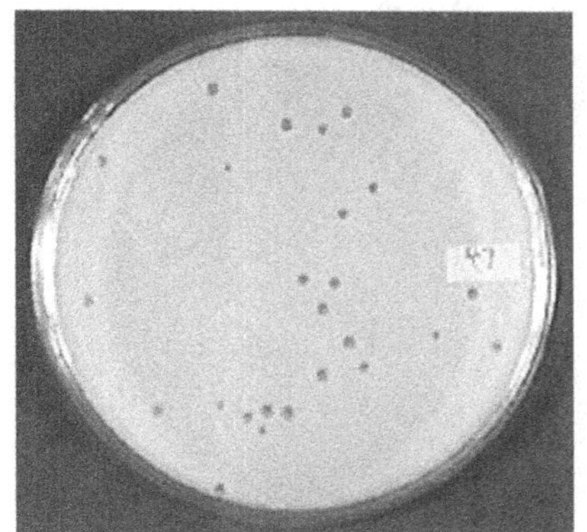

24. The clear areas in the plate indicate

    a. viral plaques

    b. Lecithinase activity

    c. Gas production

25. Which of these is Kirby-Baner Technique.

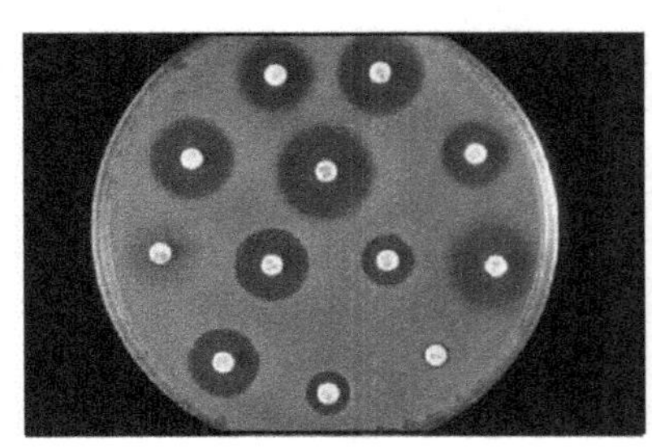

(i)

(ii)

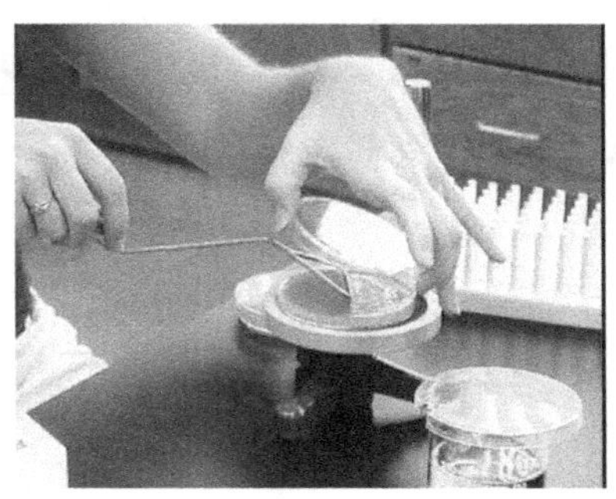

(iii)

26. Name the organisms Shown in the following pictures

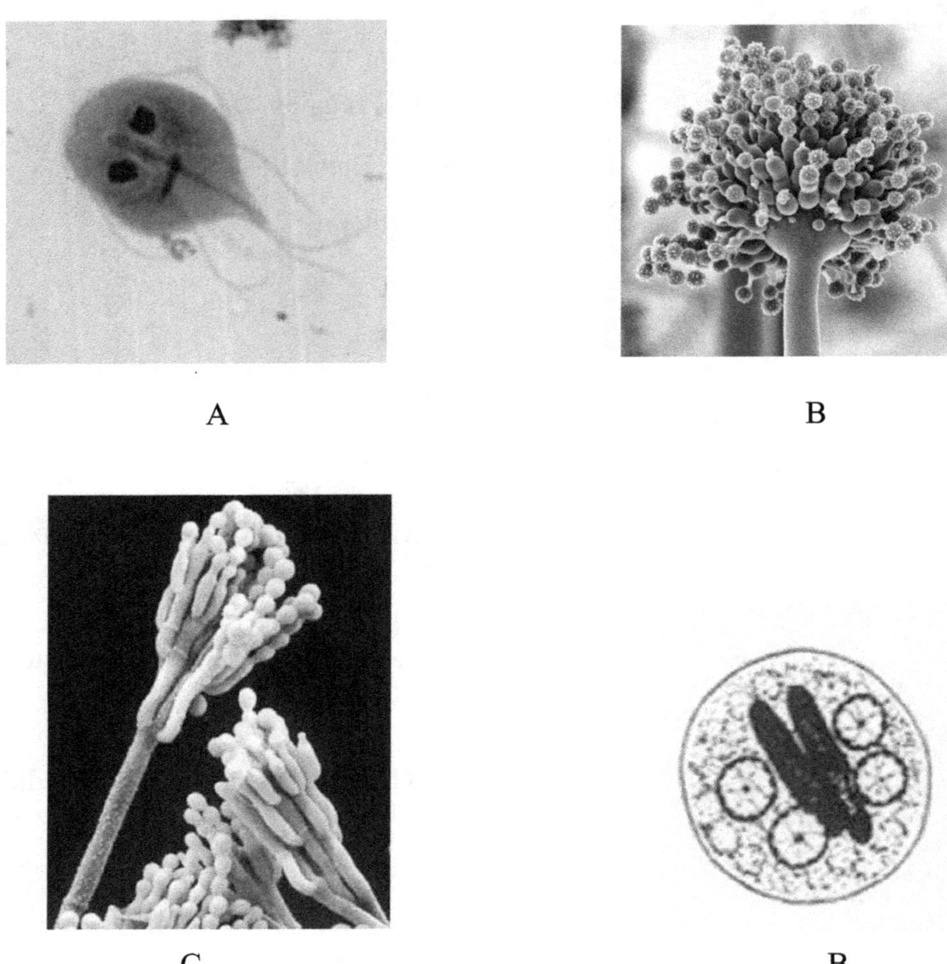

A

B

C

B

27. Which of these pictures is that of *Fusarium*

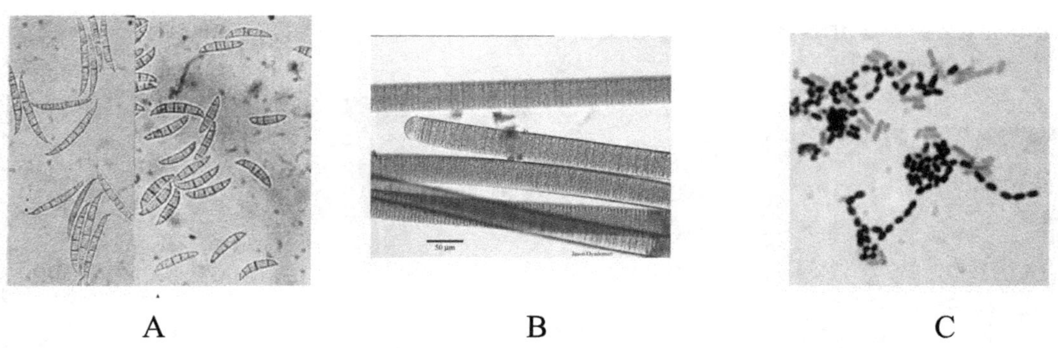

A

B

C

28. Match the organism with its correct picture

a. *Spirogyra*

b. *Acetabularia*

c. *Volvox*

d. *Nostoc*

i

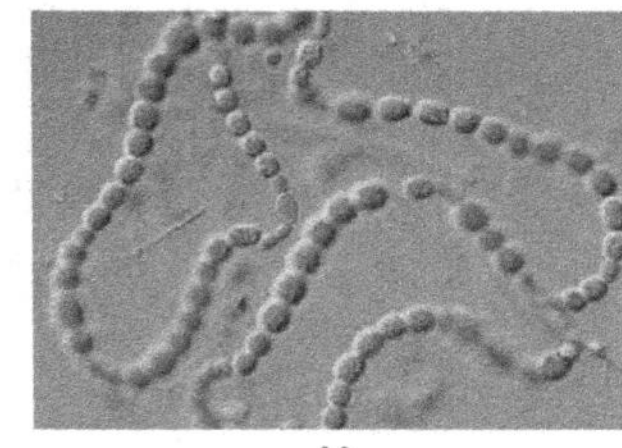

ii

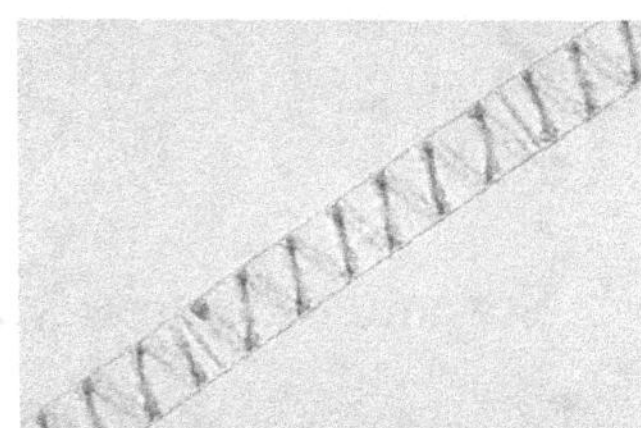

iii

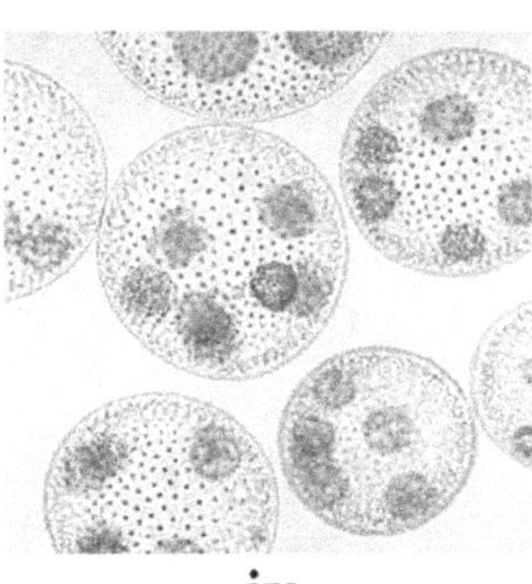

iv

30. Match the staining technique employed in each case

a.   Fluorescence Microscopy

b.   Transmission Electron Microscope

c.   Scanning Electron Microscopy

d.   Phase contrast Microscopy

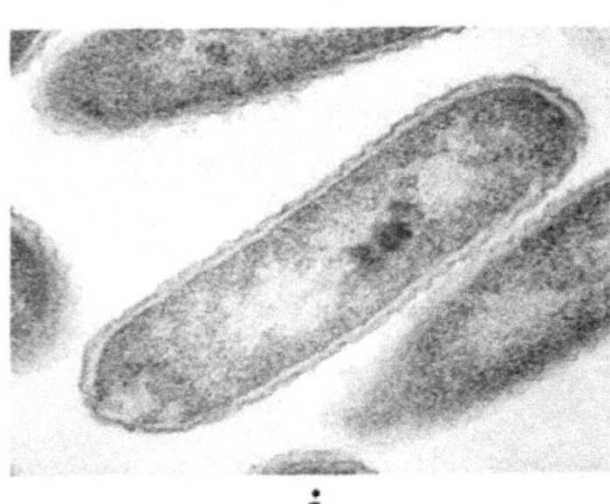

i

ii

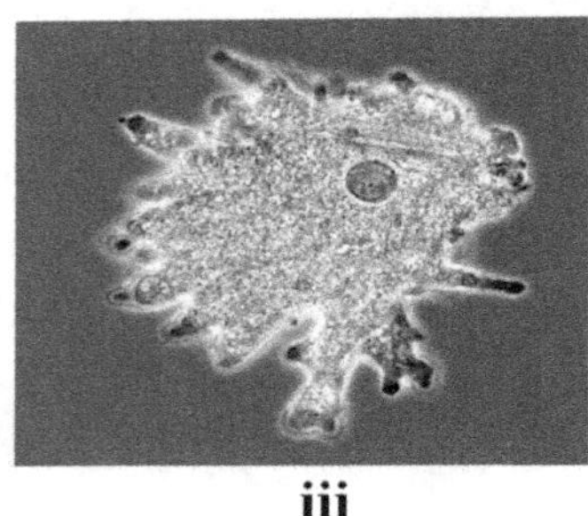

iii

iv

30. Match the staining technique employed in each case

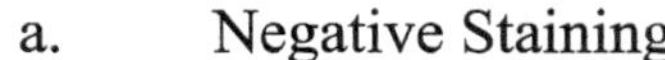

a. Negative Staining

b. Gram's Staining

c. Capsule Staining

d. Acid Fast Staining

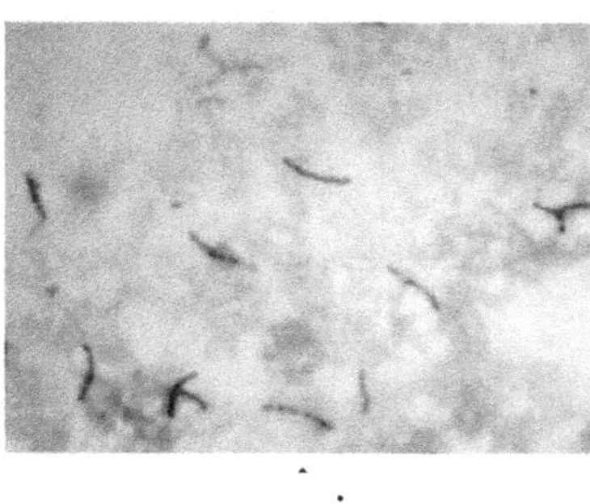

i

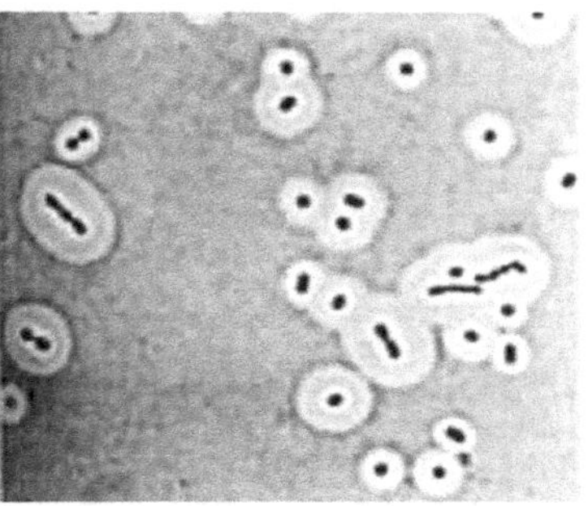

ii

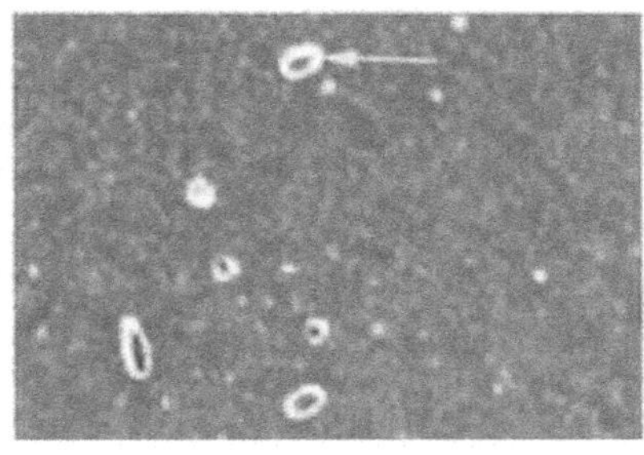

iii

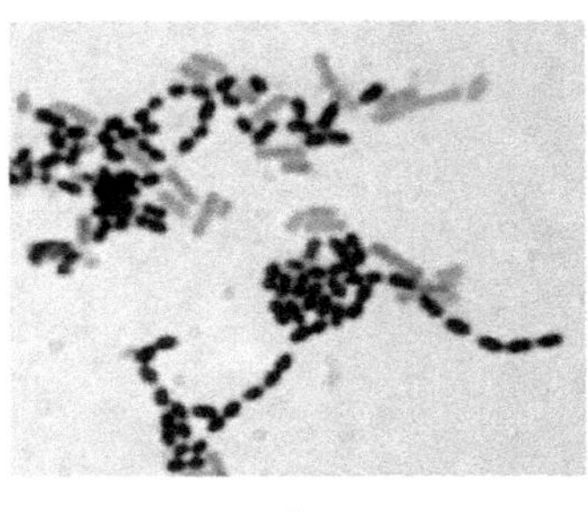

iv

31. Match the Plant disease with the picture in each case

a. Smut

b. Blight

c. Rust

d. Canker

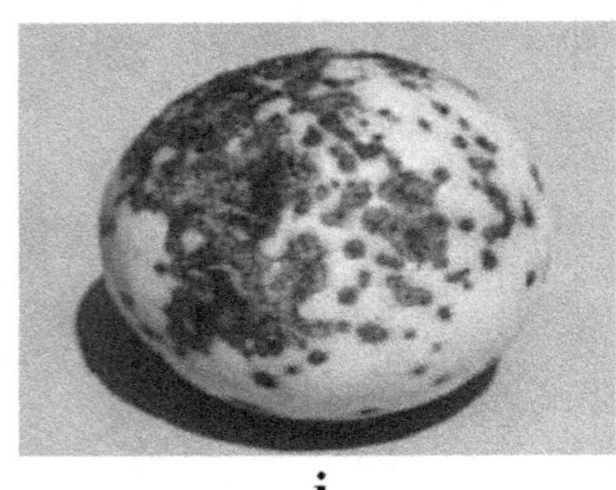

i

ii

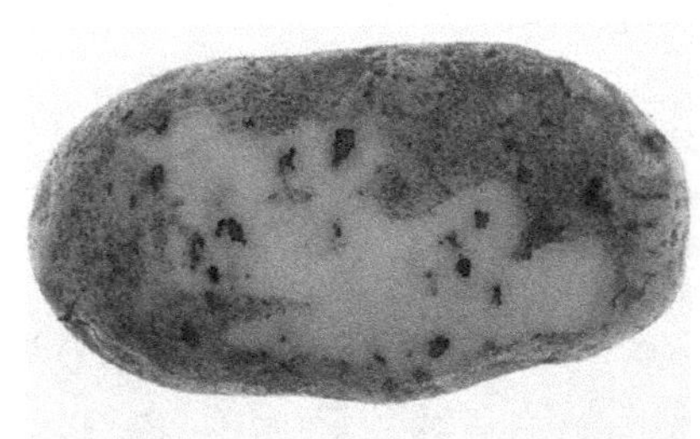

iii

iv

32. Which of these is a leaf rust

A

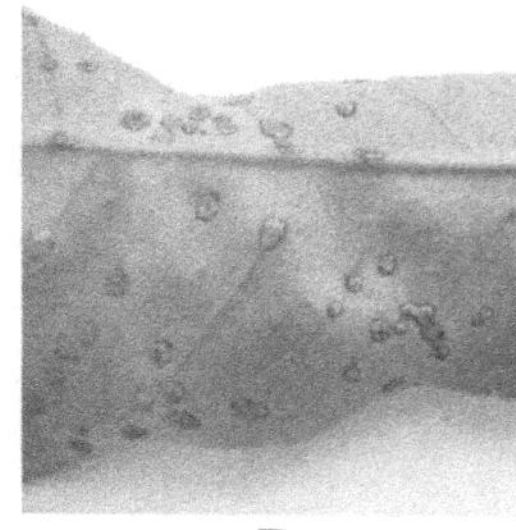

B

C

33. Identify the pathogen causing this disease

    a.   *Puccinia*

    b.   *Xanthoonas*

    c.   *Ustilago*

    d.   *Pyricularia*

34. Which part of the structure should be labelled on "HOOK" ?

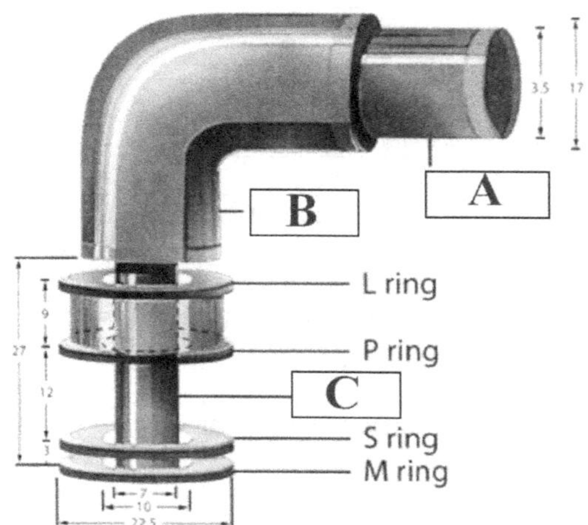

Answers to Pictorial Quiz

Q	Answer		
1.	A. iv B. i C. ii D. iii		A. Beijerick +Ivanowsky B. Louis Pasteur C. Robert Koch D. Winogradsky
2.	A		Edward Jenner: *The idea of Edward Jenner's vaccination against smallpox was criticized by many.*
3.	A. iv B. ii C. i D. iii		
4.	A = Staphylococci B = Streptococci D = Rods E = Coccobacilli J = Filamentous rods	C = Diplococci, F = Tetrads G = Spirilla H = *Vibrio* I = Slender rods	
5.	b.		*Clear zone around the colony indicates complete hemolysis which is called beta hemolysis*
6.	A. B. iii C. D. i		
7.	a.		Sugar fermentation test: *The tubes contain an inverted Durham's tube to collect gas if produced by the fermentation.*
8.	c		*Lecithinase positive Staphylococci, produce clear zone on Baird Parker agar due to consumption of lecithin in the medium*
9.	a.		MacConkey agar: *The medium contains lactose and neutral red indicator. Lactose fermentors produce pink colonies whereas lactose non-fermenting colonies are yellow to colorless*
10.	d		Potato Dextrose Agar: *The medium is selective for growth of Fungi. Hence a fungal mycelium is found on the plate.*
11.	A. B. iv C. D.		
12.	c		Tempeh
13.	B		*Sauerkraut is produced by lactic fermentation of shredded cabbage. Picture A is beer and Picture C is Wine. Both involve fermentation by yeast.*
14.	b		
15.	a		*Idlis are South Indian fermented food that involve mainly lactic fementers.*
16.	b		Clumping of particulate Ags is agglutination
17.	d		IgM is also called the Millionaire IG because of its molecular weight that is one million
18.	d		A is Rocket immunoelectrophoresis (RIA) and B is Single Radial Immunodiffusion (SRID). Both are quantitative techniques used to determine the titre of the Ag/Ab
19.	(a)		
20.	A. B. i C. D.		The pictures show various essential parts of a typical aerobic bioreactor
21.	A. B. iii C. D.		

Table *Contd...*

22.	b	The spinal spasms in tetanus are so intense that at times, only the head and the feet touch the bed.
23.	c	Granulomas are very much diagnostic for the disease tuberculosis
24.	a	The tubid background on the agar plate is the lawn growth of the host culture. The number of plaques can be enumerated to determine the counts of phage particles
25.	A	Antibiotic sensitivity test by agar diffusion method was devised by Kirby-Bauer
26.	A = *Giardia,* B = *Aspergillus,* C = *Penicillum,* D = *Entamoeba histolytica*	
27.	A	*Fusarium* is a characteristic spoilage causing fungus often isolated from raw vegetables and fruits
28.	A. B. i C. D.	
29.	A. B. i C. D.	
30.	A. B. iv C. D.	
31.	A. B. iii C. D.	i = Citrus canker caused by *Xanthomonas axonopodis* (bacterium) ii = Smut of maize caused by *Ustilago maydis (fungus)* iii = Potato Blight caused by *Phytophthora infestans* (fungus) iv = Rust of wheat caused by *Puccinia graminis(fungus)*
32.	A	A = Rust of wheat caused B = Citrus canker (leaf) C = Tobacco Mosaic Disease caused by Tobacco Mosaic Virus
33.	d	The disease is Blast of rice caused by *Pyricularia oryzae(Magnaporthe grisea)* (fungus)
34.	B	

www.ingramcontent.com/pod-product-compliance
Lightning Source LLC
LaVergne TN
LVHW080851240726
843527LV00052B/301